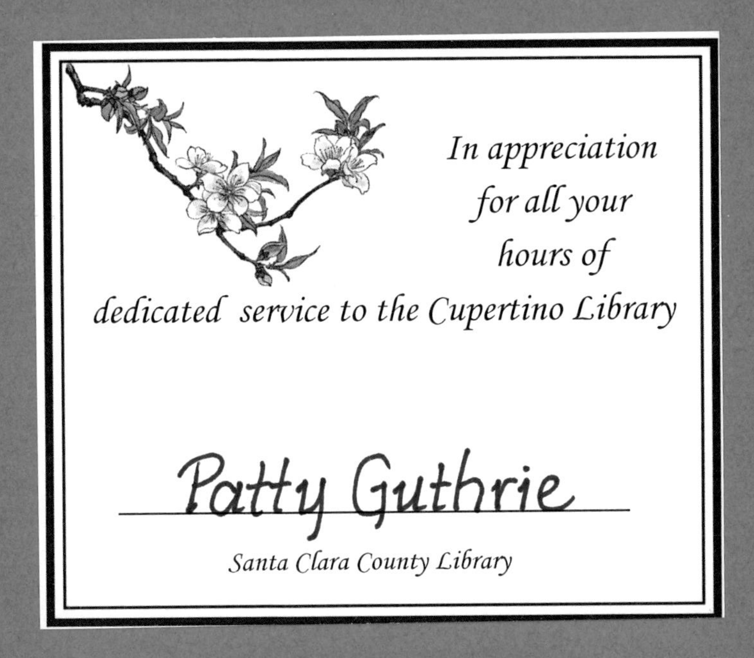
In appreciation
for all your
hours of
dedicated service to the Cupertino Library
Patty Guthrie
Santa Clara County Library

GRAVITY'S KISS

GRAVITY'S KISS

THE DETECTION OF GRAVITATIONAL WAVES

HARRY COLLINS

The MIT Press
Cambridge, Massachusetts
London, England

This book was set in Adobe Garamond and Berthold Akzidenz Grotesk by The MIT Press. Printed and bound in the United States of America.

Library of Congress Cataloging-in-Publication Data

Names: Collins, H. M. (Harry M.), 1943-
Title: Gravity's kiss : the detection of gravitational waves / Harry Collins.
Other titles: Detection of gravitational waves
Description: Cambridge, MA : The MIT Press, [2017] | Includes bibliographical references and index.
Identifiers: LCCN 2016035193 | ISBN 9780262036184 (hardcover : alk. paper)
Subjects: LCSH: Gravitational waves--Research. | General relativity (Physics)
Classification: LCC QC179 .C647 2017 | DDC 539.7/54--dc23 LC record available at https://lccn.loc.gov/2016035193

10 9 8 7 6 5 4 3 2 1

CONTENTS

1 THE FIRST WEEK
We Have Coherence

It is September 14, 2015, and I am in my study sitting on the sofa with my laptop on my knees—the way I spend most of my waking life these days, as I prefer to work from home. It's evening and I am idly reading the subject lines of the dozens of emails that come in every day. The largest number come from the gravitational wave community, and I usually delete them without reading. I've been working with this community for over four decades now, longer than all the active gravitational wave scientists except one. Once every six months or so I might save a short string of these emails in a folder, in case they turn out later to be of interest. I study how science is actually done in real time.

One email catches my eye. The subject line is "Very interesting event on ER8." It had come in around noon, British Summer Time. The email says that the interesting event has been seen by the cWB analysis "pipeline." The abbreviation "cWB" stands for "coherent wave burst" and the "Burst Group," which developed the pipeline and is tasked with finding bursts of gravitational waves without basing the search on preconceived waveforms. The cWB pipeline, then, has just detected something that looks like an interesting burst of waves. A pipeline is a mathematical/statistical procedure that is applied to the torrent of data that pours out of the interferometric detectors; different groups operate different pipelines based on different principles—they may have more than one each—to monitor the data streams and automatically search for anything that looks like a gravitational wave signal. When the machines are running smoothly there could be a few alerts per week.

The email I have noticed is addressed primarily to the CBC Group but also to other groups. "CBC" stands for "compact binary coalescence." This is the group tasked with analyzing gravitational wave signals that come from binary stars at the end of their lives—the signals emitted in the last mad second or so of "inspiral," when the pair of stars, as with a skater drawing in her arms, start to rotate faster and faster around each other until they merge into one heavier star. The theory says these systems will emit a burst of gravitational waves that, in the last moments of rotation, as the new star that is born out of the merger reverberates with cosmic passion, might be strong enough to have a chance of being sensed by the exquisite machines that humans have placed on Earth to catch them. These inspirals and mergers can be modeled, though the exact waveform depends on the mass of each component in the inspiral and the way they are spinning; there is an almost infinite number of possible combinations, and when the CBC pipeline, as opposed to the cWB pipeline, looks for a gravitational wave it looks for close similarity to one of the 250,000 possible models in its "template bank." The template bank divides the indefinite world of potential into detectable actuality. Here is that first email, which comes from Marco Drago:

> **Mon., Sept. 14, 2015, 11:56** (all times are British Summer Time until October 25, when they move to Greenwich Mean Time)
>
> Hi all,
> cWB has put on gracedb a very interesting event in the last hour.
> https://gracedb.ligo.org/events/view/G184098
>
> This is the CED:
> https://ldas-jobs.ligo.caltech.edu/~waveburst/online/ER8_LH_ONLINE/JOBS/112625/1126259540-1126259600/OUTPUT_CED/ced_1126259420_180_1126259540-1126259600_slag0_lag0_1_job1/L1H1_1126259461.750_1126259461.750/
>
> Qscan made by Andy:
> https://ldas-jobs.ligo.caltech.edu/~lundgren/wdq/L1_1126259462.3910/
> https://ldas-jobs.ligo.caltech.edu/~lundgren/wdq/H1_1126259462.3910/

It is not flag as an hardware injection, as we understand after some fast investigation. Someone can confirm that is not an hardware injection?

"GraceDB," by the way, is the "Gravitational wave candidate database." It is the web page that shows events of potential interest, and it mostly logs things automatically so long as it is triggered by the events extracted from the pipelines that surpass a certain threshold. But it logs a lot of events, none of which has ever turned out to be a genuine gravitational wave.

To be reading these emails you need to be part of a select group, and to access the URLs you need a password, which, as far as I know, outside of the scientists, only I possess. The select group is known as the "LIGO–Virgo collaboration" or "LVC." It is the group based around the Laser Interferometer Gravitational-wave Observatory (LIGO) and Virgo. LIGO has two four-kilometer interferometers in the United States, while Virgo is a French–Italian collaboration with one three-kilometer interferometer near Pisa (which was not online at the time of the putative signal). But there are 1,000 or more scientists involved located all over the world.[1] I look at the websites and see certain traces that appear to be large and

1. Actually, about 1,250, but I'll use the round figure of 1,000 throughout the text. The LIGO–Virgo collaboration contains within itself elements of competition. Virgo was originally an entirely independent operation intended to compete with the US effort but it has been less well funded and, as a multicountry consortium (initially Italy and France), has more organizational difficulties to overcome. Perhaps as a result of these problems its interferometer has always lagged behind LIGO in terms of sensitivity, and Virgo scientists will be very disappointed that they were not able to contribute to the first detection of gravitational waves. The British-German GEO600 was also first conceived as a competitive instrument, but it was relatively starved of funds and it soon became clear that its most important role would be to contribute technically to LIGO, which it has done very successfully. That said, the scientists from all groups have worked as a seamless and pretty well frictionless team when it comes to analyzing the meaning of the signal detected by LIGO and writing up the results. So they are a fairly solidaristic group when it comes to handling and presenting the data (with some residual rivalry between LIGO and Virgo that seemed to show itself occasionally in the Big Dog incident (see chapter 2) but is nowhere in evidence in respect of the event. When it comes to building apparatus, they are much more independent and competitive.

appear to coincide in time on the two great interferometric detectors—L1 and H1—located 2,000 miles apart in Louisiana and Washington State, L1 therefore being southeast of H1. But I'm not that excited, because stuff like this comes up all the time: the detectors are so sensitive that they are always jiggling about and creating false alarms. We've been waiting fifty years to see a gravitational wave and it is pretty unlikely that this is it.

The next email, which comes twenty minutes later, though it again does not get me very excited, at least makes me think I probably will save this string of emails in its own folder. This email says: "this is a nice inspiral with Mchirp = 27 Mo." The implication is that the analysis of the shape of the waveform—its relatively low frequency and "ringdown" profile—indicates that combined objects with a total mass of considerably more than 27 of our suns have been involved. This means that at least one of them, and probably both, must be a black hole, since the largest possible mass for a compact star other than a black hole is around 2.5 that of our Sun (stars can be bigger if they are still burning fuel to fight the pull of gravitational collapse). If they are both good-sized black holes, then the theory says that a lot of gravitational wave energy will be emitted when they merge, and this makes it more plausible that this could be a real detection.

Two minutes later I receive:

> According to the LLO injection logs, the last successful injection attempt was 1125400499 (Sep 04 2015 11:14:42 UTC). I looked at the LLO injection schedule around the time listed in Andy's Q-scan: 1126259462. Here are the nearest scheduled (burst) injections:
>
> 1126240499 2 1.0 hwinj_1126240499_2_
> 1126270499 2 1.0 hwinj_1126270499_2_
>
> Both are > 3 hours away

What this is telling us is that no one had officially and deliberately injected false signals into the detectors within three hours of the event; at least, it shows they had not done this in a way that was visible to the scientists.

Injecting false signals is something the scientists do to make performance measurements and calibrate the instrument.

Seven minutes later: "Very interesting indeed! Looks like a high-mass inspiral?" But this writer inquires why the event had not been seen by other non-cWB analysis channels—so-called low-latency channels, specifically designed to spot things quickly—that have been put in place to alert the scientists to events like this. The answer comes a bit later: these channels currently have a cutoff such that they do not register high-mass events, so they would not be expected to see an event generated by such massive black holes. One senses a sigh of relief among the community, insofar as a sigh of relief can be sensed over email.

Incidentally, so you know how this business works, though the detectors that are involved are located in Louisiana and Washington State, the four emails I have discussed so far came from Hanover in Germany, Florida, Melbourne, and Paris. The physical location of an instrument is, nowadays, almost irrelevant, and so is the physical location of those who analyze the data. As it happens, it is nighttime in the United States and most people are asleep, which is why the scientists who notice it first are located in Europe—in Hanover.

That the location of the analysts is irrelevant does not mean that they do not rely on a dense history of face-to-face meetings that has created a trusting community that can now do its business remotely. Nowadays, with this history behind them, perhaps two face-to-face meetings a year is enough to maintain a remote working relationship so that most of the work can be done from wherever they happen to be, so long as they are plugged into the Internet.

I have my own long and dense history of face-to-face interactions with the community: from the early 1990s to the mid-2000s I traveled to pretty well every meeting of these groups and frequently visited the sites of the detectors; I went to at least half a dozen meetings a year all over the world. In the last few years I have been merely maintaining my presence by going to only about one meeting a year, but I was with the community in Budapest only a couple of weeks ago. I was greeted warmly and am still close to being an ordinary member of the group, though the field has grown and I know a smaller proportion of people than I once did. Nevertheless, just sitting here at home on my sofa, I am right in the middle of things.

It has taken me decades of personal interactions not only to reach this position of trust but also to reach a sufficient level of understanding to make some sense of the emails (I'll discuss my expertise at greater length toward the end of the book). Note that because I can sit on my sofa and be in the middle of things does not mean that what is going on would be comprehensible to "the public" even if they had the sofa and the passwords—far from it.[2] And though I can make some sense of the drift of the emails—enough to see that something interesting is going on and where the discussion is going—I cannot understand a lot of the technical details because I have never gone deeply into the techniques of data analysis, taking most of it on trust. I can still get a good sense of what is being said, however. Luckily, my friend Peter Saulson, a physics professor at Syracuse University in New York and a senior member of the gravitational wave physics field, is endlessly willing to answer my email queries. An anthropologist would refer to him as my "trusted informant" (see "How the Book Was Written and Those Who Helped"). I try to limit my questions to things I mostly understand, so our interactions mostly have the quality of conversations, not interviews.

By the following day, if I had known how to look, I would have found out that the very initial analysis indicates a total mass of around 50 suns with no component smaller than around 11 suns; but I don't come to understand this for another day or two. (It would eventually be concluded that the component masses are around 36 and 29 of our suns.) I do not

2. I stress this because some social scientists and some members of the public believe that science is far more openly accessible than it is (see Note I in "Sociological and Philosophical Notes," 351). All notes that deal with the more wide-ranging social-philosophical questions have been gathered together into a section called "Sociological and Philosophical Notes," which starts on page 349. This has made it possible to treat the content of those notes in a less constrained way, allowing those who want to pursue these matters to read the notes as a self-contained chapter. Readers interested mainly in the scientific story can ignore these notes.

have an exact record, but I think it is on Tuesday that I decide to start shoveling these emails into their own folder. At the time, there are lots of reasons to doubt that this is a gravitational wave signal, but the marked and almost unprecedented tendency in this particular case is for the doubts to lose credibility as the emails accumulate rather than for skepticism rapidly to overwhelm the proto-signal, as has almost invariably been the case in the past.

Since I am trying to give a sense of my dawning acceptance of the extraordinary possibility that this forty-three-year-long part of my life is finally drawing to a close, let me say that it is not until now, Thursday morning, after two full days have elapsed, that I decide I must start to write this book: right now it is 10:00 a.m. British Summer Time on Thursday, September 17, 2015.

DAWN OF BELIEF

During the first two days I have read and stored about 140 emails, and I now sense that this intensely skeptical community is beginning to believe that this is a real event: the world of terrestrial detection of gravitational waves has shifted on its axis. Here is the first email that truly begins to transmit this sense of the impossible happening:

Sept. 16, 2015, 8:00 p.m.

I think it is essential we devote time to hardware injection as close as possible to the time of the event in order to establish safety of the DQ/vetoes we might consider (although, to zeroth order data is so clean around it, but it might play a role in reducing further the background events/FAR estimate) and inject using the HW system the very waveform extracted for our candidate event using some scale factor and over few instances in time. Beyond that, I think we should take an hour or two to perform the burst ER8 injections I proposed to the group a while ago:

https://wiki.ligo.org/viewauth/Bursts/ER7O1HWInjections#A_42Alternative_proposal_for_ER8_HW_injection_run_and_periodic_injections_over_ER8_47O1_42.

Such injections will allow a reasonably complete dress rehearsal of our online searches (although, apparently, Nature brought us straight to stage for the performance), including the validation of our protocol for EM follow up.

What this email is talking about is what the community has to do to start to confirm the reality of the event. But do not worry about the technicalities; it is the throwaway parenthetical in the last couple of lines that reveals how belief is changing: "although, apparently, Nature brought us straight to stage for the performance." The "N-word"—the sacred name, "Nature"—has been spoken. That never happens!

There follow more formal indications of the change. The first is the decision of the Burst Group to announce that they have sufficient confidence to demand that the mechanism worked out long ago to handle real gravitational wave candidates be set in motion. This procedure has been invoked only twice before, and then for realistic blind injections (cases where the scientists "inject" fake signals into the data, to see whether the devices can detect them; more on these in chapter 2).

Sept. 16, 2015, 9:30 p.m.

The chairs of the Burst group would like to formally initiate Step 1 in the process for claiming a first detection (M1500042) for the event labeled G184098.

This event has an estimated false alarm rate of better than one in 200 years and appears significant as viewed in three burst pipelines. Initial crosschecks with the detector characterization team have found that both L1 and H1 were working well when this event occurred. We have initiated the Burst detection checklist and other studies to better understand this detection candidate.

We look forward to working with the full collaboration to learn more about this very interesting event!

Best wishes,
Ik Siong Heng, Eric Chassande-Mottin, and Jonah Kanner for the Burst group

L1 and H1 are the huge interferometers, each with a pair of arms 2.5 miles long, located in Livingston, Louisiana, and Hanford, Washington.

Half an hour later comes the following from a senior member of the collaboration, the spokesperson of the LIGO–Virgo collaboration:

Dear Jonah, Siong, Eric,

Thanks for this good news. These are indeed exciting times—welcome to the Advanced Detector Era!

Gaby.

And the next day, from a member of the community:

Congratulations, bursters!!!

This shift in reality has happened surprisingly quickly: fifty years of disbelief has turned into readiness not to reject in only two days. Seen in terms of the human condition it is quite frightening; we humans tend to make our political and financial decisions assuming that big changes take a long time, but this is a reminder, on a very small scale, that the fragile world we live in can be transformed into something very different almost overnight. Usually those changes are of a horrific nature; here something good has happened.

HOW IT WAS FIRST SPOTTED

Incidentally, I am a typical member of the community in terms of my initial skepticism. A month and a half later, in another context, one of the physicists harks back to this moment:

> **Oct. 29, via teleconference:** So I guess many of you had the same experience I did which is that you woke up early in the morning on Sept 14th and heard that we had something interesting in the data and you probably immediately assumed that this was an injection either intentional, blind, gone wrong, software bug, something like that. And since then I think we've all been asking ourselves, you know, how could this signal which appears so perfect and so loud, could have appeared so early in the run. And it all seems a bit improbable.

A little more than two months after the first announcement I arrange a teleconference with Marco Drago, who sent that very first email, and ask him how it happened. He explains that he has an email alert set up that automatically tells him if GraceDB has seen anything. It takes about three minutes for a pipeline to place a record on GraceDB and he gets the alert immediately after that. So he knew he should have a look at GraceDB within a few minutes of the event happening. Notice that this is not something that the community as a whole is doing—they know, or they knew on September 14, 2015, that even if GraceDB said it had found something it would not be a gravitational wave. So people do not, in general, sit around excitedly monitoring the data stream; all the alerts will get looked at but, in the normal way, not with any sense of urgency. Furthermore, the Laser Interferometer Gravitational-wave Observatory was not supposed to be detecting any events. The "advanced" version of LIGO—aLIGO—had not long been "online" and was completing the last of its engineering runs, which are part of the shakedown before observing proper begins. On September 14, we were still in engineering run number 8—ER8—with the first proper observing run, "O1," not due to start for about week, so the chances were that anything seen would be a test injection, noise, or too compromised by noise to count as a clean signal. Marco explains that in the course of ER8 he had already seen two GraceDB signals that turned out to spurious and would see two more before O1 started, but this one was unusually clean looking. So he found a colleague of his, Andrew Lundgren, and asked him to check if there had been a test injection at the time of the alert. The answer was "no." They then contacted the control room at the sites in the United States using the standing telecon connection and asked the operators on night shift if they were aware of any test injections. Again the answer was "no," and they were told that the detectors were running well. So almost exactly an hour after the initial alert, at 12:55 p.m. Central European Time or 11:55 a.m. British Summer Time, Drago sent the email to the community, the one that I noticed a few hours later.[3]

WHY BELIEVE?

We Are in "The Advanced Detector Era"

One reason for thinking this could really be the first terrestrial sighting of a gravitational wave is that, as the email above puts it, this is the beginning of the "Advanced Detector Era." At the end of 2010, the large US interferometric detectors were switched off so that they could be rebuilt, within the same housings, to increase their sensitivity and range. Many improvements were planned: installation of seismic isolation down to ten cycles per second, which seems to have made a huge difference; better, larger, and heavier mirrors with better coatings; multistage mirror suspensions with better suspension fibers; lasers with twenty times the power to fill the detector with light (still to be run at full power); thermal compensation of the mirrors to avoid distortion through heating; a new configuration called "signal recycling," which allows the most sensitive frequency of the devices to be tuned if desired but which is not yet online; and an extraordinary technique called "squeezed light" that takes advantage of quantum theory to decrease the

3. See Note II in "Sociological and Philosophical Notes," 354, for discussion of Garfinkel, Lynch, and Livingstone's 1981 paper that covers the first sighting of an optical pulsar.

fundamental uncertainty in certain dimensions of the measurement at the cost of increasing uncertainty in the complementary but less interesting measurements, but which won't be installed for a few years. Some of these techniques have been refined in the smaller "development" interferometers that are located around the world, notably the German-British GEO600, which is located on the edge of a field in Hanover. Five years later in 2015, the interferometers at Hanford and Louisiana, the Laser Interferometer Gravitational-wave Observatory, are back online. While in 2010 the devices were known as Initial LIGO or iLIGO, they are now known as Advanced LIGO or aLIGO.

The range of these instruments is counted as the distance at which they can detect, with reasonable confidence, the final inspiraling and merger of a binary star system comprising two neutron stars, each with a mass of around 1.4 suns—this is the "standard candle." In 2010, LIGO's "BNS range" or, more simply, "range," was around 17 or 18 megaparsecs. A megaparsec is around 3.26 million light years. In spite of a year or more of observation, no gravitational waves had been seen. What this means is that, though the instruments were probably affected by many gravitational waves—at least if the physicists' theories right—their impact never rose sufficiently far above the background noise in the instruments to stand out. We tend to say that LIGO was unlucky in that no impactful events occurred within its range, or maybe some did but they happened when the instruments were offline. These machines are very delicate and their combined duty cycle is only around 50 percent—they are offline for maintenance or as a result of environmental disturbance or breakdown, half the time. The refurbished instruments are intended to have a range of 200 megaparsecs, but this won't be accomplished for two or three years of gradual introduction of all the new innovations and the inevitable periods of shakedown. In the meantime, the first stage of refurbishment has been a great success, completed ahead of schedule and within budget, and the instruments currently have a range of about 60 or 70 megaparsecs.

If we think of sensitivity as likelihood of seeing an event, increased range buys a proportionally much greater increase in sensitivity. An increase in range of a factor of just over three, which this is, means that in 2015 the detectors can see to the boundary of a sphere of space that is just over three times the radius it had in 2010. If we take stars to be uniformly distributed in the heavens once we are beyond our "local" area, then a sphere of three times the radius contains 3 × 3 × 3—roughly 30—times the volume. This means we are 30 times as likely to see a BNS inspiral or anything else that emits strong gravitational waves in that kind of frequency band—but the apparatus is very much more sensitive than previous editions to low frequencies so the increase in sensitivity for low frequency gravitational waves is still greater. Other things being equal, one day of observation with LIGO at the end of 2015 is worth a month of observation with LIGO in 2010, and the first ten days are worth all the years of observations that have been made to date, in terms of the likelihood of seeing an event. So that is why the scientists think the half-century of effort might finally have paid off—though that it should pay off quite so quickly and cleanly is very strange, and people are suspicious.

The Waveform and Its Coherence

Figure 1.1 shows a first glimpse of the event in something like a comprehensible form. Two lines are overlaid corresponding to the output from the two detectors at Louisiana and Hanford, L1 and H1. Time runs from left

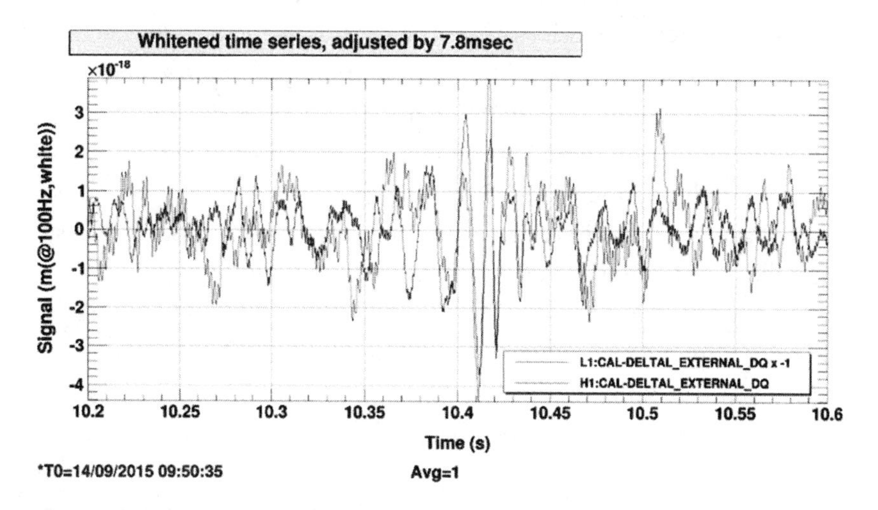

Figure 1.1
A first glimpse of the event.

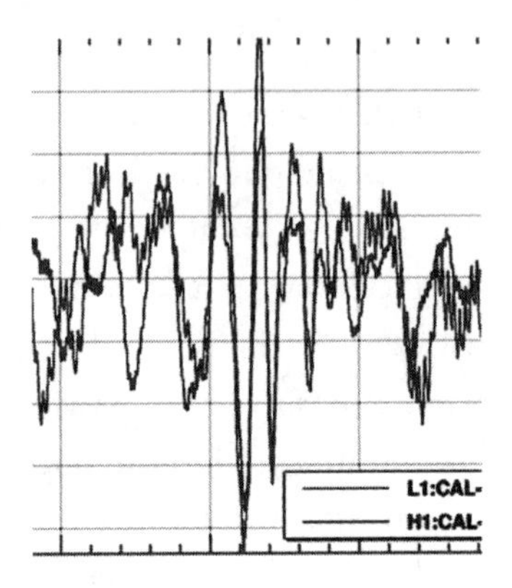

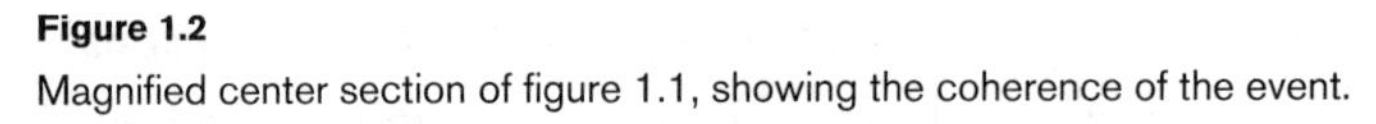

Figure 1.2
Magnified center section of figure 1.1, showing the coherence of the event.

to right. In this figure one trace is a slightly lighter gray than the other. The energetic section that starts around 10.38 seconds is "the signal."

One of the traces has been time-shifted by 7.8 milliseconds so that it sits more neatly on top of the other. The necessary time shift is assumed to correspond to the different time it took for the signal, traveling at the speed of light, to reach the two sites. These are 2,000 miles apart with the gravitational wave encountering the Louisiana interferometer before the Hanford device (it will turn out later that the correct time shift is 6.9 msec rather than 7.8). Either side of the signal the two traces are rather different, peaks on one sometimes coinciding with troughs in the other, and with both traces looking like random noise. But from around 10.38 seconds to around 10.45 seconds, in the region of much more marked activity, the peaks and troughs correspond closely. Figure 1.2 is a magnified version of that central section; here the correspondence in the energetic section can be seen still more clearly. (It is even more obvious in the original that circulated around the mailing lists, where the traces are colored red and blue.)

There are four cycles in each detector that overlap almost exactly. Here the two signals are not only energetic but "coherent" between the two detectors. This makes it very unlikely that the traces represent uninteresting noises that are accidentally correlated—noise in two detectors is very unlikely to be coherent (see *Gravity's Ghost*, 75). In the case of "Big Dog," the blind injection of September 2010, it was the coherence of the signals that caused the scientists to believe in it even though it was the statistical unlikelihood of its being caused by chance, as measured by time slides (see below, chapter 2), which was given as the official reason. In figure 1.2 we can see that the chance is very remote that two signals in detectors so far apart would fit over each other with such exactness by accident. Also, as we will see, the number of cycles, their frequency, and the way the signal rises and diminishes all fit the same kind of template, corresponding to the inspiral and merger of a pair of black holes at both sites. I find this graph completely convincing.

I write to Peter Saulson:

> **Fri., Sept. 18, 8:59:** Why is it that I find that whitened time series so convincing-looking whereas no-one else is mentioning it.

Peter is staying with friends from gravitational wave circles for the weekend, both very senior. He replies:

> **Mon., Sept. 21, 12:25:** Re the time series: You aren't as alone as you fear. [Very senior gravitational wave scientist] told me that that graph was the single most important piece of evidence that convinced him that this was real. I'm in that camp, too. As for the question of why no one is mentioning it: Ah, that is a sociological question ;-) But let me suggest a possibility: We don't have much in the way of techniques for investigating the time series in and of itself to ask whether it matches theory, or how much is corrupted by noise, etc.
>
> When we look at that graph, we are using our intuition. For whatever reason, we've developed those techniques more in other ways. For example, the second piece of evidence that was very important in convincing me that this was real was how well it matched very similar templates in the BBH [binary black hole] search ... at the same moment, a small cluster of triggers in the same short-duration portion of the template space rang off at both detectors. (Not precisely the same ones, but you don't expect that given that there's some noise.) The NewSNR [signal to noise ratio] of the event

... is considerable: over 10.5 for the event in each case. That says that the signals look like our templates. That is exciting! By contrast, glitches look like "towers" of modest SNR events across the whole template bank.

But let me come back to your suggestion. I do agree that, given that there's sufficient SNR to see a portion of the signal directly in the whitened time series, we should be sure to draw the "winning template" superimposed on the data in the time series. That will be very convincing to all, us and readers of the paper alike.

2 RESERVATIONS AND COMPLICATIONS
Malicious Injections?

ANTIPARALLEL

Nothing is simple, least of all science. It would be nice if these traces were simply the raw fingerprint of a gravitational wave impressed by a giant hand on the two interferometers, but they are not. Nothing is visible without processing. This is already obvious because one trace has had to be time-shifted by several milliseconds to correspond to the other, but much more manipulation has had to take place before that time shift could reveal anything interesting.

The interferometers each have two arms at right angles joined at what is called the "center-station" and each stretching out to an "end-station." Thus, if you stand at the center-station, you can visualize the arms as the axes of a graph—the x-axis reaching out horizontally to the right and the y-axis straight ahead. Hence the labels "X-arm" and "Y-arm." When a gravitational wave affects an interferometer, first the X-arm will shorten (say) while the Y-arm will lengthen, and then this deformation will be reversed as the wave passes through half a cycle.

The geographical layout of the two devices is shown in figure 2.1, Hanford on the left and Livingston on the right, with North at the top; as can be seen the orientations are almost opposite. So if we imagine the X-arm shortening and the Y-arm lengthening at the Hanford interferometer, the Livingston interferometer will react in almost opposite way: X-arm lengthening and Y-arm shortening. Thus the waveform produced by the

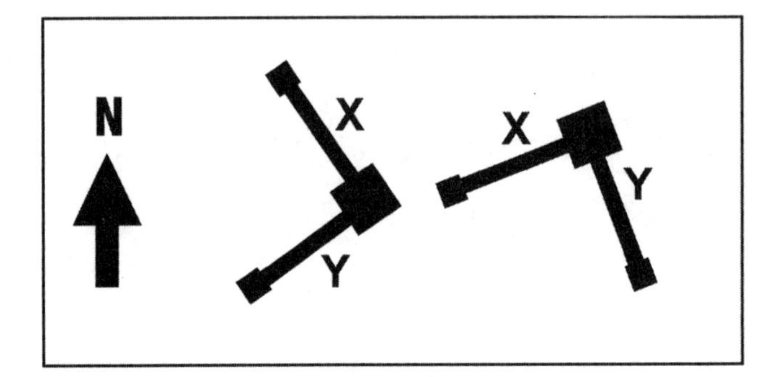

Figure 2.1
Orientation of Hanford and Livingston interferometers.

two interferometers will be almost complementary, and to get them to match even roughly the output of one has to be multiplied by –1.

But that is only the beginning of the adjustments. The simple factor of –1 will have to be adjusted because they are not exactly antiparallel. Beyond this, however, is the fact that these are devices in the real world not on paper. For one thing, the Earth is a sphere not a flat plane, so their horizontal orientation will also be slightly different. There will be many other things about them that are slightly different because they contain real mirrors and real actuators and real vacuum tubes and real electronics—and all these difference will have to be understood by calculation and careful calibration of the detectors, by inserting signals across a range of frequencies and strengths and measuring exactly how each device reacts. A few weeks later, I hear in the course of a teleconference:

> So the ESDs have different actuation functions at H1 and L1. This is really because at some point [XXXX] changed some resistor [at Livingston] and he put it in the a-log. It was the right resistor to change but it didn't happen at Hanford.

Only after all these differences have been measured can the right "filters" be designed to adjust each signal to make it look, at least roughly, as though the same device were seeing it at each site. The traces we have been looking at have been collected after most of this filtering has been done, so they are already the outcome of much careful work. There is no such thing as a "raw" signal—the data as it initially comes out of the detectors looks like a mess of noise (see *Gravity's Ghost*, 74).

A BLIND INJECTION?

On the other hand, we have many reasons for doubt and caution. The first of these is the notorious "blind injections." Blind injections are fake signals injected into the detectors in secrecy. A team of one or two researchers create false signals and insert them in such a way that they look like real signals. The idea is to keep the team on their toes and force them to rehearse the detection procedure in readiness for a real signal; they cannot do other than proceed as though the signals are real because only the team who inserted the signals and the director know if a signal is a blind injection or potentially the real thing. My previous book, *Gravity's Ghost and Big Dog*, describes two blind injections, one in mid-September 2007 and one in mid-September 2010; now here we are in mid-September 2015![1]

Blind injections are a mixed blessing. They have been great for me because I have been able to write two books in which I have carefully analyzed the process of detection—the process through which scientists reach a conclusion about whether some proto-signal is ready to be announced to the world as a discovery. Because no one knows the difference between a real signal and a blind injection, what I have been describing is supposedly identical to what would happen in the case of a genuine discovery. So I have rehearsed this moment just as the scientists have rehearsed it. It took scientists eighteen months of analysis to decide if the Equinox Event, the blind injection analyzed in *Gravity's Ghost*, was something that could be reported to the world of science. It took one-third of the time, six months,

1. *Gravity's Ghost and Big Dog* (Collins 2013) is a paperback, but *Gravity Ghost* (Collins 2011) was originally published alone in hardcover.

to complete the process in the case of Big Dog. The promise is that the next event (i.e., this one) will take only three months to analyze, so one can see that these rehearsals are worthwhile in terms of training. As for me, my account of the Big Dog analysis rehearses the technicalities of discovery in fine detail and relieves me of some the duty of repeating it here; should readers find this book's account too thin in terms of technicalities, I recommend *Gravity's Ghost and Big Dog*.

Because of what had gone before, everyone's first thought is: "Oh no—another blind injection!" There is a certain amount of bad feeling around since plans are in place, over the objections of quite a vocal group, to continue with the blind injection program. The scientists would get *something* out of a blind injection, but to analyze a blind injection is a lot of work for nothing—at least, nothing in terms of scientific kudos. I am told that at least some of the scientists were sufficiently resentful of even the past exercises to have cheated and looked into analysis channels from which they were banned, to avoid what they saw as a waste of their time. The feeling is growing that the days of blind injections should be over so that if there is a proto-signal, everyone knows it is really worth analyzing. For me there is *absolutely nothing* to be gained from another blind injection, not even another useful rehearsal; I don't need another rehearsal. I can't write another book about a blind injection. So for me, it would be a total waste of time even more so than for the scientists.

But there is an immediate reason to think that it is not a blind injection, and this was quite simply because on September 14, LIGO was not officially switched on! It is only switched on today (it's now Friday, September 18). What do we mean by "not switched on"? Well, since August 17, LIGO has been operating within an engineering run known as ER8. Engineering runs are part of the shakedown, and in an engineering run the scientists can change things and try things that they would not be allowed to do in "science runs," or as they are now to be known, "observation runs." In an observation run the instrument is kept in as stable a state as possible to increase the likelihood that any disturbance is the real thing, so that the noise background against which the reality (i.e., statistical likelihood) of a signal can be judged can be established as completely and surely as possible.[2] The first observation run, "O1," is scheduled to start today, September 18. If LIGO was not switched on at the time of the signal, it makes it improbable that the blind injection program would have begun.

From the point of view of many of those involved, and definitely including me, that the observation happened so early is a great stroke of luck. If it had happened a few days later, at the beginning of O1, then it would have taken place at a time when it could have been a blind injection. For me that could have been disastrous, because I simply could not spend another several months of my life on a blind injection, having written two blind injection books already. The scientists would have been faced with a similar problem, because taking these events seriously means a significant disruption to their lives.

As it is, the mere fact that the event happened in ER8 rather than O1 is not in itself enough to assure people that this is not a blind injection of some strange kind. So a flurry of emails goes around the system in an attempt to reassure the scientists that they are not wasting their time.

> **Mon., Sept. 14, 13:55:** Of course, if this is the blind injection test planned for ER8, and we burn a ton of time that could be spent on actually preparing for O1, then David's concerns about blind injections are again validated
>
> **Mon., Sept. 14, 16:31:** We were under the impression that blind injection infrastructure was *not* ready which leads us to believe that this should not be a blind injection.
>
> **Mon., Sept. 14, 16:38:** Agreed. We have to understand if we are in a blind injection test phase before expending much in the way of resources on this. We have Rapid Response Team people who are able to work on follow-up, but they are busy getting the detector ready for an expected Friday O1 start. We don't want to pull people off those tasks unless it is a real priority.

2. I give an extended discussion of how this noise background is calculated in *Gravity's Ghost and Big Dog*.

Mon., Sept. 14, 18:10: There were NO Transient Injections during G184098 Candidate Event. Other than continuous wave injections which were on-going in L1, there were NO hardware injections—blind or otherwise—during the event candidate G184098.

Tue., Sept. 15, 11:10: yes: there was no injection performed, blind or otherwise, at this time.

Wed., Sept. 16, 17:54 [from a very senior figure in the LIGO administration]: It is important to emphasize that this was not an injection; we were not in blind injection test mode, all injection channels were clean.

[But, on Thurs., Sept. 17, in a personal email to me]: Most of us have had paranoid feelings that this might possibly be a "secret blind injection" by some cabal of highly placed people in LIGO who might want to make some point about the LSC not being ready, etc. I spent one sleepless night picturing this possibility, and [XXXX] has mentioned that feeling, and also mentioned that someone had (partly in jest?) contacted him to ask if he himself had done it. The trouble with this thought is that, if it were true, it would represent a complete breakdown in trust. So I think everyone is shaking it off and forgetting the paranoia.

It is interesting and slightly disturbing that a scientific community can have doubts like this but, as the writer says, there is nothing to do but forget it. On the other hand, if one of the advantages of blind injections is to provide deniability to journalists—any sign of activity can be sidetracked into potential blind-injection-land (more on this below)—then it makes no sense for senior people to put out the denial, because if they did no one would be in a position to use the "possible blind-injection" gambit without lying. And these scientists do not like to lie, only to deceive (a distinction we will examine in chapter 13).

I know that if this *does* turn out to be a deliberate blind injection after all these assurances, some of them coming from near the top of the organization, I will feel so deeply betrayed that I might give up the project. This is partly because I don't believe in what the scientists are doing in one regard. I think that they are pathologically obsessed with secrecy. They are determined that no one will find out about the fact that they are searching for an event before they have completely finished analyzing it, written the paper, and had it peer-reviewed and accepted by a top journal. I think this approach is wrong and will suggest an alternative in chapter 13.

For now, let me just say that I—and I am not the only one—do not think that kind of line can be held. In war, opposing generals often get wind of an attack or the position of a secret installation by the sheer amount of activity taking place on the ground—the tracks made by many vehicles, the food and fuel deliveries, and so forth. This event is changing people's habits in the same way: there are more telecons, there are more closed-door meetings, there may be more gatherings at distant locations and changes in body language—people in the know will look at each other in a different way. I soon discover that I cannot keep this secret from my wife for longer than one day, and I ask around and discover that others' partners have had to be told: there is a moral duty to those close to you when the biggest event of your professional life is taking place, and partners will be the first to spot the changed demeanor and changed pattern of activity (I'm spending even more time on the sofa!). I hear that a social gathering that includes a past director of Gravitational Physics at the National Science Foundation is taking place over the weekend. This past director is the person who, more than any other, made it so that LIGO would be funded in the first place and then had the wisdom to ensure that it survived the near-fatal traumas of its early years.[3] I learn that special permission has had to be sought to let this past director in on the secret! Of course, everyone who is told is sworn to secrecy in turn, and, I understand, mothers, fathers, and children are not being told.

To reiterate, one reason that blind injections are loved by the senior scientists is that even if outsiders get wind of the activity it is deniable in the sense that it can be said: "We are looking for something but it might

3. See *Gravity's Shadow* for an extended account of the difficulties LIGO faced in its early years.

well be a blind injection so there is no point in getting excited." It seems to me, however, that insofar as this is a motivation for blind injections, it is a colossal waste of resources and people's time and energy for the sake of keeping a secret. Worse still, the possibility of blind injections may be counterproductive in that the scientists may not be able to generate the enthusiasm necessary to push themselves to the limits if they know an anticlimax might be awaiting them.

My view is that the scientists have some reason to try to keep the investigation secret because they don't want a fuss in the media about something that turns out not to be genuine; in gravitational wave physics, the felt need not to make such a mistake is more marked than in other fields because of the history of claims that turned out to be unsupported.[4] But this does not mean that so much energy should be expended on secrecy that it ruins people's lives and might even spoil the science. Somehow we have to educate the public into grasping that the first indication of new scientific findings is always provisional. In this case, there is the precedent of BICEP2.[5] In March 2014, the US BICEP2 team claimed to have detected the signature of primordial gravitational waves in the cosmic background radiation, but it turned out that their findings were probably the result of cosmic dust. The LIGO scientists need only say, should the level of activity leak out, that with the precedent of BICEP2 in mind they will not have anything to say until they have submitted a paper and it has been subjected to peer review and accepted for publication in a top journal. Until then, anything that shows up on their detectors has to be treated as probably caused by some kind of noise rather than a genuine signal.

MALICIOUS INJECTION?

I am now alerted to another possibility: a malicious injection. Some hackers or pranksters may have inserted a signal into the interferometers to

4. See *Gravity's Shadow and Big Dog* for more on this.

5. "BICEP" stands for Background Imaging of Cosmic Extragalactic Polarization.

cause embarrassment. This possibility was taken seriously enough in the last detector run for a special committee to be set up to consider it.[6] The possibility is being considered now just as it was in 2010. How hard is it to inject such a signal deliberately? Here are a couple of answers in emails. The first is somewhat long, but it is worth looking through it, not to understand the technical details but to get the flavor. I have emphasized certain phrases in both emails and one can see that the negative answer—it would be very hard if not impossible—rests on sociological considerations.

> **Fri., Sept. 18, 19:12:** I have been asked by several people whether I think G184098 could possibly be a malicious injection.
>
> I am coming to the conclusion that this is almost impossible. Here is why:
>
> 1) all the usual injection paths were checked—the only thing going in was the cal lines.
> 2) It is almost impossible to add such an event afterwards to the frames: the signal shows up in all sorts of extra channels (PD A & B, control channels etc). *I don't think any single person with the necessary knowledge on how to modify all frames would be able to know to which channels the signal needs to be added, and which transfer function.*
> 3) that leaves real time injection, which could be done in soft or hardware. To put some malicious c-code in the front end, you would have to work around the version tracking by [XXXX] and [YYYY]. We have a list of last mod times, as well as a history of what was running at any time and can look at the code. *Sure, somebody like [XXXX] and [YYYY] could probably find a way to sneak in the signal, but by they wouldn't have the expertise to know what shape, strength and where to inject it.*
> 4) *Similarly, to design a malicious piece of hardware in the actuation chain would be a major engineering feat, and would be hard to keep secret from anybody. Again I think it's very unlikely that any single person could pull it off without being noticed.*

6. *Big Dog*, 206–207.

5) ... and maybe most importantly, we still struggle to get regular hardware injection right after several attempts, and in the open. And now *somebody comes along, does it all secretly, and flawless in the first try? I don't believe it ...*

So in summary, to pull something like that off flawlessly and unnoticed, you would need to assemble a team with expertise in waveforms, interferometer controls, CDS computing and data archiving. That wouldn't stay secret.

Fri., Sept. 18, 20:25: I take [those numbers as demonstrating] a "high degree of coherence" [corresponding to what we can see in fig. 1.2]. One reason why this is important is that *there is little overlap between people who might know how to put a rogue injection into the data without leaving a trace, and people who know how to make coherent injections*

One of my classroom tricks when talking to scientists is to ask them how they know that the Moon landings were not faked in the Arizona desert. They always reply with technical points, such as the radio signals blanking out when the spacecraft passed behind the Moon, the way the flag did not flutter in the Moon's non-atmosphere, and the slow settling of the dust on the Moon's surface. I then point out that if one were faking a Moon landing it would be very easy to simulate all these technical features and anything else one can think of; just consider what Hollywood can do nowadays. I then point out that the reason we are sure the landing was not faked is essentially sociological: we know that if it had been faked it would have involved a very large conspiracy that would have leaked, and, furthermore, it is inconceivable that the Russians, who were watching everything, would not have pointed it out; but the Russians quietly accepted that they had been beaten to it. Faking the social is almost always much harder than faking the technical. The same, as we see from the emails, applies here. The italicized phrases indicate the difficulties of keeping things secret if more than one person was involved and rest, in turn, on estimates of the distribution of skills within the community—another social fact. Yes, the possibility of malicious injections is still in the air on Friday, but sociological considerations have almost eliminated it.

SHOULD SOMETHING HAVE BEEN SEEN ALREADY?

In the first day or so a rumor emerges that if the September 14 event was real then it is so large that something like it ought to have been seen in 2009 or 2010. It is said that there is only a 1 in 50 chance that it would not have been seen. This calculation seems to have been based on general considerations such as the threefold improvement in range and the probability that there would have been many more such events, some of them closer and still more powerful.

As explained, this event appears to be the final moments in the life of a heavy binary black-hole system, and such events emit much more energy than the inspiraling of a binary neutron star system upon which the range of the interferometers is calculated. But there are more stars in the heavens than grains of sands on the world's beaches, so if there is one such event there should be many, roughly evenly scattered, near and far, among the galaxies. Why didn't earlier LIGOs see such energetic events?

Then the calculation is done and reported over email: just how sensitive would earlier LIGO need to have been to see such an event—and the answer turns out to be quite a lot more sensitive than it was if the signal was to emerge clearly from the noise. This is mainly because of the low-frequency part of the signal to which Advanced LIGO is much, much more sensitive than the previous model than the simple increased range for binary-neutron star inspirals would indicate. So that is another worry dispatched.

One correspondent points out the crucial issue in more detail. aLIGO has undergone a threefold sensitivity improvement in respect of its ability to see the standard candle of a binary neutron star inspiral, but the improvement at low frequencies, with the new seismic isolation and mirror suspensions, is far greater than it is at the higher frequencies associated with the standard candle. As he puts it, as far as sensitivity to potential real events is concerned, the low-frequency improvement is effectively infinite: earlier LIGOs could see nothing at low frequencies, whereas aLIGO can. The new event, involving heavy black holes, emits gravitational wave in a much lower-frequency range than a BNS inspiral. To this kind of event, aLIGO is much, much more than three times as sensitive, and that is why this event is compatible with seeing nothing on the earlier generations of observatories. The difference can be seen in figure 2.2, which shows the sensitivity of the old detectors at the top with aLIGO below; the improvement in sensitivity at the low-frequency end is far greater (the lower, the greater) than is the improvement in the middle frequencies. The arrow points to a frequency range that is important in this event—around 30 Hertz (cycles per second). As can be seen, at this frequency the detector is about 10,000 times as sensitive as it was in 2010 (the values of the vertical axis, which indicates sensitivity to length changes in the interferometer arms, increase downward by a factor of 100 for each division). Incidentally, because of the huge energy output of this event, though the range of aLIGO is currently spoken of as 60–70 megaparsecs because the range is calculated on the standard candle, the scientists are saying that the event is 300–600 megaparsecs distant; that is the advantage of being able to see low-frequency events.

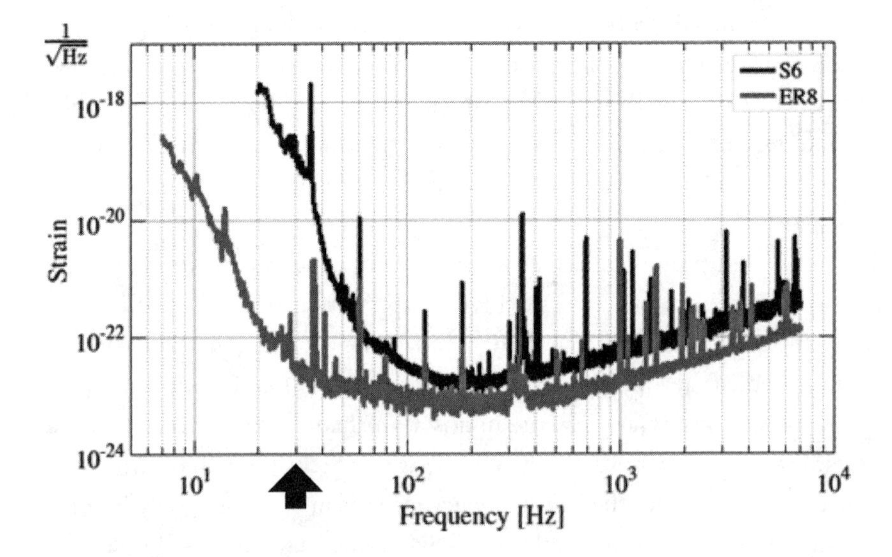

Figure 2.2
Comparative sensitivity of LIGO during S6 and ER8, showing the relatively huge increase at low frequency.

On the other hand, the sheer number of stars in the heavens means that if this is real then many more events should be seen. Indeed, some think that as aLIGO moves toward design sensitivity, there could be so many events that it will be impossible to resolve and analyze them individually—they will comprise a "stochastic background" that, if you converted the signals into acoustic waves, would sound like popcorn in a pan. The warning is that if no further events are seen in the three months of observing known as O1 that is just beginning, then that will be a cause for concern.

This event is going to take around three months to analyze to the point of being able to say it is a discovery (it actually took five months). As many of the scientists see it, if no similar events are seen during that three months it will *reduce* the credibility of this event. Indeed, the likelihood that if this event is real then other events will quickly follow has been used as an argument for those who don't want to start analysis until O1 has officially begun (more on this below). Knowing the cautious outlook of the community, I can anticipate them reaching toward a discovery claim as the analyses are completed and then drawing back as no further signals are seen. I hope that won't happen. Horribly—I feel intense guilt about this—I hope no other events do show up, because if they do it will make this discovery all too easy and therefore less yielding in terms of *sociological* interest; sociology emerges most easily from communities under strain, faced with making difficult decisions. On the other hand, if other events do turn up I will be enormously pleased for my colleagues: their half-century of effort will finally have transformed itself from the ugly duckling to the beautiful swan of gravitational wave astronomy.

Of course, calculating rates from single events is fraught with danger, as some of the scientists point out. It still could be that aLIGO has just been very lucky so that the implication that there should be many more events over the next three months is misplaced. It really is going to be interesting, assuming this event is unique. And it is very likely that the event is going to be especially interesting in terms of sociology in two other respects as well. First, the other major detector, the Italian-French Virgo, is offline, still undergoing refurbishment to a new sensitivity, and even the much smaller GEO600, which might just have been able to make some contribution but probably not, was also offline. Furthermore, if two black holes are the components, as seems increasingly likely, there will have been no electromagnetic emissions—nothing for regular astronomy to see. So, failing more events during O1, it looks as though the two US-based interferometers are going to be on their own as they face up to the question of whether this is truly a discovery.

ENGINEERING, NOT OBSERVING

That the apparent signal occurred while aLIGO was still in an engineering run may help to eliminate the idea that it was a blind injection, but this idea has a number of disadvantages. First, the detectors had not been fully prepared for observation; they were still undergoing last-minute exploration and adjustment. This hurts in three ways. One is simply that this is not what was supposed to happen, and at least one of the scientists feels strongly about it, writing:

> **Tues., Sept. 15, 1:27:** Isn't one of the selection criteria [for taking an event seriously] that the event needs to occur during an actual observation run? Unless we're planning on retro-actively declaring that the run started yesterday, that criteria was not met.

When this comment receives a rebuff, the same scientist remarks:

> **Tues., Sept. 15, 2:11:** I might argue that it is irrelevant why we weren't in an observation run. The fact of the matter is that we weren't. Or conversely, if we're willing to claim detection outside of an observation run then the words "observation run" are clearly meaningless and we should stop using them.

But this objection does not survive for long. As another email would point out, when the main instruments are down for maintenance, instruments of lesser sensitivity are on what is called "astrowatch"—they remain in readiness in case some unexpectedly loud signal arises such as a local supernova that even they could detect. In astrowatch the standard conditions for an observation are ignored; so if we are in astrowatch, it cannot be that a signal arising out of an engineering run cannot be accepted if it is clear enough.

But any signal is likely to be less clear in an engineering run because the instrument is still being adjusted and manipulated and calibrated. The problem is that the credibility of a signal depends on the extent to which it stands out from the background. As already mentioned, the instruments are always jiggling about and creating false alarms: coincident bursts of noise. What has to be shown is that a signal of this strength and this level of coherence could not come about by chance—that is, the odds of it being due to chance are at the 5 sigma level or better.[7] Making such a claim requires that the "background" is estimated, which is done by creating time slides.[8]

To create time slides, we place the continuous output trace of L1 and the continuous output trace of H1 effectively side by side, and then slide one of them along by, let us say, two seconds. Then we look for apparent coincident output from the two detectors. This coincident output cannot possibly be caused by a common outside agency, such as a gravitational wave, because the noise in each detector is not happening at the same time. Therefore, what is being gathered are pseudo-coincidences, generated by noise alone. If both detectors are producing output for a week, then we can see how many pure noise coincidences would be generated in a week by counting the number of pseudo-coincidences generated in a week when one temporally displaced output trace is compared to the other. If we slide one output along another two seconds and repeat the exercise, we will have gathered noise-generated coincidences corresponding to two weeks of time. As we will see, there is nothing special about this two-second time interval—maybe it should be three seconds. In fact, in all previous discussions of the question, it has been taken that the interval should not be too short, two seconds probably being around the minimum to prevent more

7. See *Big Dog, passim.*

8. Time slides are explained in more technical detail in each of my previous books.

than one apparent signal being generated as the same pair of glitches is slid along opposite each other. But some of the scientists will question this (more on this in chapter 3).

If we were to stop after we had done a couple of time slides utilizing a two-second interval we would probably find that there was no noise coincidence that was as loud or coherent as the proto-signal in those two weeks, so we can say that the false alarm rate (FAR) for such signals was less than one every two weeks. The trouble is that this is unconvincing. To be really convincing we need to be able to say something like the FAR is less than one in some tens of thousands of years. This requires a lot of background data and a lot of time slides. Because during ER8 the detectors were not steady for very long, it may be impossible to generate enough background to be convincing. But—and I am not sure about this right at this moment—this problem could be solved by generating background from the observation run, O1, that has just begun. So long as the devices could be said to be running in a similar fashion in O1 as they were in ER8—and remember the proto-signal happened only just before O1, so things were likely to be in the same state as they would be for O1—the O1 background should be usable.[9]

> **[From Peter Saulson, Sept, 22:]** We will surely be using O1 data to obtain enough background for estimating the significance of GW150914 (ugh, but an OK name). It ought to be valid so long as we have good reason to believe that the instruments are behaving similarly to earlier. So far, there doesn't seem to have been any noticeable change in the ifos that would lead us to believe that background has changed significantly. (This sweeps under the rug the fact that, from lock to lock, the ifos often behave differently, but we average over that and hope for the best.)

The third worry about this being ER8 is more "philosophical," we might say, and has a long history. Every quantitative science is beset with

9. For extensive discussion of false alarm rates, false alarm probabilities, and the technicalities and logic of time slides, see *Big Dog*.

"experimenter effects." This means that scientists find it immensely hard not to read results in a way that is biased by their expectations—the thing they expect to happen. It is for this reason that the efficacy of new drugs is generally not accepted unless they have been subject to "double-blind" testing, which means that the testing is done under circumstances where neither the patients nor the doctors and scientists know which patients are being treated with a real drug and which with a placebo. Only if the drug performs better under these circumstances is it accepted that the performance was not brought about by some kind of psychological effect in the patients, or in the interaction between patients and doctors, or in the way the doctors and scientists interpreted the results.

The problem for gravitational wave detection is that the detectors and the statistical analysis algorithms can be improved only by testing them against data and seeing what works best. The conundrum is generally resolved by cordoning off a small proportion of the data—say, 10 percent—and tuning the machines and the algorithms against it. Only when the final procedures have been agreed upon and "frozen" do the scientists "open the box" of the main body of data. After this, nothing in the way of tuning or analysis procedures must be changed; that way, it is certain that expectations are not biasing the way the analysis is done so as to enhance the statistical significance of a finding. In the earlier history of gravitational wave physics, we can find a number of cases where it is believed that results that are no longer considered genuine were generated by post hoc bias.[10]

Trouble arises when something of real scientific importance in respect of data analysis is discovered only after the box has been opened. Such was the notorious "airplane event," when a signal was heard on the detectors but only after the box was opened was it discovered that its almost certain cause was a low-flying airplane. Officially, the airplane could not be taken into account because the presence of airplanes had not been anticipated

10. For discussions of these effects in gravitational wave physics, see *Gravity's Ghost*, 104–107. Also see Note III in "Sociological and Philosophical Notes," 357.

before the analysis procedure was frozen. This resulted in a bizarre discussion, which I describe in my earlier book.[11] Of course, in spite of the problem of experimenter effects, the only way to proceed in the case of the airplane event was to take the airplane into account; to pretend it was not there in consequence of a methodological rule would have been close to insanity.[12] In the case of the current event, there is danger of misinterpretation because it occurred during ER8, when certain aspects of tuning and analysis procedures had not yet been frozen. This results in a series of emails filled with concern about the extent to which analysis of the new event will be done in plain sight rather than blinded, and a number of emails describing the parameters of the event—what was the likely composition of the inspiraling components, and so on—that come with warnings: "If you want to remain blinded do not read any further!" And there are worries, or shoulder-shrugs, pointing out that so much has been said over the open emails that it is impossible now to continue analysis in a blinded state: as one emailer put it, "the ship has sailed."

As far as I can see, it only really matters if the signal being analyzed is very close to the noise. If it stands out and has the clear qualities of a signal, such as the coherence that can be seen in figure 1.2, then pedantic adherence to blinding rules is not necessary so long as the dangers are borne in mind. The airplane event demonstrates this. I have worked through the nonsensical implications of double-blind testing in the case of the treatment of broken limbs, and a very funny article in the *British Medical Journal* points out that if we take blinding seriously then we should beware of parachutes since no one has ever tested them in comparison with a placebo.[13]

11. See *Gravity's Ghost*, 27–32.

12. Close to insanity *in my opinion*. Some scientists were certain that pretending the airplane was not there was the right way to proceed; one even resigned from the project in protest. I believe that the whole story is an illustration of the tension between "relentless professionalism" and judgment, which will be discussed later in the book.

13. See Note III in "Sociological and Philosophical Notes," 357.

> **Sat., Sept. 19, 21:46:** It does seem that this is no longer a blind search for an event, and we shouldn't be pretending otherwise. However, is it still useful to blind on the PE side? There seems to be a growing (possible erroneous?) consensus on the possible parameter ranges (e.g., of interesting masses or ratios). Is there a concern that the bulk of the collaboration may prematurely become convinced that the source is of a certain type, living in a narrow range of parameter space? Perhaps that ship has already sailed/is impossible to keep in harbor?

At any rate, there is now an additional reason why it is impossible to be pedantic about blinding in gravitational wave detection physics, irrespective of whether it is an engineering or an observation run. This is that the community has accepted the necessity of "low-latency" searches. Low-latency searches grab a signal and analyze it roughly as soon as possible—this is what has happened with this event. The reason is that they want to transmit as much information as possible about a proto-event to regular astronomers—electro-magnetic astronomers—or those with neutrino detectors—so that they can turn their telescopes to point to a potential source in the sky and look for the electromagnetic signature of the event—a burst of light or a burst of X-rays, or, where instruments are fixed, turn their attention to the time and direction of the event. This "multimessenger" astronomy is better than relying on gravitational wave signals alone, especially in these early days of making a first detection that people are going to be reluctant to believe. Indeed, in Budapest in 2009, Adalberto Giazotto, inspiration for and one-time director of the Virgo detector, said that no first announcement of a gravitational wave detection should be made unless it was backed up with an electromagnetic sighting. Thus, low-latency searches are inevitable, and, willy-nilly, blinding is compromised. This may relieve pressure on those who worry about the absence of perfect blinding in the analysis of this event. We will see more of this tension between low-latency searches and more thorough analysis as events unfold. A precursor of this argument can be found by jumping ahead to week 3:

> **Oct. 2, 15:43:** ... a point that several of us have been saying for a while: immediately opening low-latency boxes when the offline search has not been tuned can hurt doing the best offline search that we can do. We got away with this in Big Dog because the offline search was in a steady state when the low-latency CWB trigger came in.
>
> Of course, any CBC signal found by the low-latency burst search is likely to be so loud that it should be visible even with sub-optimal tuning in the CBC search. If GW150914 is a signal that we can claim as real, it should be there as significant in the original search with the frozen tuning. It should also be robust to any other reasonable tuning that we can do.

Here is another twist to the story. If this is a binary black hole, there will be no emissions of electromagnetic radiation or of neutrinos—that is the nature of black holes; anything that could be potentially emitted is sucked in. If, then, this event is confirmed—and it seems very likely that it is going to be confirmed—it will be without the reassurance of an electromagnetic confirmation. A bit later I will report on what Giazotto said when I asked him for his view on this. To look at it in a more positive way, the fact that if this event is confirmed it will be without electromagnetic confirmation is a great thing, because it will be an observation that *only the interferometers will be able to see*. Assuming the rough estimates of the rate of massive black hole inspirals are borne out by events, millions of such occurrences will be taking place in the heavens that are completely invisible and unknowable to anyone but this community!

Astrophysicists have long believed that there are pairs of black holes circling each other in the heavens. They have long known that if these binary black hole systems inspiral and merge they will emit gravitational waves at much higher energy levels than other typical sources. I remember being disturbed when, in 2004, I attended a public lecture by a famous astrophysicist and he said that Advanced LIGO would be such an improvement over the old LIGO that it would be able to see emissions of gravitational waves right to the edge of the universe. I knew that aLIGO was intended to have a range of 200 megaparsecs, which is far from the edge of the universe, but then I realized that, rather sneakily, as I saw it, the astrophysicist had switched the standard from the detectability of binary neutron star inspirals to black hole inspirals. Worse, I learned from another theorist at the same meeting that it was not known if there even were any black hole inspirals. The momentum in a black hole binary is so huge that takes a very long time to dissipate. It might well be, I was told, that the universe is too young for the black hole binaries that probably do exist to have dissipated enough energy for their components to have come together. Therefore, the astrophysicist was not only departing from the normal way of talking about the range of these instruments without making that clear, but was also talking of a source that may not exist.

For forty-three years I have been fascinated by the attempts to detect gravitational waves with terrestrial detectors. I have been fascinated by the way the technology has changed over the generations, by the battles over claimed detections that fell by the wayside, and by the way the communities of physicists work, but I have never been all that fascinated by the astrophysics itself—I don't care that much about what is in the sky; I just care about the means of finding out about it. But now I find myself knowing something about the sky that only another thousand people know, and among the people who don't know it are the thousands of astronomers and astrophysicists in the top universities of the world who happen not to be gravitational wave physicists. I know that, assuming this event is real, some black hole binaries have had time since the universe was formed to inspiral and merge.[14] I know a great big thing about astrophysics and about the nature of the universe that very few others know! For the time being, my wife Susan and I know more about this bit of astrophysics than Stephen Hawking. How can I not be excited about that!

14. The draft paper would include a paragraph saying (see appendix 2):

> GW150914 not only provides the cleanest evidence for the existence of stellar-mass black holes, but clearly demonstrates that *binary* black holes can form in nature, and in addition that they can form with physical properties that lead to their merger within a Hubble time.

MORE INDICATIONS OF CONFIDENCE

The scientists' confidence is not diminishing but growing, and by this time I have decided to create a new folder, called "GWdel," into which I filter the emails that I am deleting. If I simply delete them they will be permanently erased automatically after a week or so, but I decide I should keep this archive in case I miss something that I want to go back to and in case it turns out to have historic importance. Here is the sort of indication that is increasing my confidence:

> **Sat., Sept. 19, 4:40:** ... the possible detection of G184098 has important implications for stochastic searches. Earlier this week, I emailed Nelson and Tania to suggest that we look into this calculation "if the event is still alive and well in a week." However, in the past few days, tremendous progress has been made vetting the candidate thanks to the exertion of CBC, Burst, and Detchar. I think it's already time to proceed under the assumption that G184098 is real.

And here is a remark that also bears on the above discussion of blinding:

> **Sat., Sept. 19, 6:38:** Here, we have a clear signal, and it is strong. The initial estimates by the cWB group already show that its rate/probability is so low that we must take it seriously as an event candidate. ... So I think it makes sense to remove the dark glasses, and remain vigilant that our existing knowledge does not introduce unreasonable bias into our future statistical conclusions. In fact I believe that the pyCBC group has already determined the parameters that they will run time-slides with. So provided that you stick with these, or make changes that do not strongly bias the outcome in favor of this event, I would argue that the dark glasses can now be thrown into the dustbin. As you know, one of the most exciting aspects is that since this signal is quite strong, it is quite likely that the deep searches which are taking place now or will take place soon will reveal other weaker signals, at or near the detection threshold. For those, the full-blown "blind" technology developed by the group will be critical for determining if THEY are real or fictitious.

People are working hard "In this intense, exiting [*sic*], sleep-deprived times [*sic*] (Sat., Sept. 19, 9.31)." "But, now we have *real* GW data!! (Sat., Sept. 19, 9.31)."

On Saturday evening at 16:15 comes another sign of the times. One of the physicists suggests that a codename be invented so as to save everyone referring to the event as G184098 (the meaning of which I do not understand) or GW150914 (which refers to the date of detection). The two blind injection events had codenames for internal use that were easy to refer to—"the Equinox Event" (referring to the date) and "Big Dog" (referring to the constellation, *Canis Major*, in which the event seemed to be located). The emailer proposes a competition for a codename followed by a vote. Here is an early list of candidates as of Saturday, September 19 (https://wiki.ligo.org/DAC/G184098#Proposed_codenames):

- The masses are LARGE. The signal is STRONG. This event SMASHED into our detectors before we were ready. *You wouldn't like this even when it's angry.* I propose **The Hulk**.

- We **were** and **are** ready.

- I wondered when the naming game would start. From *The Free Dictionary*, **The Big Enchilada**. The most important or powerful person in an enterprise or field; Something of the highest value or importance.

- Gravitational waves + black holes: why not the "**Albert**" (Einstein) event?

- 2015, one century of general relativity. It could be the **CENTENARY** event.

- "**Dawn**"—because this may be our first detection—the true **Dawn of the Age of Gravitational Wave Astronomy**.

- I hate to be the party crasher, but I think GW150914 is easy to remember since the event happened on 2015–09–14 and looks more professional than other picturesque choices. Maybe we should stick with GW150914.

- I agree with Marco. The "boring" designations look more professional. I resist the urge for nicknames, but the best choice if we go that way would be to use eminent physicists. Einstein, Bohr, Fermi, Curie etc. This one I would call Albert and make it the only time we use a first name.

- Hydra's Head. [XXXX] and I looked for the constellation nearest the most likely sky location found by CWB, and it's the 'head' part of the Hydra constellation. The secret meaning is that seeing this now means there are more out there. There may be some who do not believe a single event. But cut off the Hydra's Head, and another grows in its place.

Summoning up my courage, early on Sunday morning I send in my own suggestion:

> **Sun., Sept. 20, 7:53:** If this is real then it will be the start of gravitational wave astronomy. So I suggest "Genesis." You know it is not a blind injection so can risk being portentous.

INDICATIONS OF DOUBT

But I soon sense that I have missed the mood. More and more emailers are saying they just want to stay with something like GW150914.

> **Sun., Sept. 20, 4:12:** This is weird and has a very high risk of embarrassment. We are absolutely not qualified to pick cool sounding names, please let us not try. There are so many ways this can end badly ... This is G184098, GW150914, the ER8 candidate, the discovery candidate, etc. Please keep it matter of fact and/or technically descriptive.

To anticipate, "GW150914" would come to be the official name of the event—the name used in the title of papers and in scientific discussion of its discovery and constitution—but in informal discussions among the scientists it would be known as "The Event." Mostly, that is how we will refer to it from here on, with occasional references to GW150914.

From the debate about the name, it is beginning to look as though I am more ready to think this in the bag than a good proportion of the scientists. And the doubts that I anticipated a day or so back are beginning to surface with disturbing strength. Those who think you can estimate a rate from a single event are worried:

Sun., Sept. 20, 22:55: ... There are many pieces of evidence that point to this being a high-mass CBC, but there are also some oddities.

At the distances we are sensitive to we expect high-mass CBCs to be distributed roughly uniform-in-volume [that means the detectors can see high-mass events to such a distance that space can be treated as uniform], meaning the prior distribution for the distance to a detected CBC increases as the square of the distance [which means you are four times as likely to see an event that is twice as far away and so on, which means the odds are one is going to see distant rather than close-up events, and this means the event is likely to be weak, not strong]. Folding in the selection effects of our network we carve out a subpopulation with the shape of the antenna pattern. So a "typical" BBH of this mass would be much further away with an SNR closer to threshold, in a part of the sky that was above the LIGO detectors at the time. [It would be above the detectors because this is the direction in which the pair of interferometers are most sensitive.] This event has an SNR well above threshold, and is far from the peak of the antenna pattern. [So it is very unlikely.]

Of course detecting such an event isn't impossible, but we should expect to see more BBHs soon. If we don't then this event is frighteningly unlikely, or maybe these oddities are hinting toward a different kind of source distribution. [Either something very unlikely has happened or this is a different kind of event that does not fit the distribution of high-mass CBCs.]

Mon., Sept. 21, 1:44: I agree with [XXXX, YYYY et al.] that one of the best tests of the CBC model will be if we see additional events soon (this all needs to be carefully quantified, but initial estimates suggest definitely within 2–3 weeks of coincident observing time). If we don't, then something is profoundly wrong, and this would have to call into question the detection. But we can burn that bridge when we get to it.

What these emailers are arguing is that, assuming a uniform distribution of such events in the sky, the odds are against seeing this one first because it is unusual in that the first one you see is likely to have been weaker and coming from a different point in the sky and, furthermore, if nothing similar is not seen within a couple of weeks (!) then we should worry that this was not a real event after all; and if nothing is seen by the end of O1, when the detectors have been online for another three months, then something is "frighteningly" and "profoundly" wrong.

I do not find these arguments as compelling as the emailers do: the whole Initial LIGO program was based on the detectors being lucky—there was this small chance of an event going off that is near enough to give rise to a detectable signal. Now luck is being taken out of the picture. There is a scientific rationale for scorning Lady Luck, which is that The Event is at a much greater distance than the kinds of things that were expected to be seen by Initial LIGO and greater distance means more uniformity in the heavens. But even then, I, as a non-astrophysicist, can imagine that these events are uniformly distributed but very rare. Could we not be at a time in the evolution of the universe when, though the heavens are full of black hole binaries, they merge only very rarely, so that even though aLIGO is scanning a far greater volume of the heavens in respect of such mergers, it is pretty unlikely ever to see one—and this is just the kind of lottery-winning luck that iLIGO hoped for? To insist on arguing backward from estimated rates to the non-existence of this event seems to me like arguing that it is going to be impossible to accept that intelligent life on Earth exists until we have evidence of intelligent life elsewhere.

Thus, it seems to me that anticipating this much doubt at this stage indicates that the old, almost pathological, fear of being wrong is still stalking the corridors (I should say, stalking the ether), overruling the joy of finding something. If no more events are seen, the next few months are going to be fascinating!

Another email comes in that presents a modified view:

> **Mon., Sept. 21, 13:48:** I put together a page investigating the implications for rates if GW150914 is a signal. ... If this is real, there's a good chance of many more events in O1, although there's still a ~15% chance that we don't get another one [in O1].

—The fact that this is loud doesn't affect the estimate of the rate at all.

If this emailer is right, then the group can find a rationale for accepting that this is a real event even if they see no other events by the time they have completed all the analyses needed to support a publication. There is a 15 percent chance that nothing else will be seen during that period.

But the doubts resurface a week later:

Sept. 30, 15:51: While of course it is essential that we move as quickly as possible—as we are doing—in understanding questions around the significance, the astrophysics and the instruments, I cannot see the arguments of rushing a publication of the first detection of GWs with data from just a few days around the event and while the run is still in progress.

In his 1984 review of the field of magnetic monopole John Preskill wrote "... as of this writing (early 1984), it is not certain that nobody has ever seen one (monopole). What seems certain is that nobody has ever seen two." [*Annual Review of Nuclear and Particle Science* 34: 461]

We wouldn't like the review of the field of gravitational waves in 2025 to say the same.

The monopole non-discovery that is haunting the collaboration was published in 1982 in the same journal as is the target for this discovery, *Physical Review Letters*. Blas Cabrera, describing what appears to be a thoroughly well-designed experiment, reported the first-ever discovery of a magnetic monopole. But no second monopole has ever been discovered and Cabrera is taken to have been wrong. He is said to have "saved his career" by making his report in modest terms, and that precedent is one of the things that argues for caution in this case (see also *Big Dog*, 233).

Peter writes to me:

Sept. 30, 21:27: Yes, it isn't clear, and we're actually somewhat confused on whether to rush to print or not, whether to insist on very well measured FAR/FAP, whether to insist on a second (or many more) detections at lower SNR, ... We'll muddle through somehow ;-)

To me it would seem like an obsession with certainty to the point of misunderstanding the nature of science if the LIGO community felt it did not have the confidence to publish a report of one event unless they saw a second.

THE EM PARTNERS

We have seen that the community is desperate to keep The Event secret from the outer world and to keep the analysis pure. But neither can be done, and the problem is the "EM partners." The EM partners are the "electromagnetic astronomers"; the "interferometeers" would like confirmation of their findings from the EM partners should they be able to spot light, X-rays, gamma rays, infrared, radio signals, or anything else coming from the point in the sky that is the putative position of the GW150914 sighting. For the EM partners to do their best in this regard, they have to know as soon as possible about any potential event. The flash of light, or X-rays, or whatever might not last long, and if the telescopes are not pointed in the right direction fairly soon after the initial evidence of a gravitational wave pulse, the electromagnetic bird may have flow the coup. So, as explained, that is why there is a low-latency search.

The news about this potential event was sent to the EM partners a couple of days back but without much detail. I think the idea was that it would be sent just as lots of other possibilities are mooted, without any real confidence. This is justified because a signal that is too weak to justify a claim based purely on the strength of the gravitational wave signal might be justified if there is an electromagnetic signal too—this is multimessenger astronomy. On the other hand, the early alerts are often vague, and the nature of the signal will not have been worked out with a great deal of confidence. A burning question is how much information to pass to the EM partners, and when. One can say, "Please have look in this direction and tell us if you see anything," but in the very early hours the direction might be wrong (as it was in this case), and something is far more likely to be seen if the astronomers have some idea what they are looking for. Moreover, telescope time is very expensive and in high demand, so if the community "cries wolf" too often the EM partners are likely to become less cooperative. The cooperation of the astronomers is enormously valued, not only because of what they can see in the heavens, but also because astronomers were notably opposed to the building of LIGO in the first place: the final "O" in LIGO stood for "Observatory," a rallying point for the astronomers who felt that physicists were trying to steal their birthright at a time when the best they could do was fail to detect anything. So the gradual increase in astronomers' interest and support is seen as a kind of victory for the foresight of the physicists, who are, at last, nearing the goal of being able to report the results of proper gravitational-wave astronomy.[15]

All of this leads to heated discussions concerning what the EM partners should be told and how regularly they should be updated as the shape of The Event firms up. In the very early days, this is a matter not just of secrecy but also of blinding: they do not want the EM search vitiated by expectancy effects, something the astronomers would not be able to avoid once they had been told too much about the nature of the hypothesized event. The usual tension arises: they need to be told enough not to miss an event because they are looking for something else, but not so much that the confidence in the purity of the search is reduced. This is the level of information that will not be sent out until a couple of weeks have elapsed:

> **Sept. 28, 15:12:** Analysis of this event is still in progress. The preliminary waveform reconstruction (if it is a real GW event) appears consistent with expectations for a binary black hole coalescence at a distance greater than 100 Mpc.
>
> **Oct. 4, 13:55:** An update is available regarding the gravitational-wave event candidate G184098, identified on 2015–09–14. Analysis of this event is still in progress. The preliminary waveform reconstruction (if it is a real gravitational-wave event) appears consistent with expectations for a binary black hole.

15. See *Gravity's Shadow* for an account of these early battles with astronomers.

Another confidentiality warning comes around:

> Dear colleagues,
> As O1 has started, you may be asked if you have received LIGO triggers or details about them. We remind you that we count on your confidentiality—you can say that you have an agreement to receive LIGO/Virgo GW triggers, and that we are exercising those agreements, but please do not provide any information about the number, times, or properties of triggers you actually receive. If your team includes others who are directly or indirectly involved, be sure they respect this confidentiality too. If you are unsure if some other bit of information is sensitive, please ask the LVC liaisons (Marica, Leo, Peter) for advice before sharing it.

As time goes on the concern shifts from blinding to secrecy, and how much to tell the EM partners will remain a bone of contention throughout the following months. We will look at this problem again in chapters 10 and 13.

ANALYSIS OF THE FIRST WEEK'S EMAILS

Table 2.1 shows the number of emails related to The Event that I receive in the first week. Of course, more emails than this have been flying about—emails between the members of subgroups, emails on lists to which I have not subscribed, and so forth. But this gives some indication of the flurry of activity and how it has grown. The decreasing ratio of those I've stored to those I've deleted gives a sense of the increase in technical detail emails—the scientists getting down to analytical labor—as opposed to discussions of principle. These numbers are not terribly accurate: for example, there was quite a long discussion about what the codename for The Event should be, but that discussion just came and went, forming a bump in the numbers that does not have much to do with science.

Thus ends the first week.

Table 2.1

Number of emails received in the first week (September 2015) of The Event.

	Mon., Sept. 14, from 11:56	Tue.	Wed.	Thurs.	Fri.	Sat.	Sun.	Mon., Sept. 21, up to 11:56	Mon., Sept. 21, all
Stored	39	65	26	50	32	47	17	10	23
Deleted	8	28	49	56	130	144	114	44	129
Daily Total	**47**	**93**	**75**	**106**	**162**	**191**	**131**	**54**	**152**
168-hour total								859	

3 HALF A CENTURY OF GRAVITATIONAL WAVE DETECTION

Efforts to detect gravitational waves with Earth-based detectors began with the work of Joseph Weber at the end of the 1950s. By the late 1960s and early 1970s, Weber was publishing papers that reported that he had detected the waves in the vibrations of aluminum cylinders weighing a ton or so. His cylinders were located at his home university—University of Maryland—with another for comparison located in Chicago. The waves should make their presence felt as simultaneous vibrations in these widely separated detectors. Since such apparatuses are vibrating all the time, his trick was to compare the number of coincidences between the detectors in real time and the number of coincidences when one of the signal traces was offset in time, so that a common cause could not be giving rise to a coincidence. This is one of the ideas that, to this day, makes the detection of gravitational waves possible. The trouble is that we cannot turn gravitational waves off, so we cannot do an experiment where the waves are allowed to influence the apparatus sometimes but we shield the apparatus from the effect at other times. This kind of on-off contrast is vital in many kinds of delicate observation or experiment: if there is a difference between "on" and "off" we know that something special was going on during the "on" period, and that "something" is the phenomenon we are looking for. But if we take the signals from two detectors and compare their output when they are offset in time, so that any effects cannot have a common cause, we are effectively turning the machine to "off"; if we compare the output in real time, when there can be a common cause, we are effectively setting the apparatus to "on." This creates the contrast we are looking for!

Weber's results were so remarkable that other groups built similar detectors, but, to simplify, they found nothing. In fact it was a lot more complicated than that, with at least one other group being puzzled by one putative signal and Weber claiming that certain others' findings would be seen to be compatible with his if they were analyzed in a "glass half-full" rather than a "glass-half-empty" frame of mind. This book is about one discovery covering about half a year, yet it is over four hundred pages long. The Weber controversy went on for a decade or more, and a retrospective summary account compressed into a page or so obviously cannot do it justice.

But this book is not about Joe Weber; it is about what he brought about. The extraordinariness of what has happened cannot be fully grasped without a sense of what went on in the previous half-century. This is the fifth book I have written that bears on the search for gravitational waves. The first, *Changing Order: Replication and Induction in Scientific Practice*, published in 1985, is an analysis of the process of replication in science using early gravitational wave detection claims as one of its studies. The fieldwork for *Changing Order* was mostly completed between 1972 and 1975, and the analysis is quite philosophical.

The next three books are entirely about the science of gravitational waves, each being a description of the science along with sociological reflection and analysis. The first was the 875-page *Gravity's Shadow: The Search for Gravitational Waves*, published in 2004, which goes back to the Weber days and carries the story from the beginning to the early 2000s. Though *Gravity's Shadow* is an account of peacetime activity, it also provides accounts of the destruction of lives and spirits and the extraordinary perseverance needed to make such a project work in the face of widespread skepticism, active opposition, and even ridicule. I once wrote to Jonathan Miller suggesting he create an opera based on *Gravity's Shadow* (he didn't reply), but I still think the creation of this field had enough melodrama and larger-than-life characters to give Puccini a run for his money. The very short *Gravity's Ghost: Scientific Discovery in the Twenty-First Century* was published in 2011, and *Gravity's Ghost and Big Dog: Scientific Discovery and Social Analysis in the Twenty-First Century* was published in 2013. *Gravity's Ghost* recounts the analysis of a "blind injection" known as the Equinox Event, while *Gravity's Ghost and Big Dog*, a paperback, reprints that story along with the analysis of a second blind injection known as Big Dog. Blind injections, recall, are fake signals deliberately injected into the detectors to keep the community on their toes; that they are not real signals is kept secret until the denouement.

I will often refer back to the earlier books for more historical and technical detail or philosophical and sociological analysis. Those who want properly to understand the way those early 1970s events unfolded all the way through to the present day, in the same kind of detail and style as is presented here, can go back to earlier books; here we will just sketch out the main turning points in the history.

By about 1975, Weber's results were no longer believed except by a very few. But Weber had started something that is now a billion-dollar international enterprise; most scientists recognize Weber as the pioneer without whose, on the face of it, crazy attempts to detect the waves, there would be no gravitational wave science today. They were crazy because, given the standard calculations—invented by Weber himself—of the sensitivity of his apparatus and the energy contained in gravitational waves, he had no chance of seeing anything. Yet he decided to try the experiment just to see what happened, and the National Science Foundation funded him with a few tens of thousands of dollars to do it. Given the extraordinary weakness of the phenomenon and the essential impossibility of the experiment, it seems probable that if Weber had not, as many physicists saw it, shamed the physics community by getting it wrong, terrestrial gravitation wave detection would never have got off the ground. What Weber did was force others to find the resources and the will to embark on the long journey needed to get it right, which has led to today's discovery. One vital part of the equation was that when Weber did seem to find evidence for gravitational waves, he was ready to say so in spite of the impossible nature of what he had found, and by the early 1970s he was publishing strong claims. Weber died in 2000 still believing he had seen the waves.

After Weber, people built devices that were similar to his but cooled to liquid helium temperatures or below—the so-called cryogenic bars. The United States (Louisiana State University) and Italy (Rome and Padua) were the most prominent countries in the enterprise, along with Perth, Australia, who were to convert to interferometry in due course. The cryogenic bars were in theory (though Weber sometimes questioned it) much more sensitive than the room-temperature devices. Weber began construction of a cryogenic bar program but he did not complete it. More than once, an Italian group published results indicating that they had seen gravitational waves with their cryogenic bars but they could not gain lasting credibility for their claims. In the meantime, the much more expensive interferometer program, the practical details of which were worked out by Rainer Weiss in the early 1970s, was building momentum. Small interferometer projects were funded, and the construction of the large-scale American devices was finally funded in 1992, spelling the slow demise of the resonant bars, the details of the bitter battle being set out in *Gravity's Shadow*. In the fall of 2015, two cryogenic resonant bars remained online, both in Italy. According to the scientist who ran one of these bars, their range was 5 kiloparsecs and the rate of events that could be expected given this range was a few supernovae or giant flares per 50 years. This scientist believed the bars were doing a good job in terms of "astrowatch" during periods when no other detectors are on air, but expected them to be shut down soon after any confirmed detection. The scientist who ran the other bar believed it was still doing interesting work detecting cosmic rays and other exotic particles that may relate to the question of dark matter. Looking back from the point of view of what the interferometers have now detected, the bars seem completely hopeless, but at that time they represented the moving frontier of optimism.

It is hard to remember that resonant bars were the dominant technology in gravitational wave detection for thirty years. It is now hard to remember the melodramatic nature of the birth of this field of science, but melodramatic it was. The events include the story of Weber, heroic instigator of what was, at the time, the impossible detection of gravitational waves; Weber making certain dreadful mistakes, such as claiming a 24-hour periodicity to the signals he was seeing when it should have been 12 hours because the Earth is transparent to the waves, and claiming to see coincident signals between his own detector and that of a skeptical rival when the two devices were running in different time zones so not in coincidence at all; an unusually violent rebuttal of Weber's findings and experimental methods by one of the most powerful physicists in the United States; Weber inventing a theory that implied that his bars were a thousand million times more sensitive than the initial theory suggested; the *volte face* of a rising theoretical physicist (now dead) who first wrote a paper saying that Weber's theories were impossible and then another saying that he was right after all, losing his own credibility as a result; Weber's credibility eroding but his dogged refusal to abandon his claims to the end of his life, leading him to feel he must reassure me that he had no intention of taking his life; the development of the first rival technology—the cryogenic bars and their making positive claims that were in turn rejected; the effective inventor of interferometry as a practical method being refused funding at the outset while the ideas were developed elsewhere; the violent battle over funding of the interferometry method—it was resolutely opposed by the astronomical community; the heat of an international the battle between technologies—Italy versus the United States this time—being turned to maximum because interferometry was under threat from astronomers, while its demand for funds threatened to kill every other kind of approach to gravitational wave detection, with criticism sometimes verging on accusations of professional misconduct and statistical massage; the painful choices that had to be made by champions of the bar technologies and the choice of an analysis method, leading to the breakdown of friendships, which prevented rivals from making claims unless all were agreed; the eventual funding of interferometry being followed by violent personality clashes between its principal scientists; paroxysmic changes in the management structure of LIGO, with a new manager brought in and rapidly departing; another, more successful leader eventually being sacked; the brilliant inventor of many of the principal features of the new interferometers being separated from the project and told he must never again enter the building containing the prototype device he had developed; LIGO being

CALTECH/MIT PROJECT

FOR A

LASER INTERFEROMETER

GRAVITATIONAL WAVE OBSERVATORY

December 1987

LIGO-M870001-00-M

By comparing the source strengths and benchmark sensitivities in Figure II-2 and in the periodic and stochastic figures A-4b,c (Appendix A), one sees that *(i) There are nonnegligible possibilities for wave detection with the first detector in the LIGO. (ii) Detection is probable at the sensitivity level of the advanced detector. (iii) The first detection is most likely to occur, not in the initial detector in the LIGO but rather in a subsequent one, as the sensitivity and frequency are being pushed downward*.

Figure 3.1
The promise in the initial funding application; stress added by the scientist who gave it to me.

rescued from the brink of death by new directors brought in from high-energy physics; the mass departure of a number of senior LIGO project scientists; and innumerable delays in the completion of the project.[1] These now begin to seem like stories from another world, a world so strange that to try to summarize it here in anything other than telegraphic points would be like trying to provide the *Reader's Digest* version of *The Odyssey*. *Gravity's Shadow* provides the full account at a length of 875 pages, and perhaps one day I'll try to pull it together in a form that requires a less full-blooded commitment, but it cannot be done in a few paragraphs; such a short book would have to be written carefully if we are to avoid parody.

If the generally accepted theories are correct, aLIGO is the first generation of detectors where the theory of what heavenly sources are emitting roughly matches the theory of what can be detected. Indeed, the failure to see anything until now is being retrospectively reconstructed as a kind

1. The story up to the completion of the first generation of large interferometers is told in *Gravity's Shadow*.

of triumph, as seen in a passage from the first research application (see figure 3.1).

But this claim is what we might call "the small print." It conceals a very different history, now being forgotten, of belief in the likelihood of detections. Succeeding ten-yearly reports on astronomy promised detection of gravitational waves within a few years even when we were still in the resonant bar era; sensitivity curves for LIGO were presented to the community that appeared to show sources within the range of initial LIGOs even while the small print intimated that these sources were unlikely to exist; two distantly located detectors were built from the outset, whereas if it was really believed that they were too insensitive to detect a source it would have been possible to do all the development work on only one, as argued by a senior critic of the project;[2] when it was being argued that an "Enhanced LIGO" should be built—a name that now seems to have been forgotten for an apparatus that has been absorbed into Initial LIGO in talk and imagination—part of the justification was that an increase in range of a factor of 2 or so would be bound to lead to a detection. When I pointed out that this was a false argument, as it was based on extrapolation from

2. I argue in *Gravity's Shadow* that while one detector would have been enough to do scientific development work, it was much wiser for various human and economic reasons to build two, but the scientists I argue with invariably tell me "we had to build two or we would not have had coincidences and would not have been able to see anything." The original claim that one detector would be enough was made by Dick Garwin:

> Nevertheless, my view remains that one should go very slowly on actually building a LIGO. One should insist that all of the automatic controls be available and demonstrated at the short length before going to the major cost of construction, and one should build first only a single site rather than build two simultaneously. Of this latter I am absolutely convinced, and it was what made our own experiment so easy—we needed only one (fortunately well behaved) detector to show conclusively that Weber had not seen anything. Of course one will eventually want to build two sites if the first one works, but even if there were no uncertainties and one were deploying a fully developed instrument, one would probably want to do it sequentially rather than simultaneously. I do not believe that the impact on smaller scale and more directly applicable Physics is justified by the benefit of rapid construction of a second site. (Garwin, letter of Jan. 7, 1993, to John Gibbons)

zero, a very senior scientist pushed me up against a wall and shouted in my face that I didn't understand anything. So, whatever it says in the small print, scientists were very disappointed that iLIGO was barren, especially as many (including this author, who lost £100.00) were willing to take on the betting firm Ladbrokes that the discovery would take place before 2010—admittedly at very good odds. Nevertheless, the willingness of the scientists and the funding agencies to press ahead on the basis of the theoretical certainty that there really was something out there to be seen, and that the next generation of detectors should see it, is a great triumph of human perseverance. But history is written by the victors, and whereas not long ago they were at the cutting edge of technology the bars now seem hopeless, and the early generations of interferometers were just prototypes leading to this triumph. That is one way in which the order of things changes.

4 WEEKS 2 AND 3
The Freeze, Rumors

SECRECY

On September 21, the LIGO spokesperson sends around an email explaining how to respond to any awkward questions from outsiders:

> Dear LVC members,
>
> We have now started O1. Welcome to the Advanced Detectors era!
>
> We have informed our astronomer partners too (see message below), and intend to send them information about triggers more significant than a FAR [false alarm rate] of ~1/month. ...
>
> About GW150914, the Burst group has now asked the Spokesperson, Detection Committee and others to consider as a GW candidate, so we are now in Step 1 of the Detection Procedure. We will communicate to LVC members when enough technical details about this event are known, to start Step 2. This may take 2 or more weeks (data to estimate the background is still being collected).
>
> We remind all members about maintaining strict confidentiality about this candidate. If you feel you need to tell somebody for any reason, even to people you trust, please consult Fulvio and Gaby **before** doing so. We again suggest below some possible answers to what may be FAQ; please tell Fulvio and Gaby if you are asked any question about this candidate by non-LVC members, or hear any comments about it from non-LVC members.
>
> Let us know if you have any questions!
>
> Gaby, Fulvio, Dave, Albert and Federico.

— — —

—Have you started taking data?
We began collecting science quality data at the beginning of September, in preparation of the first Observing Run that started on Friday, Sept 18, and are planning on collecting data for about 3 months.

—Have you seen anything in the data yet?
We analyze the data "online" in an effort to provide fast information to astronomers for possible follow up of triggers using a relatively low statistical significance (FAR ~1/month). We have been tuning the details of the communication procedures, and we have not yet automated all the steps that can be, but we will send alerts to astronomers above the threshold agreed as soon as we can after those triggers are identified. Since analysis to validate and candidate in GW data can take months, we will not be able to say anything about results in the data on short time scales. We will share any and all results when ready, though probably not before the end of the observing run.

—We heard you sent a GW trigger to astronomers already—is that true?
During O1, we will send alerts to astronomers above a relatively low significance threshold; we have been practicing communication with astronomers in ER8. We are following this policy with partners who have signed agreement with us and have observational capabilities ready to follow up triggers. Because we cannot validate GW events until we have enough statistics and diagnostics, we have confidentiality agreements about any triggers that hare shared, and we hope all involved abide by those rules.

Thus is triggered the secrecy reflex, or even "fetish," that will characterize the next five months. But there are a thousand people who know what is going on, and probably a thousand wives and partners. There are secretaries making appointments for detection-related meetings and failing to be able to make appointments with outsiders because of the unusual extent to which the time of those they work with is in demand. People's work patterns will have changed.

THE FREEZE

I learn that on September 25 the Detection Committee has decided that the detectors must be held in a steady state until enough background data has been gathered to establish a high statistical significance for The Event. Statistical significance, remember, is established by comparing the signal—which is a coincident excursion of energy in both detectors—with pseudo-coincidences, generated by aligning noise "glitches" in one detector with noise glitches in the other using time slides, in which the output from one detector is slid along the output from the other until glitches appear opposite each other. Each coincident glitch that can be generated this way is compared with the genuine coincidence that constitutes the signal. The number and form of these pseudo-glitches is compared to the signal, and that generates the statistical confidence. But to make sure confidence is high, a lot of this noisy background is needed, and the background has to be comparable to that found at the time of the signal; this means that the detectors have to be kept in roughly the same state as they were at the time of the signal for a period of time—which in this case will turn out to be sixteen days of coincident running generated over about twice that period of calendar time, so the freeze would end on October 20. The freeze means that some important maintenance work has to be delayed. Here is the message from the Detection Committee:

> The Detection Committee recommends that changes to the detectors from the GW150914 configuration be minimized until the following have occurred:
>
> 1. Five days of useable coincidence data is collected in that configuration, as requested by the CBC group. 2. The useable data are calibrated. 3. The CBC off-line analysis is performed. 4. The results are presented at least in a preliminary way to the DAC and DC.
>
> If the CBC results raise unexpected questions, then a further hold might be needed.
>
> This recommendation is made with the expectation that these steps can be completed within 2 weeks from today.

This "freeze" will be discussed again. Also around this time it is decided to extend the observation run, O1, by a month up to January 12; this would provide an opportunity for more events to be seen before shutting down for development work to prepare for O2.

SEPTEMBER RUMORS

It takes eleven days for the rumor dam to start to leak. At around 6:30 a.m. on Saturday, September 26, I open my emails and find one from *Nature*:

> I am a journalist with *Nature*'s news team in London. As you may have heard, there is a rumor that Advanced LIGO has "seen" a signal of gravitational waves. I am putting "seen" in quotation marks because of course even the people who initiated the rumor—assuming they are bona fide collaboration members and not pranksters—have no way of knowing whether it's a blind injection or not at this stage.

I note that it was sent the night before. And the rumor it refers to came from a person of substance in science, well-known theoretical physicist and public intellectual Lawrence Krauss. At 1.39 p.m. on September 25, he had tweeted:

> Rumor of a gravitational wave detection at LIGO detector. Amazing if true. Will post details if it survives.

I am away that day but write back to the journalist the following day after consulting with Gaby Gonzalez, the LIGO spokesperson:

> Thanks for the heads up on the Krauss rumour and if those other scientists have heard things independently of Krauss, anything you can tell me would be useful—as a sociologist I am interested in how this kind of rumour begins and spreads and how the community handles it.
>
> On the other hand, as a guest of the gravitational wave community I never say anything—and that, as American politicians say, includes neither

confirming nor denying rumours—that refers to the substance of the ongoing science. They've done the work and it's their business. So Gabriella Gonzales (cc field) is your person.

But if you want to talk about general principles of blind injections and how long it takes to work out whether a proto-signal can be announced as an event (months and months), I'm happy to do so—it will be the stuff that's in "Gravity's Ghost and Big Dog."

But then I decide to send a quote to the journalist, just in case he can use it.

"The detectors are spider's webs a couple of miles in diameter. The scientists must sense a tiny gnat passing through at the speed of light. The webs tremble in anticipation, shaken by winds, by falling trees, passing trains, waves on distant shores, earth tremors on the far side of the world and anything else you can or cannot think of. So it takes months of calibration and statistical analysis for the spiders to decide that what woke them came not from earth but from heaven."

He does not use it.

One of the most reliable features of Western societies, and it is what we rely on for our freedom, our safety, and our defense against conspiracy theories, is that it is hard to keep a secret. I really do not like the obsession with secrecy that characterizes the community's way of working. I don't think scientists should keep secrets. This is not to say that scientists should be forced to surrender their raw data for anyone to analyze—a view given foolish support from many directions—but that is not the same as saying that, outside of the defense industries, they should keep secret the very fact that they are doing work and the fact that a promising finding has arisen.

Why do they want to keep it secret? First, because they will have to field endless questions from journalists—but surely it is just a matter of diverting them to a press officer. And second, because of the fear that they may appear to have found something only to have hopes dashed, and for the scientists to look, as they see it, foolish. But they look foolish only because they are trying to maintain the image of science as infallible. That game, however, is long over; everyone knows science is endlessly fallible. What we need to establish is that science is the best we've got and one aspect of it being the best we've got is the honesty of the scientists. Scientists cannot "show their work," as is demanded by some philosophers and thinkers about science and democracy, because that work is not really understood by the scientists themselves—they cannot describe the subtle interplay of measurement, argument, and trust that leads them to one conclusion rather than another. But what they can describe is the faltering growth or disappointing collapse of publishable certainty that goes with making a discovery. Just stop pretending that's not what happens. I love the community of gravitational wave physicists, but this obsession with secrecy is not good. We will return to it in chapter 13.

HOW SECRECY AFFECTS RELATIONS WITH ASTRONOMERS

One way in which secrecy hurts the field is in its relationship with its electromagnetic partners—the astronomers. A huge body of emails discusses what the astronomers should be told, a problem made still more difficult by the fact that some astronomers are members of LVC and so know what is going on as a matter of course, and some are not. It is what to tell those who are not members that is the problem. Very rapidly they are asked to turn their telescopes, but what the astronomers do not know is what to look for. This is because for them to make the right judgment about how to expend their efforts would require them to know as much as the gravitational wave scientists, and that is because no one quite knows what it is that has been sighted. All the initial indications are that it is something very heavy, and if it is heavy enough, there will be nothing to see. This is because two black holes merging will suck everything into themselves—that's why they are called black holes. Still, the calibrations have not been completed and the analysis has not been done that would allow this to be pinned down with certainty, so it is still a matter of judgment. And the community does not want to put out information for others to judge. But here is an example of the kind of concern felt by some of the scientists:

Sept. 25, 14:26: First, I'll bet that everyone following the analysis developments [through the community's websites and emails] believes by now that if it IS a real signal, it's almost certainly a BBH merger. I understand the point of view that that has not been formally stated as a finding or "reviewed" by the DAC. ...

That basic picture has implications for EM follow-up observations. It's not a question of being certain (even that the event is real), but of making informed guesses to plan both initial and subsequent observations; quantitative measures and parameter uncertainties can come along in due time. Honestly, I'm pretty sure the BBH hypothesis implies that observers should not bother looking for a counterpart appearing at this point, 11+ days later. But THEY don't currently know that; some might still be ... spending resources unwisely.

So let's tell our partners what we CAN tell them, being clear to avoid rumors or misunderstandings and being frank where there are uncertainties. If we can't give them any better information about sky position yet, fine—let's tell them that, and why, rather than remaining silent. The attitude and tone of our messages are as important as their content. Refusing to share information we DO have because of a fear that they would leak it would go against the spirit of cooperation we've worked hard to create and the framework of trust we established with the MOUs. Anyway, we'll have to share sooner or later. ... We can hold our detailed results private until the endgame, but should keep our EM collaborators in the loop on the general picture and communicate whatever is relevant for their observations.

This is about more than just this event candidate; it's about our relationship with astronomers and their respect for us as we try to establish GW detection as an astronomy resource. ... We can hardly ask them what they've learned from their data if we're not forthcoming with what we're learning from ours.

WEEK 3

Monday, September 27, and my prevailing sense is one of anticlimax. As I remarked above, after only a couple of days there was a new normalcy, and now the normalcy of life is palpable. Yes, a gravitational wave has been seen. Two weeks ago I was joking that I would write book number four if I lived long enough. Now, all being well, I will write book number four and complete my forty-three-year project—and it all seems terribly routine, not something I probably wouldn't complete before I was senile or dead. Back then it seemed just so unlikely that a gravitational wave would ever be seen by LIGO and the other interferometers: we are creatures of inductive tendencies, and having been promised "jam tomorrow" for forty-three years[1] with no more than a fruit-perfumed sniff of nothing, it has been impossible to believe that the jam jar would not be empty once more. Now that we have swallowed that exquisite mouthful of intensely concentrated fruit flavor it's all gone a bit dull. You can only taste something for the first time once.

Big Dog is part of the problem. The excitement of Big Dog extended over the whole six months of its investigation. The smaller reason is that there was going to be a climax when the envelope was finally opened and we found out whether it was real or a blind injection. The more profound reason is that, in the case of Big Dog, whether it would turn out to be an event (or proto-event) that was significant enough to report as a discovery was in doubt right up to the end and, indeed, the subject of dispute right up to the end: was it a "discovery" or "evidence for"? The rehearsal of discovery that comprised Big Dog now makes it obvious that unless something has gone badly wrong, this is going to be real. Now there is nothing to look forward to but three months of scientists doing calculations to make it so. We've just shifted to a new form of life. And it is relatively mundane, because there is nothing unusual about The Event given the hugely increased sensitivity of the detectors: with Initial LIGO, any discovery would have been a cause to question previous assumptions about the brightness of the gravitational wave sky; though GW150914 tells us that black hole binaries have had time to collapse, that's not unexpected, just unknown. If I were an astronomer or astrophysicist it would be different because a new world of professional activity would be opening up; but for

1. By the time you are reading this, of course, it will be more like forty-five years.

me the world of gravitational wave physics might well close down once this book is written. I suppose the hope for a bit of excitement is the question over whether there will be more sightings during O1. If there are none, that could create some nice tension. If there are some, that will be nice for me for a moment, but it will be mainly exciting for the astrophysicists.

The lava flow of retrospective accounting is pouring out and forming a new world with frightening speed. "Of course, we never expected to see gravitational waves with those early detectors; we always knew we would see them once they reached the required sensitivity. And now we have built the sensitive machines and are in the world of gravitational astronomy. All those cryogenic bars had no hope: they could see only one event in thousands of millions of years, and Joe Weber was a thousand times less likely to see anything—not that factors of thousand make any difference when it's already thousands of millions against."

THE RUMOR DIES

Surprisingly, the rumor begun by Lawrence Krauss does not have legs. Here are the first three tweets we find on the morning of September 28:

1. Marco Piani, @Marco_Piani, Sept. 26
   @LKrauss1 amazing if true, but as scientists shouldn't we avoid spreading rumors, especially in a public space, and wait to know the facts?
2. Harry Bateman, @GeoSync2016, Sept. 28
   @Marco_Piani @LKrauss1 Ditto! This is almost surely just one of the calibration signals that they insert electrically to check themselves
3. Lawrence M. Krauss, @LKrauss1, Sept. 28
   @GeoSync2016 @Marco_Piani I reported it as a rumor ... it is ... suggested no more.

Nature's journalist has written his story and somehow gained the impression that there is nothing worth following up because it could all be a blind injection. I'll return to this.

NOT SO ROUTINE AFTER ALL

On September 28, I have a long telephone call with scientist Peter Saulson that indicates that things are a bit less routine when you start to understand the detail of what's going on—the kind of detail that I can't extract from the emails without help. One of the reasons it is not so routine is that it is far less obvious than I thought that the amount of background could be built up by running O1 until whatever level of statistical significance desired was reached. This turns out to be wrong because the instruments have certain known faults, and one group of scientists wants to fix them. Unfortunately, if the faults are fixed then the instruments are no longer "the same" instruments as were in operation at the time of The Event, and therefore the background they generate has no bearing on the background at the time of The Event—this seems to be taken for granted. There is strong pressure by some to take the instrument down and strong pressure by others to keep it running. Here is an email that, once one knows the context, makes the competing pressures plain:

> **Sept. 28, 21:43:** At its meeting last Friday, the Detection Committee discussed the current configuration hold on the detectors. We agreed to the following recommendation, which has been forwarded to the run coordinators, who are responsible for making final decisions about any changes to the detectors.
>
> ***********
>
> The Detection Committee recommends that changes to the detectors from the GW150914 configuration be minimized until the following have occurred:
>
> 1. Five days of useable coincidence data is collected in that configuration, as requested by the CBC group. 2. The useable data are calibrated. 3. The CBC off-line analysis is performed. 4. The results are presented at least in a preliminary way to the DAC and DC.
>
> If the CBC results raise unexpected questions, then a further hold might be needed.
>
> This recommendation is made with the expectation that these steps can be

completed within 2 weeks from today.

****

Pressure from one direction is coming from the fact that The Event occurred during ER8 and they have only a few days of quiet coincident running from which to generate time slides. What is worse, because this is a low-frequency event, and noise is always worse at low frequencies, the background noise could be quite high.

Confounding things further is a continuing argument about how to do the time slides. One group has proposed that slides be 0.2 seconds long and another that they should be a minimum of 3 seconds. Given the same stretch of coincident running, if the first group wins the argument there will be 15 times more time slides than if the second group wins, and that means 15 times more background data supporting a statistical conclusion. The trouble is—as I understand it, but its consequences are disputed by the first group—that very short time slides tend to produce more than one pseudo-event from a single pair of glitches because glitches can be a couple of seconds long; this means that the background data will be inflated, lowering the statistical significance. The second group has done an analysis to prove this, and it certainly fits with what seemed to be the consensus during the earlier blind injections.[2] The first group has new technical arguments that go against this view. I don't understand them, and it is clear that I am not alone. It seems, however, that the first group is winning the battle. The second group has generated graphs that clearly show how apparent background does increase as the time-slide interval shortens, though the increase is not great. But the analysis has been done for higher-frequency glitches than GW150914. As Peter Saulson puts it to me:

> We are going to be tempted to do something that is risky with two-tenths of a second time slides and we're pretty sure this event is going to be found in

2. See *Gravity's Ghost*, 125.

> the mass bin with the poorest background statistics, so it's not going to be smooth sailing. (Peter Saulson, telephone call, Sept. 25)

So that makes it less routine than I thought.

I sense, though no one has said this explicitly, that there is pressure from the other direction because the scientists are what we might call "gravity-wave happy." They are sure they've seen a wave, and they think this means they are going to see others, so that makes them care about this one a bit less. Back in the days of Big Dog, I became convinced it was a blind injection because I was sure the detectors would not have been taken down for the start of the next upgrade when they could have been kept running to provide more background diagnostics, if the director had not known it was a blind injection (I am still inclined to believe this in spite of denials). That was because everyone knew that if Big Dog was real it was a huge stroke of luck and another gravitational wave was not going to come along any time soon. Now, because the sensitivity of the detectors fits with theory, the scientists think other events will come along soon, and this makes them value and nurture this one a little less, being more ready to contemplate the readjustment of the instruments that is the equivalent of shutting them down during the Big Dog incident.

> A lot of people are saying "darn it," we're going to get another event soon if this is any indication and no one will care in two months because the second one will come. Two and even the poorest confidence level that we might end up being stuck with will surely [be good enough]. (Peter Saulson, telephone call, Sept. 25)

I hope they are not living in a fool's paradise but we will soon know.

THE 40 HZ ANOMALY

Another thing that emerges from this phone call with Peter Saulson is that one scientist seems to have found something unusual in the waveform of The Event. It can be seen in figure 4.1, which shows the filtered trace from

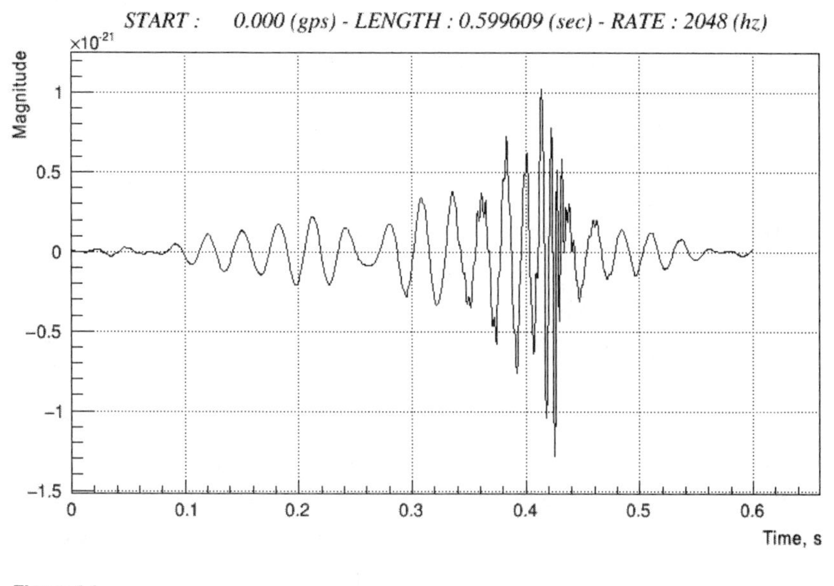

Figure 4.1
The 40 Hz anomaly.

the Hanford interferometer (H1). An almost identical trace from the Louisiana interferometer (L1) is also presented on the website.

Here everything up to 0.42 seconds fits the model of an inspiraling binary black hole system with the "ringdown" that immediately follows merger shown as the dense high-frequency region at around 0.41 seconds. Following this, however, is a decaying low-frequency section that goes on for four or five cycles. Such a thing is quite unexpected and not predicted by any of the astrophysical models. Peter explains that since the person who discovered it is someone who has to be taken seriously, this is going to have to be explained before they can reach the point of publishing a paper. So GW150914 is much more exciting that I thought it was, because not only will it be the first ever "direct" sighting of a black hole, not only will it be the first ever sighting of an inspiraling black hole binary, something that no other technology could detect with all the astrophysical consequences that follow, not only will it be the start of gravitational wave astronomy—all these things have already been relegated to the routine bin—but it could require a complete reformation of our understanding of the dynamics of merging stars—something that is bound to cause controversy.

WHAT ELSE IS GOING ON?

It may all seem routine to me, but the community is a beehive of intense activity. I hear of people whose marriages are being wrecked. I hear that people are working eighteen hours a day. These people would hate to hear what is going on described as routine.

What are they all doing, if this event is so obviously a gravitational wave?

> Are we prepared to accept this?—yeah. That has become no longer the news. So it went from roller-coaster panic and paranoia the first week, to real excitement last week, and this week, yeah, maybe we've just got to buckle down and put one foot in front of the other, do a good job. and be ready to go. (Peter Saulson, telephone call, Sept. 25)

What does putting one foot in front of another consist of?

> We get a signal ... that is proportional in some sense to [strain over time]. But the proportionality is actually a frequency dependent function. [The devices have a greater response at some frequencies than at others so that the initial trace that comes out of the machines, even after gross noises have been filtered out (see *Gravity's Ghost*, 74) does not give an accurate picture in terms of the size of the peaks and troughs of the waveform.] And knowing what it is involves understanding lots of details of the interferometers ... things about the actuators that push on the mirrors, things about the various filters that are used as part of the control system, and it's based on a tremendous amount of measurement and then fitting and seeing if whether the model you have put together gives an adequate representation of the measurements. And then, at the end of the day, that has turned into a time-domain filter [something that standardizes all these different response strengths] where you can run in this recorded [initial strain trace] and produce a new time-series that is accurately within a few percent. (Peter Saulson, telephone call, Sept. 25; square brackets indicate my gloss)

The whitened time series we have looked at (figure 1.1 above) does not show the amplitude at each frequency; it just brings out all the features as clearly as possible.

I suggest that maybe having an accurate trace of the strain is more important to the estimation of parameters than the demonstration of the existence of the signal. But Peter says that is not true, because to demonstrate existence to the level of significance that can be obtained by comparing the signal with the CBC Group's template banks, one must have a signal that is accurate in amplitude at different frequencies.

> The signal-to-noise ratio [SNR] also depends on match with the template and since we are going to use a high SNR to deal with the problem of trying to establish that it's not a chance event from noise, if we throw any of that away it will make our job that much harder. For establishing that this is a signal, calibration is crucial even though [luckily] it was not crucial for the event to be noticed. (Peter Saulson, telephone call, Sept. 25; around 17 minutes into the call)

BUILDING CONSENSUS

But measurement and calibration is not the only thing going on. What takes time is turning people's heads around. Yes, this event has been accepted as a true event in a remarkably short time, but the fear of being wrong is still there, and so is the sense of trepidation associated with putting this before the world:

> How fast can we actually do it [i.e., firm up the detection to a sufficient level of statistical significance]? ... Last year we said if we really rush and we have all our ducks in a row before the run starts we think we can do it in three months. And people are saying well maybe, given how loud this is—dada, dada, dada—maybe we can do it in two because *we don't have to do that much consensus formation—the consensus has already been formed* and, you know, a fair fraction of that is true, but even just doing the bare minimum of steps, let alone taking a deep breath when we confront whatever surprises we have, it's going to take a while. *And we are going*

> *to have to debate all over again, this time for keeps, you know, how do we describe this? What degree of confidence do we either put forward or hold back: people are feeling very shaky about the example of the BICEP2 experiment.* That was an even more amazing Nobel Prize discovery if it was true—and boy they ran with it. (Peter Saulson, telephone call, Sept. 25, stress added)

I hear the term "ghosts" being used to describe the BICEP2 scientists—the team who claimed in 2014 to have found gravitational waves, only for them to turn out to be cosmic dust.

By the way, one normally says "emphasis added" in these circumstances, but "stress" is probably the right word here.

So, of course, this is a social transformation we are looking at—a matter of building consensus and a matter of risk versus adventure with the many examples of failed adventure in the history of science before us: the monopole from a quarter of a century back, and BICEP2, scarcely cold in the ground, whose ghost now shuffles shamefaced in the anteroom of science.

We can see now that there are at least three ways of thinking about the reality of The Event.

(1) What makes the scientists (and me) believe in it—a decisive range of things with the coherence of the signal being crucial
(2) What the scientists present themselves as believing when they want to fend off journalists: "nothing yet"
(3) What the scientists feel they have to do to reach the standards of belief in the wider world of physics—generation of statistics

A REQUEST TO OPEN THE BOX

By the end of week 3, the scientists feel they have done all they need to do to the machine and to their pipelines. They are ready to "open the box" and do the time-slide analysis, and they send out requests to this effect. So far they have gathered only five days of good quality coincident data—periods when both interferometers were in lock and observing and the data quality was acceptable. Five days' worth will not produce the final statistical significance—more noise background would be needed—but this will be enough to know whether The Event is a discovery as far as the collaboration is concerned and indicate whether it could eventually turn out to be acceptable as a discovery to the wider community.

> **Oct. 3, 17:36:** On Friday, DAC requested that the CBC group consider "opening the box" on Monday in their "BNS/BBH/NSBH/Waveform call" at 8am PDT in TeamSpeak. Opening boxes should be done only *if* and *when* the group considers all checks are done on the closed box results. Those results are very recent, so it is possible that questions may arise and need time to sort out—I trust there are many eyes looking at these results already.
>
> I ask that we all respect the CBC experts in their opinion on readiness for opening boxes. ... We are not blind, and we are all very anxious to see the results of what we expect are the most sensitive analyses to the candidate seen by cWB—but we need to make sure we do things *right*—that is of paramount importance for what will likely be the most important paper we write in our academic lives.

The box would be opened in week 4.

5 WEEK 4
The Box Is Opened

It is now Monday, October 5, 2015, twenty-one days after The Event; it is the beginning of the fourth week, and "the box is opened." What does this mean? Up to now the scientists have been working eighteen-hour days, feverishly calibrating the apparatus and refining their statistical analysis procedures. The calibration changes the signal from a coherent trace to a coherent trace the magnitude of each part of which faithfully inscribes the strength of each moment of the wave, undistorted by the characteristics of the detector—the way it is more sensitive in its response at "this frequency rather than that." This means that the signal can be matched accurately against the huge bank of templates—calculated waveforms corresponding to the inspirals and mergers of pairs of stars of varying masses at various orientations. The statistical analysis procedures have been refined to make them as sensitive as possible to these waveforms and as insensitive as possible to the various noises—the glitches—that are ever present in apparatuses as sensitive as this. Though the procedure cannot be made perfect, the scientists have done everything they can not to be biased in their tuning and statistical refining by what they know about The Event: they must not "tune to the signal"—that is, make their analysis adjustments by reference to the best way to enhance the signal they want to see. Tuning to a signal renders all statistical analysis flawed, and everyone nowadays believes that, for example, Joe Weber's original claims were generated by tuning to a phantom signal. Thus the tuning and refining has been done without looking at The Event and looking only at a subset of noise against which the right procedures for these brand new interferometers in particular can be refined. Only "after" it has been agreed that everything has been checked and every refinement made will the box containing all the data be opened and the new procedures used for generating as many time slides as possible from the whole week's worth of data that includes The Event. Incidentally, the argument between 0.2-second and 3-second time slides seems to have dissipated, with 0.2 seconds being accepted as good enough.

I put "after" in scare quotes because what has to be done takes time. What has to be done is run as many time slides as can be run on the week of data. Readers of *Big Dog* will remember that it took weeks to generate all the time slides necessary to bring the statistical significance of that "event" up to near discovery status. This time things are easier, first because the programs have been improved, second because computing resources have been markedly increased, and third because there were three-months' worth of Big Dog data whereas here there is only one week. For this reason running the time slides is a much quicker operation—it takes a day or two. It is, nevertheless, a huge amount of computing—each "glitch" generated by one interferometer is matched with around 250,000 templates and with the glitches generated by the other interferometer. If there is a match with a template and a match with the other interferometer, then there is an alarm. If the match has occurred in real time, there is delight—we have detected a potential signal; if the match has occurred between the interferometers when one of the signals is offset in time, then there is misery, because this shows that such matches can arise from noise alone and it weakens the statistics of matches in real time—probably fatally.

In this case, the job of analysis has already been completed before the box is opened; it is just that the outcome is sealed from view. When, on October 5, the scientists finally agree that everything that needs to be done before the box is opened has been done, what will happen is that they will "break the seal" on some files and look at the results. If they don't agree that enough has been done then the seal will not be broken until more work has been completed, and the time-slide analysis would have to be run all over again under the new conditions.

October 5 comes to pass and the team, dispersed across the world, assembles to hold a teleconference, which starts at 8:00 a.m. in California, 11:00 a.m. in New York and Florida, 4:00 p.m. in the UK, and 5:00 p.m. in Germany, France, and Italy, though participants are far more widely spread than that. It will last for just under two hours and be attended by more than 150 people or groups—one never knows how many people are present at each node that is visibly contributing to a telecon.[1] Listening to the voices, the whole thing is very low key. Most of the two hours is spent polling each person or group in turn who have been involved in refining the search and asking if they are satisfied that the box can now be opened. Once opened, further refinement would be banned in case it should be inadvertently biased by the scientists' knowledge of the signal.

About ten minutes before the end of the teleconference the last group agrees that the box can be opened. The seals are broken and the signal turns out to be loud and clear, statistically unsullied by noise coincidences in the time slides. There was a possibility that further real-time signals would have been found lurking in the data, which would have given rise to great delight, but this has turned out not to be the case. That aside, the result is the best that could have been hoped for. Nevertheless, the voices remain low key—as though nothing out of the ordinary is happening. When I first hear about this, I intend to write something about how careful physicists are not to get carried away in such a way as to make it more difficult to concentrate and continue to unearth possible mistakes; but in fact a celebration has been going on under the surface of the spoken word. As Peter Saulson would tell me:

> **Oct. 6, 21:00:** There were definitely cheers. The thing that you have to understand about TeamSpeak [the telecon software in use] is that almost all

1. It was 12:30 p.m. in Port Rexton, Newfoundland, where I was, to my regret, without the means to take part in the teleconference. I learned of the results immediately, however, and the next day was able to listen to an audio recording as well as access the transcript of textual contributions.

microphones are muted almost all of the time. But I can assure you that the group of 15 people here at SU were cheering and laughing in amazement at the wonderfulness of it all. Then we went right out for beer.

I didn't see the textual transcript until later, and that, as one can see, is indeed celebratory. Here is an edited section in Central European time with identities anonymized.

<18:37:38> "[AAAA]": Good luck everyone!!!!

<18:39:17> "[BBBB]": I see only one significant event in the Hanover set, is that right?

<18:39:24> "[CCCC]": yes [BBBB]

<18:39:28> "[DDDD]": That's what I see

<18:39:32> "[EEEE]": Yes

<18:39:42> "[FFFF]": champagne!

<18:40:01> "[EEEE]": [FFFF], i need something stronger

<18:40:04> "[GGGG]": Yes, all events are the same.

<18:40:05> "[BBBB]": So, no weaker events.

<18:40:10> "[HHHH]": And just one

<18:40:17> "[EEEE]": sordid?

<18:41:35> "[AAAA]": We are opening the bottle here in IIII's flat!

<18:48:28> "[DDDD]": Remarkable!

<18:48:32> "[JJJJ]": yes

Figure 5.1, from which I have removed some confusing features, shows what the scientists were looking at. What is seen is the result of 11,974 time slides of the week's worth of data from the two detectors with "events" comprising coincident matches with one of the other thousands of templates. All the gray lines join matched events generated by noise; we

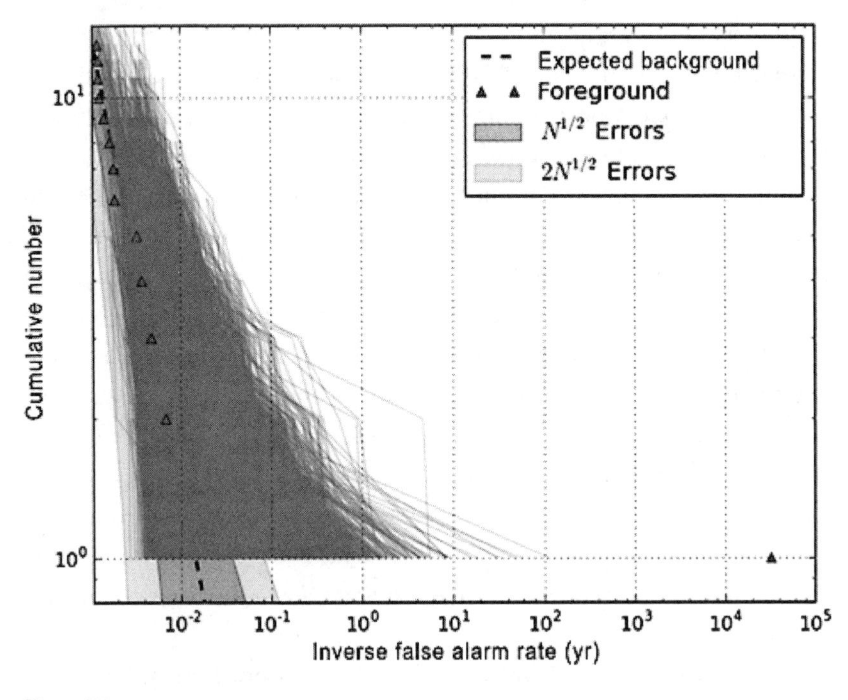

Figure 5.1
What was seen when the box was opened.

know they must be generated by noise because they are not coincident in time. The triangles are "events" where there are matches in real time, and so each triangle is a potential genuine gravitational wave or just a noise coincidence in real time. Small noise excursions are always more frequent than large noise excursions, so there are many small noise excursions and fewer large ones; there is nothing to tell us whether they are gravitational waves or noise coincidences but one has to assume they are noise. The horizontal scale sets each event in the context of the noise, showing how often a coincident excursion of such a size would occur by chance. The whole pattern slopes down to the right because, as explained, the larger events occur less frequently and will show up by chance less frequently. As can be seen, though there is a slope, nearly all the coincidences that are not really coincidences (because the signal from one detector has been delayed in relationship to the other) are pretty frequent, being expected to appear once every ten years or more; there are just one or two that might be expected to occur only once in 100 years. Likewise with all the triangular coincidences except one: the signals or noise-coincidences they represent are small and might be expected to occur ten times a year or more, so even though they are coincidences in real time they, to repeat, cannot be read as anything but noise coincidences. But the single triangle shown at the bottom right-hand corner is GW150914—"The Event" is not small. Such a coincidence would be expected to occur by chance less than once in 10,000 years. The odds against it occurring by chance—the odds against it not having been caused by an external disturbance such as a gravitational wave—are therefore huge, and it is those huge odds that make The Event *a discovery*. That is why the champagne is flowing now, three weeks after The Event was first sighted. It might have been intuitively obvious from looking at those first overlaid plots that this was the real thing, but only now does the statistical analysis show that it is what the *wider world of physicists* will be willing to call a discovery.

Notice that this has involved luck. It has depended on the detectors being in a relatively stable state for a week of ER8 so that enough time slides could be generated to give this statistical backup. If the detector had been less stable with more data quality flags posted, there might not have been enough stable background to generate enough time slides to give rise to solid confirming statistics, even though GW150914 would still have looked like a real event.

So we can see that this "discovery" works on a kind of ratchet principle. First we saw an unusually big event, then we made sure it was not a deliberately inserted test coincidence, then we saw the coherence between the two detectors, then it was made to seem unlikely that it was a malicious hoax, and now it is shown to be something of statistical significance strong enough to be presented proudly to the outside world. There is no single "aha" moment—it is more like "A—ha—ha—ha—ha—HA!"

The last "HA" is an especially big one.

Oct. 6, 13:22

Dear Harry,
You might be interested that yesterday's box opening is bringing out the sentimental side in lots of people. I immediately emailed Rai [Weiss] with congratulations. Last night I got a phone call of congratulations from my former student [XXXX] (... recently "retired" chair of [ZZZZ]) and then a congratulatory email from former SU grad student [YYYY], now postdoc at [PPPP]. People are really feeling this as a life-changing moment.

Peter

SOCIOLOGY VERSUS PHYSICS

On October 13, I write a little note to myself. All the physicists want another event to show up so they can be reassured that The Event is something real, not an artifact, and so that they know that gravitational wave astronomy has really begun. But, guiltily, I note that I would prefer it if another event did not show up. That is because sociology is easier when the people being observed are under strain; it will be easier to see the bones of the detection if there is only one. If there are two or more, everyone will be relaxed, and they will talk about their uncertainties less. I feel a bit mean about this, and the other side of me hopes there will be more detections for the sake of my friends and colleagues.

What we are seeing here is what Peter Berger, in his book *Invitation to Sociology*, called "alternation." The sociologist must be able to alternate between the worldview of those being studied and the worldview of the distanced or "estranged" analyst living in a different "taken-for-granted reality."

THE ANOMALOUS TAIL OF THE SIGNAL AND TACIT KNOWLEDGE

Figure 5.2 shows the signal with an anomalous tail—a clear 40 Hz ringing that should not be there according to the templates. But it has now gone away, its highly credible champion agreeing that it is an artifact—some ringing in the interferometer that can be discounted. I learn this from Peter via email on October 13: "BTW, I spoke with [YYYY] yesterday about other matters, and he confirmed that he no longer believes that there is any genuine 40 Hz feature in the signal." The very credibility of the signal's champion means that if he has agreed it is to be discounted then it is to be discounted.

But now another not dissimilar anomaly has shown up. This time it is a 30 Hz ringing at a point in the signal when there should be nothing, according to our understanding of the black hole merger process.

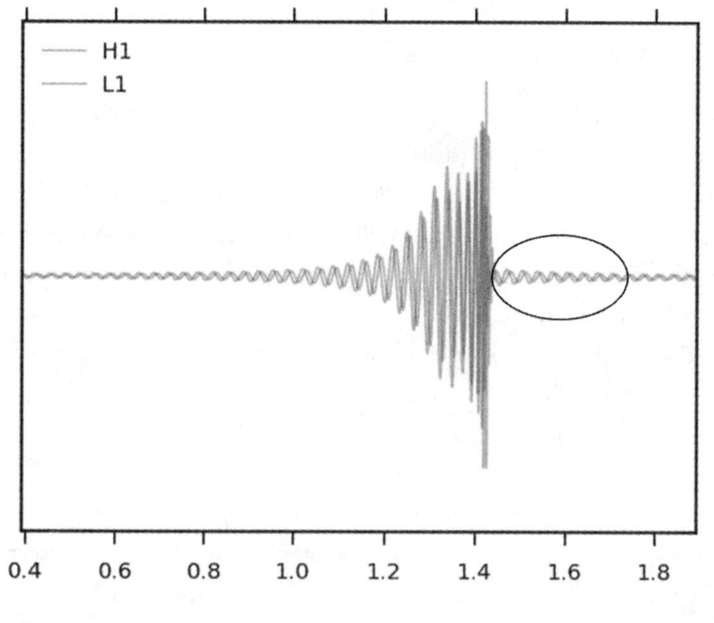

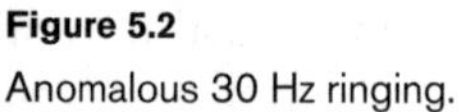

Figure 5.2
Anomalous 30 Hz ringing.

We can see the anomaly in figure 5.2 within the oval outline. Immediately to the left of the oval is the dense ringdown, lasting only a fraction of a second but containing the information some feel is needed to nail this down as a black hole merger—the signature of the "QNMs"—the "quasi-normal modes" that characterize the momentary "ringing" of the merged black hole. After that, all we should have is a new and larger black hole sitting immutably in the heavens; but what we can see here is a further ringing. (Incidentally, by the time the whole game is played out it will have been agreed that the signal-to-noise ratio is too low to allow the ringdown to be actually seen in the waveform, and that all this talk of the bit of the waveform that marks the ringdown is wishful thinking.)

Let me say I would never have noticed that 30 Hertz tail had someone not asked a question about it in the emails. By pure chance, the person who asks the question is Peter Saulson:

> **Oct. 12, 19:21**
>
> Dear [XXXX, YYYY, ZZZZ, et al.],
> Could I ask a question about physical interpretation/heuristics of the waveforms that we're looking at? ... We see a chirp waveform sweeping up in amplitude and then rapidly in frequency. ... At peak amplitude we see the highest frequency oscillations, which then damp very rapidly—am I right that this is the quasi-normal mode oscillation?
>
> Am I on the right track so far?
>
> Now, here's the question where I know that I'm "at sea": What is the lower amplitude, lower frequency, lightly damped oscillation that occupies the last 0.5 sec of the time series graph?
>
> Thanks in advance for your help.

In response to this Peter receives a number of replies explaining the source of this anomaly. For example:

> **Oct. 12:** Isn't the waveform plotted inverse FT of the frequency domain ROM? That would explain the oscillations following the ring down that Peter is talking about, presumably caused by imperfect windowing in the Fourier domain.

Here is his reply to one of the explanations:

Oct. 12, 22:14

Dear [AAAA],

Thanks for offering an explanation. So, do I understand you correctly that the feature I'm most confused about is not genuine?

If so, is there any good way to get a more faithful time-domain representation of the waveform? Or do people just know not to pay attention to that part of the h(t) time series, since it is obviously bogus?

While we are at it: Is this the only feature that is dominated by problems with the Fourier transform back into the time domain?

Since Peter and I have fallen into conversation about this event, he also offers a fuller explanation to me:

Oct. 13, 1:43

Dear Harry,
So, my innocent question provoked a nice set of explanations.

To make sure it is clear: the short answer about that post-waveform ringing is that it comes from having rather abruptly truncated the Fourier transform of the waveform (used in the frequency domain calculation of the cross-correlations between data and templates) at 30 Hz, the low frequency cut-off of the search. When you do that, and then inverse-FFT back to the time domain, your time domain waveform acquires spurious features that look like sinusoids at the cut-off frequency, which was what we saw: that post-waveform ringing was at about 30 Hz.

Perhaps something like this was at work in cWB's spurious feature in its waveform reconstruction?

Best regards,
Peter

For once, Peter has fed me more than I can digest. My understanding of this field has always been a matter of principles rather than techniques.

But I can get a sense of what is going on. It seems that in order to conduct the template matching searches the signal is subjected first to a "Fourier transform." A Fourier transform is a mathematical trick to decompose a complicated waveform into a set of simpler waveforms that are much more tractable in respect of mathematical analysis. It is the numbers that describe this nested set that are used to look for correlations. When something of interest is found, it can be transformed back into the ordinary "time domain"—the way the signal unfolds over time. But, if this is done in an abbreviated way—as it has been here—it can result in spurious elements. The 30 Hz tail is such a spurious element.

The interesting thing, and the reason for spending so long on a spurious element, is what it tells us about the community. Peter, who is just far enough outside of the specialist group of analysts to have trouble completely reconstructing what has happened, has noticed the problem. The specialists, as with me, have not even noticed it. But the reasons are different. I had not noticed it because I don't know enough for my perception to be sharp enough. The specialists have not noticed it because their perception is too sharp: they know so well what is going on that they pay no attention to it—it is just a typical result of spurious Fourier transforms that sits within every specialist's tacit knowledge. So here we have three groups with three different levels of expertise seeing (or not seeing) the same thing in different ways: me seeing nothing because of my lack of expertise; the experts seeing nothing because, as we might say, "familiarity breeds contempt"; and Peter seeing a problem because he sits somewhere in between (though, of course, much closer to the experts than me).

What Peter is concerned about is what the community is going to show to the outside world, whose understanding is not going to be as sophisticated as that of specialists and, if they have a bit of astrophysical knowledge, are going to wonder what crazy stuff the scientists are claiming to discover. He asks them to redo an analysis with a lower-frequency cutoff so that the community can reveal to the world a waveform that is more "physical."

Oct. 13, 1:32

Dear [XXXX, YYYY, ZZZZ, et al.],
Separate from tutoring me (I was halfway there, but clearly needed help), we ought to think about the best way to explain this to the various audiences that we'll be speaking to when we present our detection claim. One language/representation might be best suited for a PE paper (for example), another for an FAQ page for our scientist colleagues, and perhaps even a 3rd for the general public/journalists. Clearly, the inverse FFT of our frequency domain template (and/or the whitened time series) might be valuable to the more sophisticated readers of a PE paper.

But to explain what we've actually discovered to the general public, we probably want to have the most precise time-domain representation, even if that isn't precisely what we used in the search. And for there, I'd suggest extending the time series to substantially earlier than 30 Hz, since that is what we think really happened, with some indication that we only use the last few cycles for the search because that's where our noise lets us look. And please: in those, be sure that things that are unphysical are also invisible.

The crucial lesson is that one can never understand scientific data without the interpretative help of the scientific community.

6 WEEK 5 TO THE END OF OCTOBER
Directness, Black Holes

Oct. 16, 00:37

Dear all,

Please see the good news in the attached document, also in LIGO-L1500147-v1. We copy the conclusion here:

Based on the documentation presented, we decide the detection case for GW150914 is strong enough to proceed with Step 2 of the Detection Procedure.

The "Detection Procedure," a rather long document, can be found in appendix 1. It will be referred to from time to time as the discovery unfolds. The interesting thing to note about it is that it describes the unfolding of a *social* procedure—the discovery moves from one group or committee to another, each making their judgment and passing it on. As will be seen, the procedure has four steps, the last being a joint meeting of the collaboration tasked with deciding whether the final draft of the discovery paper, which will have been written by that time, is ready for submission to the journal of choice.

The Detection Procedure is important because it will guide the discovery process. Its institutionalization means that there can be no short cuts, which will turn out to be significant because, deep into the discovery process, the leadership will be able to state that no discovery has yet been made because the scientists have not yet passed through all its steps.

> We look forward to the consolidation of the detection case for GW150914 as we make progress in the Detection Procedure, but remind all collaborators that there is a significant amount of work involved in the review of this candidate before we can consider making a final Collaboration decision on this candidate.

Discovery is here being defined as the completion of the Detection Procedure. The Detection Procedure, then, not only is a social process but also provides a useful bureaucratic definition of discovery to head off those who think it has already happened.

A much shorter text than the Discovery Procedure is this timeline, which was projected at the beginning of October:

> **This is just a possible roadmap**, will be adjusted as the detection procedure progresses and we have more information from the analyses groups. The dates are not deadlines, each step should take as long as it needs to. There is an implicit assumption that the event will be significant enough to guarantee a paper and a press release, that we will use a fraction of (not all of O1) data in the paper, and that there will be no other significant events. We should revisit this plan whenever any of those assumptions becomes invalid.
>
> Event was on 9/14, number weeks from then and use ~1-week time scale:
>
> * Week 3 [Oct. 5–11]: A detection case (and other deliverables from Step 1) is presented to spokespersons. It proposes a paper with calibrated ER8 and X calendar weeks of O1 included (I will assume X=4, so this is ~ Sep 12–Oct 16). Spokespersons finish discovery plan, interact with EPO for press release and FAQ. Detection committee reviews checklists.
> * Week 4 [Oct. 12–18]: Spokespersons (consulting with appropriate people) agree with move ahead with Step 2, appoint a paper Coordinating team and announce a LVC-wide telecon for following week. DA chairs and review committees estimate how long obtaining and reviewing results will take (Assume it is ~4 weeks from here, ~Week 8). Detection committee asks questions, reviews checklists.
> * Week 5 [Oct. 19–25]: An LVC-telecon is held with latest information. Detection committee asks questions, paper coordinating team begins to

prepare skeleton paper and assignments.
* Week 6–8 [Oct. 26–Nov. 15]: DA groups, review committees and EPO finish their job, paper coordinating team circulate two drafts (~skeleton beginning of week 6 and ~a draft end of week 7), detection committee asks questions, reviews the information and the paper.
* Week 9 [Nov 16–22]: Outcomes of Step 2 are gathered in a meeting convened by LSC/Virgo leadership. Assume a favorable decision to move to Step 3, and Detection Committee is charged to review the detection case. Another paper draft is circulated. PRL is contacted about preparing a submission and accelerated review.
* Weeks 10–11 [Nov. 23–Dec. 6, including Thanksgiving week]: Detection Committee review the case and the paper. DA groups begin drafting other collaboration companion papers that may appear soon afterwards. Review committees review results for those papers. DA groups and reviewers answer questions from Detection Committee.
* Week 12 [Dec. 7–13]: Detection Committee present evaluation (assume it's favorable), final paper draft is circulated. An LVC-wide meeting is held to discuss and vote on the submission of the paper.
* Week 13–14 [Dec. 14–27]: Paper is submitted for review, deals with referee reports. Press releases are prepared.
* Week 15 [Dec. 28–Jan. 3]: Material is prepared for press conference.
* Week 16 [Jan. 4–10]: Press conference is held.

As we will see, the roadmap will slip by a month.

IS THIS THE "DIRECT" OBSERVATION OR EVEN THE OBSERVATION OF BLACK HOLES?

Now something happens that I did not anticipate. For decades it has been promised that more and more sensitive versions of LIGO or LIGO-like devices would surely detect gravitational waves. On the one hand, we've had the increasing reach of the detectors, and on the other hand, we've had the increasingly sophisticated understanding of heavenly objects. In 2005, a theoretical problem that had proved so difficult that the search for a solution was known as "The Grand Challenge" was solved. This problem was to model the inspiraling of binary black holes. For years and years, however the problem was approached, it would not crack: exact solutions would not work, and numerical models built on computers encountered "infinities" and crashed. In 2005 a scientist from Canada showed how to write a code that would work.[1] Solutions could take months to calculate given the computing resources available, but it was now possible, in principle, to build a bank of models for black hole inspirals against which interferometer signals could be matched. This made it easier to undertake the search that, in 2004, the famous astrophysicist claimed would extend LIGO's range to the edge of the universe (see p. 36) and allowed more information to be extracted from the signals.[2] As explained above, it seemed to me that The Event had resolved the problem that had not been mentioned by the famous astrophysicist: Was the lifetime of the universe long enough to allow for enough dissipation of the energy of binary black holes so they could reach the point of inspiral? Suddenly, or so I believed, my wife and I, and a thousand or so others, knew that it *had* been long enough—and we knew the famous astrophysicist's claim about LIGO being able to see to the edge of the universe was true, even if he couldn't be sure it was true at the time he said it. As one of the astrophysically oriented members of community would put it in an email sent a couple months later:

> **Nov. 4, 15:59:** The source clearly demonstrates that binary black holes can form in nature, and in addition that they can form with physical properties that lead to their merger within a Hubble time.

But what I do not anticipate is that the community, or at least some members of it, would decide that it is not so certain after all that this is a binary black hole.

1. Frans Pretorius, "Evolution of Binary Black Hole Spacetimes," *Physical Review Letters* 95 (2005): 121101.

2. That black holes would be the first objects detected by interferometers had been predicted at least as early as 2000.

To me it seems crazy: the binary black hole mergers had been put forward as the most promising source of gravitational wave detections because they were so energetic compared to the inspiraling of lighter objects such as neutron stars. The promise that it would happen had put reputations on the line. A huge effort had been put into solving the difficult technical problem of making a terrestrial detector sensitive at low frequencies so that these highly energetic but relatively sluggish events could stand out above the noise, and all this had borne fruit. So, we have the templates thanks to the solution of the Grand Challenge, we have the sensitivity, and, triumphantly, we have an observation that fitted a template and stood out miles above the noise fall into our laps even before we had really started to observe properly. Yet some physicists still aren't sure that this is really a binary black hole!

The argument, which begins around October 16, starts with the question of whether the black hole claim should be made in the first "discovery" paper or merits a separate paper on its own. The argument is that to prove it was a binary black hole would require a lot of refined calculation and this would take too much space to be included in something like a *Physical Review Letters* paper, which, in the normal way, would be limited to four pages. In any case, it is said that the black hole claim deserves a paper on its own.[3] This argument then transmutes into being more about whether the evidence actually proves that this is a binary black hole. One scientist, heavily engaged in modeling such things, claims it does:

> **Oct. 16, 20:20:** I do think that numerical relativity establishes that what we see is a BBH, and then forms a final single BH. We are observing the latest stage of the inspiral, merger and ringdown. When we superpose a purely NR [numerical relativity] waveform to the signal with a single set of parameters, it represents a unique "finger-print" of the BBHs.

3. I am not sure to what extent this argument reflected the desire of the theoreticians to stake their own claim separately from the experimentalists. There was certainly a lot of discussion about LVC scientists wanting to get astrophysics papers out before non-LVC astrophysicists grabbed the result and started interpreting it. This will come up again.

I do not know of any other solution to GR that would have this specific waveform.

The next day, however, an email from another scientist casts some doubt on that view:

> **Oct. 17, 6:56:** While I could say this is almost certainly a BBH merger... I am not so sure that the evidence that black holes exist is too much stronger than existing evidence. There is a ringdown to be sure, but it is not clear a boson star or whatever would not also produce such a ringdown. I think we would not claim "proof" of a horizon until the ratio of at least two QNMs were shown....
>
> So my sense is that this comes down to the differentiation between "this is consistent with a BBH system" and "this proves that BHs exist."

QNM, recall, stands for "quasi-normal mode" and refers to what happens in the fraction of a second after the two bodies have merged. Think of the merged object as a huge spherical blob of very stiff jelly that quivers in one direction then another before it settles for eternity. This quivering has certain characteristics that pertain to black hole jellies alone so, it is being claimed; it is the quivering alone—the QNMs—that can prove that the object at the end of the inspiral is a new, single, black hole.

One can see the logic here, and I have already discussed the matter with Peter under a different heading. Earlier in the book (p. 37) I talked about how exciting an event this was and included among the reasons for the excitement that "it [will] be the first ever 'direct' sighting of a black-hole." But this was hasty, since what is meant by "direct sighting" is heavily disputed. The argument and its history is set out in *Big Dog*, 197–199, with me being surprised at how reluctant the scientists are to claim a first "direct" detection. Ten days earlier I had opened the topic with Peter, asking him if he is going to count The Event as a direct detection so long as all goes well. He had replied:

Oct. 5, 20:49

Dear Harry,
Yes, I'll be prepared to use the "d" word ;-) The reason is this (if it stays true): It seems possible that our Parameter Estimation team will be able to show that the waveform contains a section at the end that corresponds to the ringdown of the merged black hole. My standard for DIRECTLY proving that we've seen a black hole is to have seen its event horizon vibrate. So, if I heard correctly and it remains true, we'll have done something truly amazing.

Even without that, it now seems virtually certain that we'll have demonstrated that we were looking at a binary system of two black holes. That weaker claim will be based just on the inferred masses and on the agreement of the signal with the calculated waveform. It will only be wonderful, not (wonderful)^2 ;-) [^2 means "squared"]

So here we see Peter saying that "direct" will apply only if we see the horizon of the merged object vibrating even though just fitting the waveform is good enough for saying one has seen black holes; whereas the earlier respondent considers that only if we see the horizon vibrating according to the model can we say we have really seen black holes.

I feel still less stupid when a very senior analyst writes as follows. I will include most of the email, as it nicely covers the full range of technicalities:

Oct. 17, 20:47: My question was referring to your statement... that having detected a GW from a BBH doesn't imply that we have also proved that BHs exist. You seemed to say that we will prove the existence of BHs only (or mainly) from the merger remnant (i.e., the final, single BH) through QNMs [modeled vibrations of the merged object] and I don't understand why. I am raising this point because other people seem to share your opinion.

By showing (as many people are already doing in PE [parameter estimation: working out the masses of the objects and their distance etc.] studies)... that the signal we have detected is consistent with a binary black hole as predicted by numerical-relativity simulations in GR [often shortened to "NR," these are computer models of the inspiral and merger that initially took months to run on computers], we will prove, within the precision of our

> instrument, that black holes exist. This comes from the fact that it is "unique" the gravitational waveform emitted by two black holes, and this uniqueness is there during the inspiral, the plunge, the merger and the ringdown, i.e. throughout the coalescence process. …
>
> If the two bodies weren't black holes, the phasing, frequency and amplitude evolutions during the inspiral, the merger, the plunge and the ringdown would be different. To continue focusing only (or mainly) on the QNMs of the remnant to prove the existence of BHs is, in my opinion, an old way of looking at this problem, which doesn't take into account the fact that we have solved the binary black hole problem in 2005 and that we spent the last 10 years working to understand the signal throughout the inspiral, the plunge, the merger and the ringdown. You can of course do extra studies for the QNM stage only, but, in my opinion, the proof that we have discovered black holes can come already from having discovered a signal that is consistent with a gravitational wave emitted by a BBH.

All current theories say that anything compact and more than 3 times as massive as our Sun must collapse into a singularity—a black hole. Estimates of the parameters of The Event show we are dealing with objects around 30 times the mass of our Sun. So, one might think, we must be observing black holes. But not if the theories are wrong! Not if there can be, say, boson stars that are more than 3 times the mass of our Sun, even 30 times the mass of our Sun, but do not collapse. No one has any evidence of the existence of boson stars, but that does not stop them being invented. After all, black holes were invented before anyone had any evidence of their existence, and gravitational wave experiments were begun before we had any evidence of their existence.

Boson stars are made of things like photons and gravitons, which aren't even as substantial as the components of subatomic particles. Neutron stars weigh as much as two or three of our suns but are only 10 kilometers in diameter. A lump of matter a centimeter high on the surface of a neutron star would take as much effort to climb as Everest here on Earth. But boson stars make neutron stars seem like dandelion seeds. The trouble is that, as far as I can see, the astrophysicists can carry on inventing new stuff forever—we already have multiple universes, wormholes, and the anthropic principle—and that means you can never prove anything. Maybe even the QNMs are consistent with something else just waiting to be invented.[4]

So, I begin an email exchange with Peter with the subject line "tearing out of hair." This is my hair that is being torn out because I am fully in participant mode and am finding it difficult to see "my" black holes being wrenched from my grasp by an argument that turns on science being mistaken for something logic-like. In the course of this debate, Peter explains what we already know about the existence of black holes:

> **Oct. 18, 13:07:**... There are degrees of discovery for black holes. We've had multiple discoveries: in X-ray binaries (Cygnus X-1) there was a star with around 10 solar masses that didn't emit light: a black hole. In the cores of galaxies (esp., for example, M87) things with millions to billions of solar masses, packed into very small volumes, that didn't emit light: supermassive black holes. The coolest thing so far was the inference drawn by Ramesh Narayan about two sub-classes of X-ray binaries: one showed X-rays from accretion onto a neutron star, the other evidently had accretion onto something, but no X-rays, leading one to think that the accreted matter was falling INTO a black hole. (I think that is fabulous "simple" but beautiful reasoning.)

So the existence of these other "proofs," or pieces of evidence, is one of the things that is causing this community to pause before saying they have made the first direct detection of black holes; they feel they need something special because the existing evidence is rather nice and you don't want to look as if you are claiming too much in the face of it.

4. Later, the claim about being unable to distinguish between a BBH and a boson star was withdrawn:

> Oct 27, 6.37: Earlier I had said that I thought you would need more than one QNMs, but I forgot that you get both the frequency and the quality factor, so it really would be quite remarkable for some non-black-hole system to produce this waveform.

But this is a change of mind for technical reasons so the philosophical questions that the stronger claim gave rise to still stand.

Peter also points out the technical difficulties of accomplishing what the other scientist counts as proof.

> Well, people (i.e., [ZZZZ]) are pointing out that you can invent objects that have a similar fundamental mode, but that the distinguishing thing is in the spectrum of (relationship among) multiple modes. … That's the trouble with "proof": depending on your standard of rigor, you can always insist that you need more proof to nail a case. … Reaching [ZZZZ's] standard is going to be REALLY hard, because the higher modes in the spectrum are only excited weakly, and we'd need amazing signal-to-noise ratio to see them.
>
> … Maybe we'll have to find a different word than the "p word" [proof], and describe it that way, finding a less contentious way to highlight an amazing accomplishment. There probably ways to say it that will make us all happy. And [ZZZZ] won't be able to police the shorthand way we'll describe it to each other, which will be peppered with the word "proof."

Let us switch registers and try to bring some sociophilosophical order into this. As Peter says in an email, the words at stake are (i) "detection vs. evidence for"—though this one is not on stage yet, it soon will be and the argument is endlessly rehearsed in Big Dog—(ii) "direct" and (iii) "proof."

"DEPENDING ON YOUR STANDARD OF RIGOR, YOU CAN ALWAYS INSIST THAT YOU NEED MORE PROOF TO NAIL A CASE"

What we are seeing are philosophical problems working themselves out as sociological choices in the day-to-day practice of physics. This has been the theme from the very first days of sociology of scientific knowledge. Standards of rigor are social agreements, and that means that whether something meets an appropriate standard is a social agreement, and that means that whether something is a "discovery" is a social agreement. In this respect, the cultures of the different sciences are markedly different and, within the sciences, the culture of physics is different from place to place and decade to decade. A quick and easy way to see this is in varying standards of statistical significance: most of the sciences require only that there be fewer than 5 chances in 100 that an observation could be due to random error to count as a publishable finding—the "2 sigma" level of significance. In the 1960s, physics required 3 sigma—only 1 chance in 1,000—but its standard has evolved over the years to 5 sigma—1 chance in 3,500,000.[5] But not all sciences use statistical analyses and statistical standards. Most of my work has been a matter of what I call "participant comprehension"—that is, embedding myself into a community until I understand how it works and then reporting the results. It is in one sense a "subjective" method but, I claim, extremely revealing and powerful, and entirely "scientific" in the sense that if anyone else repeated my work they would come up with the same findings.[6] But my methods, even though I claim they are even more reliable than physicists' methods, could never be considered acceptable within physics.[7]

On the other hand, there was a time when something like this was more acceptable in physics. Consider Robert Millikan's famous oil-drop experiment, which he took to prove that the charge on the electron was not infinitely divisible but came in discrete units. Careful reexamination of his experimental notebooks reveals that on several occasions Millikan had observed oil drops that seemed to violate his hypothesis but he chose to ignore them, feeling that his instinct as a scientist allowed him to discount them as noise rather than signal.[8] Nowadays physicists are still separating noise from signal in their experiments but not in Millikan's way.

5. See *Gravity's Ghost*, chapter 5, for an extended discussion of such statistical matters.

6. Other social analysts see themselves as working in a more humanistic style with their personal interpretations being more important than the robustness and repeatability of their findings. Here I am speaking only for those—probably a minority—who consider participatory methods robustly scientific. See Note IV in "Sociological and Philosophical Notes," 357.

7. My claim goes back to a discussion I had at LIGO Livingston with Gary Sanders who was ragging me about my methods. I said that we could each be more certain than we would be about the first detection of gravitational waves that we could not go into Baton Rouge, get on a bus and ask for two tickets, the second to reserve the seat next to us, even though neither of us had ever ridden on a bus in Baton Rouge; that is what it means to understand how a society works.

8. See Note V in "Sociological and Philosophical Notes," 360.

As far as terrestrial detection of gravitational waves is concerned, physicists are also no longer separating signal from noise in Joe Weber's way. Let us ignore the question of whether or not Weber was inadvertently massaging his data so as to produce results and simply note that his methods were nowhere near as refined as today's methods. Once more, this is a *cultural* change in physics' practices. But thank heaven for the cultural change—I do not mean because methods are so much better now, I mean because they were so much less demanding back then that Weber could found the field of gravitational wave detection physics; had he been bound by today's standards there is reason to think that there would be no interferometers and no book like this.

To bring another sociological focus to bear, scientists have a choice about the "evidential significance" they claim for their findings. Trevor Pinch first noticed this in his analysis of Ray Davis's detection of solar neutrinos.[9] In Davis's experiment, evidence of neutrinos was the transmutation of chlorine atoms, in a large tank of cleaning fluid, into argon atoms. Pinch pointed out that this could have been announced as the discovery of some argon atoms in a tank of cleaning fluid, or as evidence that the argon atoms had appeared as a consequence of mutations caused by energetic particles, or that the mutations could have been caused by neutrinos, or that the mutations could have been caused by solar neutrinos. Each choice had increasing "evidential significance" and each involved an increasing level of scientific risk, because in each case there were more things that could have been wrong: more subhypotheses that needed to be accepted.[10] We see the same going on here: is this the discovery of gravitational waves, a sighting of black holes along with evidence of their evolutionary history, or is it

9. Pinch first sets out these ideas in his 1985 article and follows up with a full account of the story of Ray Davis's experiment in his 1986 book, *Confronting Nature*.

10. See *Gravity's Shadow*, chapter 5, where I discuss the idea in terms of Joe Weber's attempts to cling to a conical island as the sea of skepticism arises around him. He can cling but only so long as he is ready to sacrifice more and more of physics' taken-for-granted world and thus push himself farther and farther from the mainstream. See Note VI in "Sociological and Philosophical Notes," 360.

also the first "direct" sighting of black holes? We'll return to a more general discussion of the sociophilosophy of discovery in chapter 13.

THE LVC-WIDE TELECONFERENCE, OCTOBER 22

On October 22, there is a collaboration-wide teleconference with 330 active nodes.[11] To start, various members of the collaboration walk through the discovery process to date; nothing is said in the first hour or so that has not already been discussed. After this, it proves to be an opportunity for scientists to raise any doubts and, most animatedly, to discuss the publications that are going to be prepared. It is made clear that, in a break with past practice, no early results will be promulgated on arXiv, the physics preprint server, prior to peer review and acceptance of the publication and the press conference.

Very few people express doubts about the reality of The Event. Nearly everyone seems to accept that this is close to a gold-plated discovery, and it is now just a matter of going through the predetermined checks, decision-making committee stages, and writing team assignments. Such doubts as there are lie, as anticipated, with the fact that only one event has been seen, whereas with an event of this strength, some people at least would expect to see more. It has now been five weeks since September 14, and no more stand-out events have been seen and this is beginning to worry people. There have been a couple of automatic alerts with signal-to-noise ratios such that the false alarm rate is less than one per month—the threshold at which alerts are sent to the EM partners—but this is run-of-the-mill stuff and cannot count as a sighting of any real event. Strangely enough, one of those alerts has gone out today with the ugly identifier, G194575, and discussion of it is sufficiently out of the ordinary for me to start a special folder called "22 Oct alert" and divert around twenty-five emails into it by the end of the day. I notice, however, that there is almost no mention of it

11. Many of the nodes represent small groups, so we never know how many individuals are attending a teleconference.

during the telecon, so it's much less exciting for the community than it was for me when I first saw it—another bit of my tacit knowledge that doesn't quite gel with that of the community. Indeed, by the end of the day, this one seems to be going the way of the rest of such things, leaving The Event in sole possession of the field and the worry about the uniqueness of The Event as pressing as ever.

Here, incidentally, is a typical report regarding these "near events":

> **Oct. 25, 11:02:** There was an oLIB trigger this morning with estimated FAR 1.6e-7, low enough to start the alert process, but it turns out that an obvious instrumental problem had appeared in H1 right around that time, so we rejected the trigger. I've summarized this in a log message at https://gracedb.ligo.org/events/view/G195294.

TeamSpeak, as we have seen, has two modes of communication, written comments and spoken. Here is a written comment reflecting the same concern that was raised very early on and the same comparison with the famous unique and never-to-be-repeated detection of a magnetic monopole.

> **17:07:58**
> "[XXXX]": My main question: Since the event showed up so early, everyone outside LIGO will ask "why haven't you seen a second event?" Is this the magnetic monopole all over again?

And repeated a little later:

> **17:20:08**
> "[XXXX]": [@YYYY]: my impression, from rumours on the timing, is that we already should be starting to get a little nervous that we have not seen a second event. The question needs to be addressed quantitatively, maybe in the papers you reference? For a discovery paper, a second detection puts a nail in a dozen coffins.

The response is as follows:

17:21:27

"[YYYY]": [@XXXX]: Yes, the question is addressed quantitatively by a number of groups (incl paper I linked). The answer depends on priors, but essentially there is *no worry* until we are very far into O1; near the end of O1 the significance of "seeing nothing else, given GW150914" becomes near 0.05/0.01.

The community is being told there is nothing to worry about until the end of O1, around the time the discovery paper will be published; the argument is also being made that there is nothing to worry about even then because you cannot generate an expected rate from a single observation and therefore cannot predict in any strong sense what ought to be seen. But everyone agrees that if nothing else is seen during O2, the putative period of running with a much more sensitive pair of detectors, then there really will be something to worry about—the thing that my more ruthless sociological self hopes will be the case. In the meantime, the scientists live in hope that the next event will show itself before the end of O1, which, as was mentioned earlier, has been extended from the middle of December to the middle of January.

The real heat in this telecon, however, is generated by a discussion of proposed publications.

17:26:54

"[ZZZZ]": Amazing trove of information describing the signal and event—we are absolutely privileged by Nature to have all these for first detection!

Where can we find the list of planned papers that actually address the fundamental physics uncovered and the surprising astrophysics we discovered as the direct consequence of this detection? (e.g., we need to write papers that substantiate the >1G$ expenditure to taxpayers.)

I am quite surprised at some of the positions taken over publication and at some of the concerns expressed; I am also surprised by the degree of animation. To try to disentangle what is going on, one needs to understand that this is more than a "point" discovery; it is also a new astronomy and a new astrophysics. The backbone of the discovery is the first detection of gravitational waves by purpose-built Earth-based detectors—the end-point of fifty years of effort. But we can see that the detection is intimately tied up with identifying the nature of what has been detected. And this is no surprise, because in the great battle between the interferometer technology and the resonant bar technology that took place around the mid-1990s, and which now seems to belong to a remote geological epoch, one of the great selling points of the interferometers was that they there were broadband," which meant they could see not only the energy from a wave but also its waveform. Being able to read the waveform also promised to increase the effective detection sensitivity.[12] This has now come to pass: the "plate" on this gold-plated event is that fact that its waveform matches the template of the inspiral and merger of a binary black hole, so astrophysics and confidence in a detection-claim go hand in glove. Parameter estimation—how massive are the black holes and what are their orientations and spins, etc.?—is one of the fingers of the glove, while the astrophysical consequences, which are the consequences for the number of such objects in the sky and the rate that can be expected, is the handprint. The question of publication has the potential to pull the dense collaboration apart into its separate groups, since parameter estimation and the astrophysical and astronomical consequences are in the domains of different groups of specialists, which are again separate from the groups whose main task is detection.

One can see the future, by the way, which is going to be a bit like high-energy physics or astronomy. In these fields, those who build and run the accelerators and telescopes, equivalent to the people who have been at the apogee of the whole fifty-year enterprise as far as gravitational waves are concerned, are relegated to the status of something not far from engineers or technicians (not counted as "physicists' physicists," according to one resentful scientist who ran a particle accelerator),[13] while the true science is done by those who interpret the data rather than create the conditions

12. See *Gravity's Shadow* for a full account of the victory of the interferometers over the bars.

13. *Gravity's Shadow*, 486.

for its existence. The same is beginning to happen to "the true science of gravitational wave detection"—a topic we will treat at length in a later chapter—and this is beginning to cause some strain.

By this time, it seems to have been agreed that the detection paper will be submitted to *Physical Review Letters*, the same *PRL* that was the target for the proto-papers put together in response to the Equinox Event and Big Dog blind injections. This journal is where gravitational wave astronomy has settled down as far as its big claims are concerned, rather than the "newsier" *Nature*. It seems clear that the first paper that is going to be ready to submit will be the detection claim paper, but what is less clear is what will be said in that paper. *PRL* has a strict limit of four pages, and it looks as though that is not going to be enough space to include the details of the way conclusions about the parameter estimates—the details of the masses, spins, and so forth, of the stellar object observed—are supported. Estimates of what this means about the distribution of objects in the heavens—rates—are also not going to fit. So for this reason—and I cannot work out how much of this is entirely a matter of space with no hint of specialist professional pride—there are going to be two, or even three, papers (it will turn out to be twelve) rather than one. The specialists seem to think that their papers are going to take longer to write than the detection paper and so they are leaning toward delaying the main paper until they can catch up.

I find it really strange that they are so concerned about not being "scooped." They think that as soon as the news is out—and the collaboration is committed to releasing the data that supports the observation as soon as it is in the public domain—other astronomers and astrophysicists, not tied in with the collaboration, will jump on it and put out their own papers interpreting the meaning of the Event. Worse, those papers will be based on a less technical understanding of the data and so might draw wrong conclusions. As one analyst puts it on the aural channel:

> **[About 1 hr. 40 mins. in]** If the collaboration wants to have an astrophysics paper, then it has to be short so that it can get out with the detection.

Because if we don't do it with the detection then we'll be writing the tenth paper a few months later in astrophysics.

What I don't understand is why these groups think there is danger of being scooped given that they have the data at least three months ahead of anyone else. Nevertheless, there is pressure for delay so that all this material can be published together.

Another source of this pressure for delay comes from a very persistent contributor, who makes a number of interventions, and wants a delay of several months so that the collaboration can do every kind of analysis to ensure that no mistakes are made.

> **[Around 1:33]** I'm somewhat concerned with our very aggressive timetable. When we are talking about producing papers within 4 months of the initial discovery of the event. For a discovery of this magnitude that's quite unusual—very ambitious—almost unprecedented—especially in a case where we cannot be scooped. And so I would advocate ... that we wait an additional few months and ... allow time to fully develop the papers that we want to produce and that the announcement include papers fully describing the instruments, fully describing all the work we've gone through, and why we believe in detail that this case is solid and we have the first detection. ...
>
> I fundamentally don't understand what the rush is. There is not pressure from the external community to produce a result right now and adding a few months and producing a much more solid case, and making the detection announcement a much, sort of, rounder announcement—first of all there's much precedent for that in terms of other collaborations ... and I think there is a good case to be made in our case.

Selecting from the responses on the "chatline":

> **17:34:10** "[PPPP]": [@QQQQ]: I disagree. We need to get this announcement out as soon as we can. Others should write to the short time scale—we shouldn't delay the discovery announcement because we can't get other papers written.
> **17:34:33** "[RRRR]": [@PPPP]: why does it have to get out so quickly?

17:35:11 "[PPPP]": This is something that matters, and the discovery is solid. Let's tell the world.
17:35:23 "[SSSS]": Yes
17:35:26 "[TTTT]": yes
17:35:38 "[UUUU]": agree
17:35:58 "[ZZZZ]": [PPPP] is right. There is a huge astrophysics community outside of LIGO and we need to enable them to pitch in with papers if we cannot finish our papers in time.

One of the most senior members of the collaboration adds in the aural channel:

[Around 1 hr. 45 mins. into the meeting]: The detection paper itself I think has really special importance and we shouldn't lose sight of it. This is, if we do it right it's going to be a classic paper in physics. And so we have to make sure that it's written extremely well and that it has all the points in that we want in the one paper that'll stand above the others. I'm not degrading anything—the astrophysics of black holes binary discovery—but still most of all the observation of gravitational waves stands above the rest. So we have to write a paper that's going to be a classic paper. I think that can be done on this time scale—I'm not worried about that—I think the comments about not seeing another event we are trying to pay attention to. How we include or don't include that in the paper is a question [to bear in mind]. But delaying I see no point in at all.

But the debate is not over:

17:39:21 "[QQQQ]": [@PPPP]: what is the rush? why wouldn't we want to produce papers about both the instrument and the science at the same time as the detection announcement? i don't undertand why waiting a few months would be a problem? we simply cannot do a thorough job of this with the current timetable, and with ~3 short papers.
17:40:36 "[PPPP]": [@QQQQ]: I don't understand why theory papers should hold up the announcement of the discovery of the decade. Why can't those papers be written promptly?
17:40:43 "[VVVV]": [@QQQQ et al.]: I don't see how we can defend

keeping this a secret from the world for 6 months. The detection case won't get firmer—the data is what it is.

17:44:05 "[QQQQ]": [@VVVV]: the "world" would much prefer having us do a thorough job (as presented), than hearing about it 6 months earlier. they've waited ~100 years, i think another 6 months would be fine. but we need to produce an emphatic case. we can't do this in a single short paper.

In spite of the argument, from which I have selected only some of the turns, *I think* this is where the matter will settle. This is because, on the whole, those going for publication of the discovery paper as soon as possible are the more senior, more experienced, and more powerful members of the collaboration. *I think* that what is going to happen is that there will be one or even two more papers out at the time of the discovery announcement, but the separate groups will have to take responsibility for getting them ready even if this means making them short papers rather than long.

I learn a couple of days later that a senior committee has met and voted to try to publish only one discovery paper that will include the claim about the source being an inspiraling binary black hole and its parameters. Only if it is felt that not everything can be squeezed into the one paper will a second paper be contemplated. It is intended that the paper, or if really necessary, papers, will be published without delay. The interesting thing is exactly how much will be claimed and how strongly. But all this will be overtaken by events with many more papers prepared for simultaneous publication.

The competitive side of the collaboration is beginning to show itself with the argument over publications and who is going to write them. A problem arises because of the publication rules the group has set for itself. The rule is that for at least a year, all publications drawing on data generated by the collaboration, must have a collaboration-wide author list—they belong to everyone. The knock-on effect of this is that they must be carefully internally reviewed before submission, and this will slow them down. Since the collaboration is committed to releasing all the data that supports The Event claim as soon as the paper is published, members fear that they will be scooped in spite of their three-month head start because of the delay in producing collaboration-wide papers, compared with outsiders who can take the data, analyze "overnight," and be ready to publish "in the morning." Jumping ahead to the end of November, I find myself involved in an interesting exchange over this matter.

I write to one of the collaborators—Peter Shawhan—asking if he would be willing to modify a figure he has prepared for the discovery paper specifically for my book. I make a mistake that is very easily made: in pressing the "reply" button to an email that has come from Peter Shawhan, I accidentally copy my request to the whole collaboration. At that point I get a friendly warning from someone I have known for years who tells me to watch out or "you may have torches and pitchforks outside shortly." He points me to the heated argument over what you are allowed to publish. I write back:

> I cannot imagine the republication debate kerfuffle applies to me. I'm not trying to do new physics just report on yours.

But he replies, making me laugh out loud:

> Rationally of course you're right. But folks have lost their heads, like pirates dividing coveted treasure.... [I'm] Just concerned about tossing more chum in the water.

MALICIOUS INJECTIONS REVISITED

In chapter 2 we noted that the possibility that The Event was a malicious injection was considered but dismissed after careful consideration. It seemed impossible that the conspiracy required to manage such a thing could be organized. But the community is not to be entirely satisfied until this conclusion has been fully explored and the results formally presented. An extraordinary feature of this discovery is the time spent ensuring it is not a hoax. I send around an enquiry (Nov. 1) to a few people including some very experienced senior scientists and the person who led the team looking into the possibility of fakery, asking if they knew of any precedent. The answers that come back indicate that this is something completely new. So The Event represents not only the discovery of a new feature of the natural world but has also given rise to a new feature of scientific method: fear of hoaxing.

It is true that the possibilities of hoaxing had been considered in the past. For example, in the case of the notorious Cabrera monopole experiment, it was thought possible that someone had entered the laboratory at night while the detector was running unattended and faked the signal; when experiments were set up to confirm the original result more precautions were put in place in consequence. There were other experiments where the idea of inserting fake data was considered and, of course, fake data is an integral part of the search for gravitational waves, not only in respect of the blind injection challenges but also the loop of data with fake starting points created by the Louisiana resonant bar group. In these cases, the scientists wanted control over the way the data was analyzed: if others could not tell true from false they could not use the data to reach conclusions without the agreement of the scientists who generated them, but none of this is malicious; the fakery is part of the method.[14] Here the methodological innovation is the huge effort expended into eliminating malicious fakery. Why has such effort been put into eliminating fraud in this case? There seem to be three types of reason, the first to do with the nature of the discovery, the second to do with the technology, and the third to do with changing times.

Beginning with the nature of the discovery, the detection of gravitational waves is a long-lasting, expensive, high-profile experiment—something that might well tempt a hoaxer. Also, this is a single "point" event and is much easier to hoax than the slow buildup to statistical significance that typifies the discovery of new elementary particles such as the Higgs boson. Furthermore, in the case of the Higgs, two very different detectors were involved, even though they were both sighted on the same beam tube—so

14. See *Gravity's Shadow*, 426–427.

there was something close to replication taking place, albeit a collocated replication. As Barry Barish explains to me, high-energy physicists are conscious of the need to check in this way:

> **Nov. 2, 16:31:** Having the extra expense of two experiments at accelerator facilities has always been justified on the basis of an independent check on each other (not just for rogue discoveries, but for different hardware, software, analysis techniques, etc that help defend against wrong results). Another interesting approach has been by Sam Ting for all his work, but certainly for the J/Psi discovery. He divides his group into two independent data analysis efforts that have no communication between each other or even common programs to analyze data. They come together in the end as two independent efforts almost like two LHC experiments.

With GW150914, all the time is being spent on analyzing the statistics of this one event, not collecting other events to back it up. Of course, the scientists hope that something else will show up during O1, but they have to at least start the analysis and write the first drafts of the discovery paper on the assumption that it won't. When it comes to replication, gravitational waves can be close to indistinguishable from noise, and that is why the method of coincidence between widely separated detectors is fundamental; traditionally it has been accepted that there is no signal unless it is present on two detectors at once. Here the waveform of the signal is exceptionally clear and fits a well-understood model; I think a case could be made for saying the *same* signal has been seen on two detectors rather than that one signal has been synthesized from the output of two detectors, but this is not what the scientists are saying. So, there is, to a first approximation, only one event, and that is easier to fake than two events.

Turning to the technology, the experiment is almost totally dependent on computers, not only for extracting and analyzing the data but even for generating it: the data *are* the computer-generated feedbacks needed to hold the mirrors stationary against the force of the gravitational wave. This keeps the output on a "dark fringe," which is the best place to measure tiny changes. As Rai Weiss writes to me:

Nov. 2, 14:27: [at the time of our early microwave background detections] the computer was already the way to store and manipulate data. It is the device that allows the problem, it is much harder with notebooks and paper. Erasures and cross outs are pretty obvious in a notebook. … the computer makes data analysis on the scale we are doing possible and without it we could not carry out LIGO both in the instrument control and the data analysis. But, it comes with some hazards.

Thus, it is because computers are so intimately involved in this result that the experiment is potentially vulnerable to hackers.

Third, and I think most significant, the times have changed. Physics and other scientific experiments have been dependent on computers for decades, but it is only recently that hacking has become used for intelligence gathering, for negating security systems, as a weapon of war, and for large-scale financial crimes. Computer hacking is now on everyone's mind; hackers can build careers in security by first demonstrating their skills in well-publicized scams. A couple of decades ago, even in the case of an experiment identical to this one, no one would have thought it needed an extended effort to make sure the signal was not a malicious fake. Here we are probably seeing a change in the way science is done; henceforward all high-profile experiments are probably going to have to take security seriously. I note, however, that nothing about the possibility of a malicious hoax will appear in the discovery paper so it is not yet *that* much of a regular part of science.[15]

We have a teleconference on October 29 that is dedicated to disproving the possibility of malicious injections. It comes to the same conclusion as the first look at the problem discussed in chapter 2. The participants in a teleconference have access on their computers to the equivalent of Powerpoint slides. One of the slides presented in this telecon makes clear:

15. Peter points out that there are also reasons specific to this discovery that make hoaxing more salient: the mindset that comes with earlier blind injection, and surprise at the strength and clarity of the signal, which also leaves nothing much to argue about except hoaxing.

> We can't say that faking GW150914 was impossible, but we can say that faking it would have required an internal conspiracy of our most knowledgeable people.

In other words, it is the required social organization that cannot be envisaged. But a good part of the reason that faking would be such a complex social achievement is that, if the malicious insertion were done early in the detection process it would have to be done at the two sites simultaneously, and the two sites differ in subtle ways. Furthermore, these differences were not even fully known before the instruments were calibrated, so it is hard to see how anyone could fake two signals to fit two devices that differ in unknown ways. As the analyst in charge of eliminating malicious injections puts it at the teleconference:

> **[Around 58 mins. in]** You'd need to know that the filtering is different between the sites. So the ESDs have different actuation functions at H1 and L1. This is really because at some point Rana changed some resistor and he put it in the a-log. It was the right resistor to change but it didn't happen at Hanford. So again you'd need to know some pretty detailed information to get this right. At that point you're going to need to travel to both sites, get access, to end stations at both sites. Probably if you are one of the very few people who travel to both sites we know who you are. If you are not one of those people then you must have recruited some conspirators.
>
> **[Around 1:09]** I think the analogue hacks—I think we can rule them out because you would need to have some sort of internal conspiracy to pull this off, and I don't think you can easily imagine any way to pull off an analogue hack that does not require at least a few people, probably a few people at each site and that restricts the number of people significantly and also limits you to people who have the motivation to do such a thing.
>
> **[Around 1:15]** The only conclusion we can give is that we can't say that faking the event was impossible but we can say that if you wanted to fake it you would probably need to get a fairly large group of insiders to conspire to make this and that's probably as high as you can put the bar in any sort of scientific endeavor.... We are not talking about everyone on the planet,

> we're talking of the few who are insiders—people who really know the system well and that gets you a relatively small number of people and the idea that you have more than one of the inside folks who've decided that it will be profitable somehow to do this strikes me as a very remote possibility.

The logic of this is interesting because it rests on the difference between the machines. It rests on the fact that because the machines are subtly different it would be very difficult to inject a signal that would look the same on both. But what that means is that we have a case of the same waveform being replicated on two separate and different devices—just as in high-energy physics—rather than one signal comprising a coincidence between two devices. I think the community is slow to recognize this.

The other feature of the conclusion has already been pointed out. We don't believe this is a hoax because we think such a thing is *sociologically* implausible. This is key if we are to understand how scientific findings come to be accepted in the world we live in.

7 NOVEMBER
Ripples, Belief, and Second Monday

In *Gravity's Shadow* I described the search for gravitational waves by referring to the logo chosen by the Laser Interferometer Gravitational-wave Observatory (LIGO)—the pair of American devices that have actually made this observation (figure 7.1).

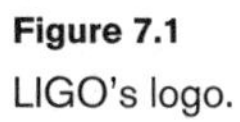

Figure 7.1
LIGO's logo.

The logo represents gravitational waves emerging from an inspiral or supernova or some such. The point I made was that the logo showed only half the story. In fact, there are two sets of waves, as shown in figure 7.2.

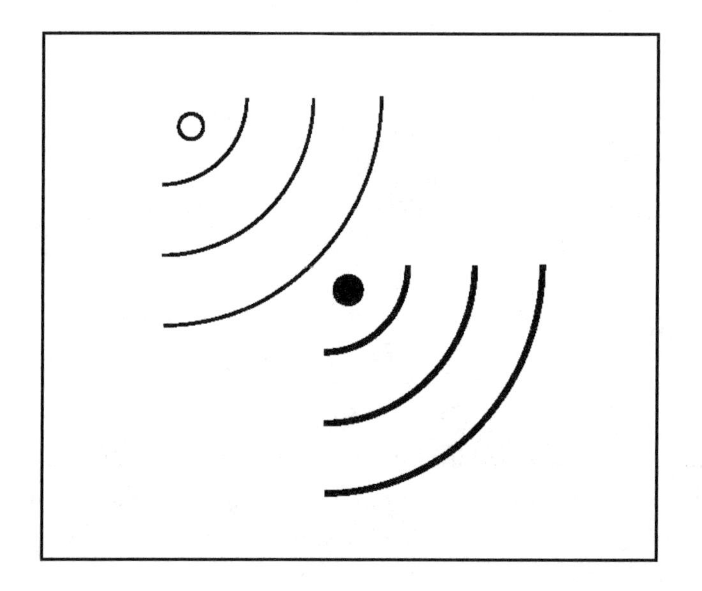

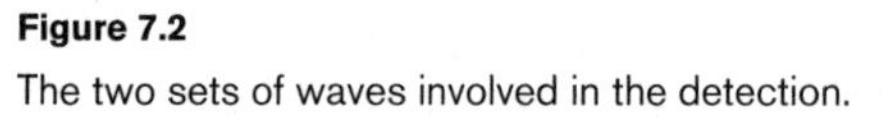

Figure 7.2
The two sets of waves involved in the detection.

The lighter ripples at top left of the diagram show the heavenly object with gravitational waves emerging from it, rippling through spacetime, and then impacting on the detectors—the solid black disc. The dark ripples at the bottom right represents what happens after that—the discovery ripples through social spacetime. That is, the scientists start talking to each other about what they have seen and try to decide whether it means anything or not—the first dark ripple. What is happening in this first dark ripple is that the scientists are trying to convince themselves that they have a finding. As we have seen in *Gravity's Shadow* and *Gravity's Ghost and Big Dog*, this is no easy task in this field given its history of claims that did not survive. In this case, however, it has happened in a few days.

If the scientists can convince themselves through their negotiations in social spacetime, they get together to write a paper intended to persuade the scientific community that their beliefs are justified. This reception is the second dark ripple. The reception in the world as a whole, including the more distant parts of the scientific community, is the third dark ripple.

The following email, albeit taken from next month's interactions, is a small illustration of the way the community is thinking about how it is going to convince the world:

> **Dec. 20, 11:58:** My proposal is that Figure 3b show the cWB background before and after the chirp cut. Like including the Little Dogs [this will be explained later in the book], I think this is a much more honest presentation of our data. Being open an[d] honest about how we've performed the analysis is the best way to reassure the reader that they can believe our results.

It is the first dark ripple that we have been looking at so far in the book—the talk that has led to the conclusion that there is something here worthy of wider notice. Now we begin to look at the second dark ripple—the preparation of a literary artifact that captures what The Event means and explains it to the community.

RETROSPECT: THE GROWTH OF BELIEF

Before turning to the second ripple let us recap on what it has taken to get us to this point. What is it that has caused the scientists to be ready to write a paper and, for the first time in the interferometer age, to try to project their findings into the outer world? Many claims—maybe fifteen—were presented to the outside world in the age of resonant bars (see *Gravity's Shadow*), but none survived. Their incredibility was made salient by the interferometer community, who were sure that their (much more expensive) way of doing things was the right one; they had, therefore, to show that the bar teams' claims were wrong. These scientists have, then, developed entrenched critical habits and it was always going to take something special to reverse the mindset. It was reversed by:

(a) *The fact that the finding matched the sensitivity calculation.* This, however, is not as decisive as it appears in retrospect because no one would have been surprised if nothing had been seen until aLIGO had reached its design sensitivity; that it did see something so soon came as a surprise—it was not an expectation.

(b) *The strength of the signal.* The signal was even stronger than Big Dog because, in both interferometers, its magnitude was greater than any noise event in either interferometer.[1]

(c) *The waveform.* The waveform was a triumph for the interferometers—a triumph for the claim made during their long battle with the resonant bar technology (see *Gravity's Shadow*, Part III) that broadband was best. The bars could only see the energy inserted by a gravitational wave, whereas the interferometers can see the shape of the signal. Seeing the shape of the signal gives confidence. It also makes it possible to do astronomy: to say not only that an event has been seen but that "the event is of this type"—in this case, an inspiraling binary black hole of such-and-such mass.

The power of the broadband is captured in a paragraph submitted for publication to the committee writing the draft paper (but excised from the first proper draft):

> Having found this event to be highly significant in searches for un-modeled transients, any hypothesis for a non-astrophysical source would be forced to explain the "coincidence" that the most significant transient found in LIGO data so far, also displays a waveform consistent with the predictions of general relativity. This qualitative evidence strongly supports the compact object merger scenario described in this letter.[2]

1. In the case of Big Dog, some noise events in one of the interferometers (because it was only in one of the interferometers we know it is noise) were bigger than the "signal"; this is what gave rise to the "little dogs" paradox: these noised combined with the Big Dog signal produced a spurious noise background if Big Dog was real, therefore making it seem less real! This is not the case here because all the noises are smaller than the signal and that is why there is no "little dogs" problem.

2. Looking back, given what has been achieved in seeing The Event, the limited ambition of the resonant bar technology now seems striking; compare what has been extracted from The Event from what could have been extracted from even the best array of resonant spheres.

(d) *The coherence of the signal at the two sites.* As discussed in chapter 1, it was the almost perfect match of the waveform in the widely separated interferometers that caused many of the scientists to believe almost from the beginning. One should add, however, that this aspect was rehearsed in the case of Big Dog, even though the signal was not quite so perfect. A good bit of the confidence in that case—hugely watered down by the blind injection probability—was the way the signals overlapped at the two sites. This was even so in the much weaker case of the Equinox Event (see *Gravity's Ghost*, 74–75).
(e) *The statistical significance of The Event.* The statistical significance was worked out long after nearly everyone had decided for themselves that The Event was the real thing. Peter Saulson said to me at some point, "We knew it was going to be significant because of the experience of Big Dog," so there was some sense of the statistics before any calculations had been done; but no one could be sure until the box was opened on October 5, three weeks after The Event. The celebrations of that day were in anticipation not so much of what had been seen, but of the fact that it was going to be possible not only to believe in it but also to tell the world about it and get them to believe—to move on to the second dark ripple.

What I am trying to explore here is how different elements of the scientific community came by their belief in The Event. I am not, at this point, discussing the power of different groups or their taken-for-granted assumptions about the physical world, or the different ways in which the existence of this discovery could be doubted—we will come to those at the end of the book. Rather, I am talking about the way belief formed within the community of gravitational wave physicists and the world of mainstream physics.

THE SECOND MONDAY EVENT

The switch from the old world of no credibly detectable gravitational waves to the ordinariness of terrestrially detectable gravitational waves is not yet guaranteed, largely because there is only one event seen by one detector (counting LIGO's two interferometers as one detector). Well, that would have been true; but, at the beginning of November, the existence of a second event becomes apparent. The event happened on October 12 but no one was aware of it until now at the beginning of November because it was too weak to be spotted by any of the low-latency search pipelines. This means it appeared only when the box was opened on data gathered between the October 8 and October 20 and template-matching applied to the whole period using the pipeline known as pyCBC, where "CBC" is, of course, "compact binary coalescence."

October 12 was a Monday, and because September 14 was also a Monday, this event, when it is not being referred to with an abstruse code number, becomes known as the Second Monday Event; I will call it "Second Monday" from here on. Because it is hard to maintain both interferometers in lock at the same time and there are some periods of data that are vetoed because they are glitchy or there are signals on the environmental monitors, it has taken five weeks to gather the sixteen days' worth of data used for the background analysis. That is why the data-gathering period ended only on October 20. That is why, even though it occurred as late as October 12, Second Monday was included in the sixteen days of double-coincident data that was subject to the "freeze" being used to establish the background for The Event. This means, among other things, that it could, if desired, be mentioned in the discovery paper. I hear about it from Peter Saulson:

> **Nov. 2, 17:11**
>
> Dear Harry,
> A "genuinely marginal" event in the latest PyCBC box opening that happened one hour ago:
>
> https://www.atlas.aei.uni-hannover.de/~miriam.cabero/LSC/O1/final_analysis3_c00_v1.2.4/7._result/
>
> ... At 3 sigma, it is going to be tougher, but this is what we expected. ...

As another emailer remarks:

Nov. 2, 23:02: this is quite a weak event by cWB standards, however it has a beautiful chirp shape both in time and TF domains.

It is retrospectively notified to GraceDB, the database, where it is given the unmemorable name "G197392." On GraceDB we find an initial estimate for the component masses of 32 suns and 14.6 suns: if it is real, it is another black hole inspiral though smaller and weaker than The Event.

This signal would never stand as a first detection and, were it not for The Event, would have been forgotten by now. But it has become worth discussing because it helps make more sense of The Event. Reality is already changing! Thus we hear the following from the chair of the Detection Committee:

Nov. 10, 20:07: ... I think it is pretty clear that this event by itself would not have launched the Detection Committee into action, but since this event will have to be included in the plots of the detection paper, and likely discussed at some level in the text, the Detection Committee will have to look at it, albeit at a somewhat less comprehensive way than "The Event."

And from another emailer:

Dec. 12, 00:40: ... I feel like the collaboration is really underselling the significance of the second event. While the second event certainly doesn't constitute an independent 5-sigma detection, if we take the rate inferred from GW150914 as a prior, the second event still has a ~95% probability of being real! I think most observational astronomers would love candidates with this degree of certainty.

A long argument now arises about what Second Monday means and how or whether it should be mentioned in the discovery paper.

Dec. 7, 05:57: I think there is some confusion. There is no "magic" at all! This event has a 3% false alarm probability, period. That means that the statement "this is a signal" is only going to be wrong in a few out of every 100 equivalent experiments. This event is a signal. ... It does not meet the

5 sigma confidence required for a first detection claim and we would ignore it if we hadn't already seen something else. But it's not a first detection.

In a world in which we believe we have built a gravitational wave detector and that the universe contains a sufficient number of sources of sufficient brightness to be detected by us—what we learned from the first detection—this is a signal and extracting astrophysical conclusions from it will lead to robust conclusions.

If people want the "detection paper" to only discuss the manner by which we conclude the first event is a detection, that's fine, but it's actually incorrect to imply this data does not contain two signals. The event rate posterior, which doesn't care about our human concern for high confidence in a first detection, which considered only what the data tell us, says there are two signals. We can't screw with the data to force a different conclusion, one we like more.

In contrast, another emailer harks back to the time of the resonant bar technology when claims were made that did not stand up. These claims were made in what I call, in *Gravity's Shadow* (chapter 22), a spirit of "evidential collectivism." I try to argue that evidential collectivism, in which speculative results are published for the wider collectivity of scientists to accept or reject, is an understandable and reasonable strategy, though it is rejected by most Americans who hold the view that the responsibility of discovery lies with the individual or individual partnership. This emailer fears the return of evidential collectivism:

Dec. 5, time unknown: I know many people have expressed the opinion that we should quote the system parameters [of Second Monday]. I am very opposed to this because it sends a mixed message: on the one hand we do not claim it as a detection, on the other, sort of underhandedly, we let the reader know that we think that it actually is, and here are the source parameters.

Many years ago a GW resonant-bar group published a paper where they did not claim any detection but they reported an excess rate of events when the detector was most sensitive to the Galactic Centre. That paper was very criticised, among other things, for its ambivalence. I think that **unless we put forward a clear case as to why we think that the second loudest event is a signal, then we should refrain from treating it as such**.

The suspicion of evidential collectivism is clear.

Thus there is conflict about Second Monday, some seeing it as so weak as hardly to be worth mentioning and others thinking that not mentioning it would be to mislead readers about what had been seen. It seems to me that it markedly reduces the probability that The Event is an artifact. This is because The Event is so strong that it is expected that there will be indications of a population of similar events, with the probability that most will be farther away and therefore weaker simply because the farther you go the more events there are: the volume of space scanned goes as the cube of the distance. Second Monday fits this model. It shows that the upper limit on possible weak events is not severe enough to make people ask "why one strong event and no weak ones?" A "replication" would be another event strong enough to be statistically significant. The Event is still a singleton in spite of Second Monday but it is a slightly more credible single event than it would be without Second Monday.

WHAT IF THERE WERE ONLY AN ISOLATED EVENT: IMAGE AND LOGIC?

Remember, at the time of the Big Dog episode it was still unclear whether a single event could count as a detection. In a conference in Hungary in 2009, Adalberto Giazotto, one-time head of the Virgo project, set out a demanding set of requirement for a gravitational wave detection, demanding enough to surprise many of the community. I reproduce his PowerPoint bullets:

- There is the possibility of nonreproducible (because rare, owing to our "bad" sensitivity) events. What are the requirements we need to accept them?
- Always "Coincidence" with Astrophysical observatories (neutrinos, optical, radio . . .)

- High statistical significance
  - To be defined in advance
  - Detection committee should have a key role
  - Strong interaction with the Data Analysis Committee
- Redundant analysis
- "Coincidence" on the GW network
  - Use all the network at the maximum possible extent
  - Event "not seen" by some of the ITF in the network: a relevant information, that should be used
  - Coincidence strategy decided in advance
  - Diversification of DA pipelines should be encouraged

Giazotto is now one of the team drafting The Event paper so I write to him asking how he was squaring what he said then with what he was writing now. He replies (Nov. 16) that he no longer feels so strongly because there is less pressure to publish a discovery claim in respect of a weak event (as was happening at the time in the case of Big Dog), and that the unsuccessful claims by the resonant-bar groups are now further in the past and therefore less salient and less of a worry. Also, this signal is so strong and matches the waveform of a BBH inspiral so well that "even a single detector could make a discovery claim" (we will discuss this point again). He also says that he thinks that the future GW astronomy will involve many weak signals and this means that attempts to confirm with electromagnetic observations, neutrinos, and so forth, will continue to be important.

Even if this signal is strong and distinct enough to be discovered by one detector, there is still the interesting question of how the statistical confirmation squares with the compelling nature of the waveform. Remember, in the first day or two Peter Saulson told me that, like me, he and another senior physicist considered the coherence of the waveform convincing but that "When we look at that graph, we are using our intuition," whereas the proof will be in the statistics (see above, p. 15).

The physicists seem able to cite only two precedents for discoveries based on single events, and one is the unfortunate Cabrera monopole that crashed and burned. The other is the more fortunate "Omega-minus," the discovery of which led to the award of the 1969 Nobel Prize to Murray

Gell-Mann. This discovery[3] was based on a single track in a bubble chamber and no statistical analysis at all.

> **Nov. 1, 16:12:** ... Historically, very few discoveries in physics have been claimed and accepted on the basis of a single event. The classic (only?) example is the discovery of the Omega-minus particle, reported in a 3-page article entitled "Observation of a Hyperon with Strangeness Minus Three" (http://journals.aps.org/prl/abstract/10.1103/PhysRevLett.12.204). As I understand it, that was accepted as a discovery by the community because there was a picture of the decay chain that made it totally obvious what had occurred, and because the discovery exactly matched a theoretical prediction—in fact, it was a "missing piece of the puzzle." We've been blessed with a similarly remarkable event, I believe, but we shouldn't take it for granted that readers will accept a discovery claim based on a single example. The need for "confirmation" is a deeply ingrained concept in science reporting, as well as in many scientific communities. If we are really claiming that this single event is a discovery, then the whole paper will have to consistently make the case that this event is undeniable—that it is like the Omega-minus, not like the magnetic monopole.
>
> As a corollary to the points above, we will need to include some text to directly explain WHY we are publishing a paper now, without waiting until we have all the data from the O1 run. (I'm personally convinced that a paper now is the right course, but only because this event is SO spectacular.)

Does the monopole provide the precedent for the single event claim being made here, or is it the Omega-minus? And is the claim supportable on the basis of the coherent waveform alone with the statistics being the icing on the cake? I write to Barry Barish, long-time director of LIGO and hugely experienced high-energy physicist, asking him to comment on a view I put to him:

3. Reported by V. E. Barnes, P. L. Connolly, D. J. Crennell, et al., in "Observation of a Hyperon with Strangeness Minus Three" (1964).

Nov. 8, 10:41: It seems to me that the community believes in this event because of the perfect waveform and excellent coherence, not the statistical significance—even though statistical significance is what is going to be needed to sell it to the wider world. That is to say, everyone believed in it after about 3 days even though the statistical calculation had not been finished. The question was solely whether there would be enough good background to generate the statistical significance required to present the event to the wider world. It seems to me that in this respect "broad band" GW (the interferometers as opposed to the resonant bars) has something beyond what most high-energy physics experiments have—a waveform that you can look at and a complex set of events that the waveform describes—you are seeing a complex event. So there is something more here than just a "single event" and that is what that paragraph expresses. As I understand, in HE [high-energy physics] you just have the theory and the statistics so you have nothing but the statistical calculation to rely on. That makes quite a difference.

This last sentiment of mine turns out to be stupid because I had not noticed that the single event being cited as a precedent dated from 1964 when high-energy physics was based on bubble-chamber photographs rather than statistics.

Barry explains his way of looking at things in a couple of emails. He thinks that my claim that "the community" came to believe in this event within a couple of days because of the coherent waveform is too simple:

Nov. 13, 21:19: There is a huge spread in what people had seen and understood after a few days, so the "experts' opinions tended to dominate. But that may well not be the case as the paper is revised and especially internally reviewed. I think for some, the statistical significance is crucial and is maybe the most important single piece of information, BUT not the total story. This comes from lots of history that people are fooled when the statistical significance is not very high, and this has led to the traditional requirement that a claim of a new discovery required 5 sigma statistical significance. For many of the data analysts in LIGO, they are very partial toward seeing what you are looking for, and in this case it is pretty distinctive ... an inspiral of increasing frequency and magnitude; and [a] merger—not well defined but presumably strong gravity; and a characteristic ring down frequency and behavior. That is a pretty convincing waveform, but how to quantify "convincing" is an issue. Every potential signal is compared with many many potential waveforms, so how significant is it if it fits one of them.

The last point here refers to the CBC method of analysis that matches a signal to something out of a 250,000 strong template bank. On the other hand, cWB has extracted a signal that looks like an inspiraling binary black hole without any template bank, that is, unmodeled—although a dispute would rumble on right to the point of preparation of the final draft of the paper as to whether cWB's pipeline is quite as free of preconceived models as is expressed by the word "unmodeled." Every pipeline depends on some preconceptions (e.g., that gravitational waves travel at the speed of light), and it is a question of just how much understanding of possible waveforms are integral to cWB. As the final draft was being prepared, the proper form of the figure representing cWB's contribution to the statistics of the finding (the left hand frame in figure 4 in the final draft) was still being disputed.

In the same email, however, Barry seems ready to agree with some of what I am getting at:

Another issue and probably what is influencing your question is not just how you convince yourself, but how you convince the scientific community. Seeing what you are looking for (e.g. wave form evolution) is a hard argument to make, while everyone understands statistical significance. So, that almost must be the lead argument in a paper, even if it isn't what convinced all of us.

In a later email the same day, Barry draws out important differences between The Event and the Omega-minus:

Nov. 13, 21:45: [The Omega-minus] event had more distinctive criteria than our waveform argument, because the mass and detailed features of the particle were predicted, the energy of the beam was defined, the target bubble chamber was hydrogen, which meant there was just one hypothesis to compare with the event. In our case, qualitatively we have inspiral, merger and ring-down but the behavior depends on the masses of the two objects and to lesser extent on spins, orientation, etc. We use many hypothesis with different masses etc to compare with the data, so the waveform, though distinctive, isn't unique, while the Omega minus was unique. That is why it is less obvious to make the argument of a discovery on the basis of one event with the observed waveform, while the coincident large bursts at LLO and LHO make it highly unlikely to be background (the statistical argument). Other HEP discoveries—e.g. the Higgs—are almost entirely made with a statistical argument. Nevertheless, in the end we could see both how well the observed waveform agreed with General Relativity waveforms and the statistical significance was more than 5 sigma. The combination made the claim of discovery based on one event convincing.

It is striking that none of these discussions occurred around the Big Dog injection—that debate was solely about statistical significance.

I next write to Allan Franklin who has written a book on changing standards in physics (see *Big Dog*, 239–240, and Franklin's book, *Shifting Standards*, 167, where he writes about the Omega-minus). I ask him if he thinks the Omega-minus discovery would stand up in today's world of physics. Of course, what I have in mind is how well the nonstatistical element of The Event would stand up, but as it is still secret I do not mention it directly. Allan writes of the Omega-minus:

Nov. 16, 18:41: My own view, which I checked with some colleagues, is that it is very hard to judge, but that it [Omega-minus] might be accepted as a discovery. The experiment had only rough estimates of background so one would not be able to estimate probabilities and sigmas. This was before there was even a sigma criterion. The measured mass did, however, fit the predictions of the Eight-Fold Way, which would make it more likely to be a real effect. My own thought is that it might be acceptable.

The Omega-minus has also been discussed by Galison in his classic history of particle physics, *Image and Logic*. Galison (p. 22) groups the

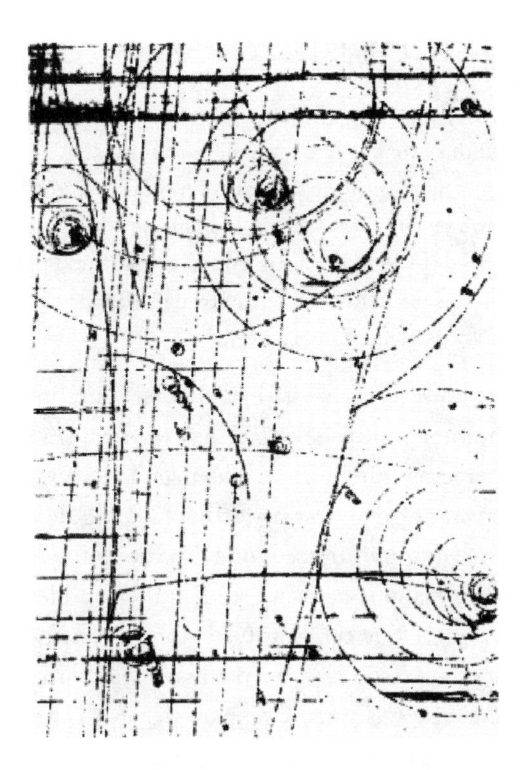

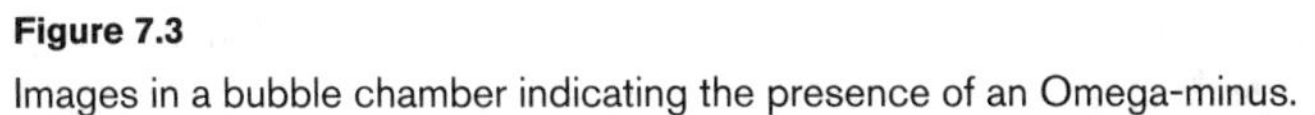

Figure 7.3
Images in a bubble chamber indicating the presence of an Omega-minus.

Omega-minus with other "golden events"—"the single picture of such clarity and distinctness that it commands acceptance" (figure 7.3).

In his book, Galison documents the shifting back and forward between *images* generated in cloud chambers, photographic emulsions, and bubble chambers and the *logic* of the statistical analysis of multiple particle signatures generated in spark chambers and the like.[4] What we are asking here is whether The Event is best thought of as image or logic. In an early draft

4. For a critical discussion of Galison's distinction, see Kent Staley's 1999 paper, "Golden Events and Statistics."

of the discovery paper we find a passage (soon to be excised) that expresses the force of the image:

> Having found this event to be highly significant in searches for un-modeled transients, any hypothesis for a non-astrophysical source would be forced to explain the "coincidence" that the most significant transient found in LIGO data so far, also displays a waveform consistent with the predictions of general relativity. This qualitative evidence strongly supports the compact object merger scenario described in this letter.

A closer analysis, however, suggests it is neither quite image nor exactly logic, while at the same time it is a mixture of the two! Let us start with the mixture: the convincing thing for the scientists, I am suggesting, has been the image of the waveform almost perfectly matching the predicted waveform of a black hole inspiral. Indeed, one deep-thinking scientist wrote to me that the 5 sigma claim is nonsense and all the confidence in the result comes from its clarity. But this clarity is not the clarity of an Omega-minus: first, it doesn't exactly match a predicted waveform, it matches one of 250,000 predicted waveforms. Compare that with the Omega-minus, the track of which matched only a single prediction! This is what Barish points out. And then again, it is not a golden event because the fact that it is so convincing comes from the fact that two such images generated by separate detectors overlap almost perfectly. And yet again, it is not exactly an image like that found in a cloud chamber, photographic emulsion, or bubble chamber, because the causal chain between event and image is so much longer and so much less direct. In particle physics images, the track is "directly" and locally caused by the passage of the particle, given that conditions have been carefully prepared to allow a particle to leave a track and that there is a well-established theory saying that these tracks are caused by the passage of subatomic particles generated in a purpose-built accelerator. In the case of The Event, by contrast, something that happened around 1,300 million light years away and 1,300 million years in the past is the putative cause rather than something deliberately generated in a local machine. Furthermore, the local effect is a force on some mirrors that, when canceled out by feedback loops, gives rise to a set of numbers that can be retrospectively converted into a trace that can be said to match a template only when the noise has been filtered out of it and the signal from one detector has been manipulated and shifted in time to match the signal from the other. So the "image" is not really an image; it is a reconstruction from a set of numbers, which is more like "logic" in Galison's terms.

And, in any case, even if we do think of it as an image, it isn't going to be allowed to be sent out there to convince the world without the "logic" of the statistical analysis that shows that such a coincidence would be likely to turn up by chance less than one time in a gazillion years. And that logic is not the logic of sparks from multiple particles building up to statistical significance over a long period of accelerator time; it is a single golden "statistical event" because all the buildup is in the generation of noise background multiplied up by time slides from a mere sixteen days of actual observation, not the collection of more and more events.

So in most ways The Event does not fit the high-energy physics (HEP) model. And The Event does not fit with HEP precedents in another very important way when it comes to credibility. HEP's credibility has built up over many decades from success with the first bubble chambers, cloud chambers, or whatever came before, to more and more powerful generations of accelerators, each discovering new particles that fit a theory. It has been a triumphant cumulative process with an increase in power and cost at every step. Gravitational wave detection is quite different: it too has a long history—five decades, or ten if you take Einstein's general relativity as the starting point. Gravitational wave detection has also built increasingly powerful and expensive detectors, generation after generation, and nearly every generation of new detectors has claimed to have seen the gravitational waves. But every claim has turned out to be wrong; terrestrial gravitational wave detection has seen nothing in spite of fifty years of trying. And it has, unsurprisingly, been the subject of vitriolic criticism by other scientists who felt their potential funding was going to this wasteful enterprise, and it has been ridiculed by those without funding worries but who still think it was tilting at windmills. Astonishingly, each new generation of detectors was funded in spite of a track record of failure, with, as I argue in *Gravity's*

Shadow, only the demonstration of managerial virtuosity and responsibility as the guarantee that the money was not being wasted. Therefore, there is nothing to go on in the way of believing this event. Gravitational wave detection has to accomplish far more in the way of creating acceptance and changing the order of things than HEP, which has only ever had to nudge things along a bit further.

8 NOVEMBER
Writing the Discovery Paper

ALLOCATING CREDIT

In the case of Big Dog, a proto-paper was also prepared, and, in the book describing it—*Big Dog*—I focused the discussion on the arguments about how the title ought to be phrased—were we looking at a "discovery" or merely "evidence for" the existence of a new phenomenon? The scientists settled on "evidence for," and *Big Dog* shows why they did this and how they set their thinking in historical context. Here I will widen my focus to the whole of the paper on The Event, not just its title.

To begin, there is a not very heated debate about how to distribute credit. Insofar as there is any heat, it goes into arguments about the companion and subsidiary papers that will be written and how to arrange things so that the team will not be scooped by outsiders; this we have already discussed. Mild argument also arises about how to distribute credit among the various national teams. In the case of the Big Dog proto-paper there was a slightly more heated discussion of this matter, but here, since the American LIGO detectors were the only sufficiently sensitive devices that were online at the time of The Event, it is clear that it was LIGO alone that saw it. One view is that "LIGO" should be included in the title of the paper, but this is rejected by the community because of its potential divisiveness. A senior member of the LIGO team remarks:

> I personally find it awkward [to have "LIGO" in the title] since the work is joint LIGO, VIRGO, GEO people are all involved with the data analysis.

An opinion remains among some that Virgo are getting get "a free ride" on the back of LIGO because the institutional authorship treats LIGO and Virgo equally. The Virgo team has made a full contribution to every aspect of the discovery except building a machine that could respond to the signal, and equal treatment does not reflect that reality. GEO600, whose members were also full participants, have contributed significantly to LIGO's hardware in, for example, designing and supplying the new mirror suspensions, but because GEO has a much closer relationship with LIGO than Virgo, its institutional visibility is in jeopardy; its existence has been absorbed into LIGO's whereas Virgo's hasn't. Of course, all the individuals from all the teams will be among the 1,000 plus coauthors of the paper and their university affiliations will be mentioned among the roughly 133 listed.

I have always found the attitude of the LIGO team extraordinarily generous in terms of distributing credit. LIGO, having two detectors, is the only experiment that could, if it wished, announce a first discovery on its own (barring something remarkable); LIGO is a bigger and more expensive project than its nearest rival and has always been ahead of the rest in implementation and in sensitivity. Its ready willingness to present what its members have made possible as a thoroughly international achievement elevates the spirit.

THE PAPER: TITLE

As the scientists start mulling over initial ideas for the paper's title, the following are suggested:

> **Nov. 1:** Direct Observation of Gravitational Waves from a Binary Black Hole Merger
> **Nov. 2:** The Direct Observation of a Gravitational Wave Event in LIGO having a Waveform Indicating a Binary Black Hole Merger
> **Nov. 4:** Coincident laser interferometer observations of a signal consistent with gravitational waves from a binary black hole merger.

This emailer adds:

I dread the comparison with Cabrera's result [the monopole]... as it was one that could have been done by hacking. I agree on the importance of being conservative in our claims.

Another emailer suggests:

Nov. 4: Perhaps a simple "Direct observation of gravitational waves" would do.

Personally, I would also add "First." While it is true that PRL may discourage that, I think there is no dispute that this is the first direct observation of a gravitational-wave signal. Let's have confidence on our achievement and tell the world! :-)

THE PAPER: ABSTRACTS

Four very senior scientists have been tasked with writing the abstract of the paper, and between November 1 and 4 they exchange drafts: draft 1 is modified by scientist 2, then modified by scientist 3, and then further modified by scientist 4. All four agree that each is an improvement on the last. Here I want to draw attention to what is in play. The target journal is *Physical Review Letters* (*PRL*), whose guidelines demand that the abstract be no longer than 600 characters (with spaces), but the scientists think that *PRL* could probably be persuaded to relax their rule for such an important discovery. Thus the fourth draft, with which everyone is initially happy, comes to 1,700 characters. I'll number the drafts 1–4.

The first sentence of all the drafts begins the same way but then varies after the first few words:

(1) On September 14, 2015 at 09:50:45 GMT the Laser Interferometer Gravitational-wave Observatory (LIGO) detected the gravitational waves from a binary black hole merger.
(2) On September 14, 2015 at 09:50:45 GMT the Laser Interferometer Gravitational-wave Observatory (LIGO) detected a large gravitational wave signal having the characteristics of a binary Black Hole inspiral, merger and ringdown.

(3) On September 14, 2015 at 09:50:45 GMT, the two interferometers of the Laser Interferometer Gravitational-wave Observatory (LIGO) observed a strong gravitational wave signal from the coalescence of a binary black hole.
(4) On September 14, 2015 at 09:50:45 GMT, the two interferometers of the Laser Interferometer Gravitational-wave Observatory (LIGO) observed a strong gravitational wave signal matching the waveform expected from the coalescence of a binary black hole system.

A useful concept in the analysis of the scientific literature is "modality." It has been pointed out that, as findings move from the laboratory to the world and thence to the scientific "taken-for-granted," the phrases used to describe a finding change—the "modalities"—are stripped away.[1] That is, instead of a finding being described as particular historic event it is described as a general feature of the world. To caricature, as a finding becomes every scientist's common sense, we no longer see "On the *n*th of September Smith found evidence consistent with the existence of Thargs" but instead see something like, "Thargs can strongly deflect Knoos." In other words, that Thargs have now become a feature of our world is indicated by the fact that accounts of their discovery—their history—are removed, just as in ordinary speech, when we mention dogs and potatoes we do not think it necessary to mention how dogs developed from wolves or how potatoes were first used as a foodstuff because, in our world, dogs and potatoes are just *there*.

We can see the difference in the following phrases:

(1) detected the gravitational waves from a binary black hole merger
(2) detected a large gravitational wave signal having the characteristics of a binary Black Hole
(3) observed a strong gravitational wave signal from the coalescence of a binary black hole
(4) observed a strong gravitational wave signal matching the waveform expected from the coalescence of a binary black hole system

1. See Note VII in "Sociological and Philosophical Notes," 361.

Phrases (1) and (3) are far stronger than (2) and (4) because they simply refer to detection and observation without mentioning the mechanism; (1) and (3) say "We saw a black hole" (like "We saw a dog"), while (2) and (4) say this is why we think we saw a black hole (like "We saw something that may have developed from a wolf").

In the light of this analysis, the almost uniform first few words of the sentence are especially interesting:

> On September 14, 2015 at 09:50:45 GMT [LIGO] detected/observed …

This clearly defines what is being reported as a singular historic event. It is being reported as the culmination of fifty years of searching—something of great historical significance. The concluding sentence of version 4 makes the historical importance crystal clear:

> This is the first direct observation of gravitational waves and the first direct observation of black holes.

This would be slightly modified in a fifth iteration to

> This is the first direct observation of gravitational waves and the first direct observation of *the dynamics of* black holes. (italics added)

This is still historic but scientifically weaker, giving priority to others' observations of, for example, the absence of X-ray emissions from one of a pair of similar stars. "The dynamics of" would, however, be removed from the final version.

We should note that, as one member of the collaboration pointed out, astronomical observations are inherently historical, so the historical flavor of the paper has this additional source. That is to say, something like a Higgs boson is not inherently historical because Higgs bosons have no identity as individuals; stars, however, have particular locations, exploding stars have dates associated with them, and some astronomical events are named—for example, Supernova 1987A. This is a binary black hole system that came to the end of its life on a particular date. That said, the flavor of these drafts is still unusually historical and modality-rich for physics.

I'm writing this part of the book at the very beginning of November, but I am not the only person to notice what is going on. One scientist writes:

> **Nov. 8, 9:39:** … It is not crisp enough for a detection paper reporting an important discovery. Its abstract starts (with a date and time, which makes sense only as a detail in text) too dramatically, at least for me. I do not want to be too critical on the great efforts of the paper preparation team, and their experience in the field is far more than mine. However, that very fact gives me a different view as a reader of the paper.

Another critic writes:

> **Nov. 15, 21:52:** We believe that the historical portion of the introduction should be shortened significantly. This should then be followed by a short outline of what we have seen, with some of the material from the Observational Results section moved to the Introduction for this purpose.

As we will see, the historical style being criticized here is carried through into the introduction to the paper and the conclusion. It is, to repeat, a very unusual paper because it abandons the normal "literary technology," which stresses the passive voice and tries to render the reader a "virtual witness" of what went on in the laboratory.[2] Here the paper is active: it reports things that the scientists did at particular times rather than states of the world that would have impacted on them whenever they looked and wherever they were; the devices *detected* the waves rather than being impacted by the waves, and it is impossible for readers to be virtual witnesses *because* the act took place at a particular time and place. In sum, this is more like a historical recapitulation than a scientific paper, and that, of course, is

2. See Note VII in "Sociological and Philosophical Notes," 361.

because the scientists in writing this paper are conscious of their place in history (just as I am in observing these events).[3]

In contrast, someone will write in reference to a later draft:

> **Nov. 21, 16:57 [referring to draft 6]:** I really enjoyed reading the latest draft. It presents a compelling story. I like the historical context provided in the introduction, it sets the tone nicely for a historic paper.

So two things are being argued here: first, that the way the things that have been discovered are being described is ahistorical, which gives them more than usual facticity for a scientific discovery paper, but second, the way they have been discovered is described as a historical event, giving the process much less passivity than is normal in a scientific paper: something exceptionally real and ahistorical has been found, and it is also a great historical event to have found it—nice!

THE PAPERS: THE EVENT PAPER VERSUS BIG DOG PAPER

Exactly how unusual is the historical tone of The Event paper is hard to say without a survey of all the papers in *PRL* but one can gain some sense by comparing it with the proto-paper written in respect of Big Dog. At

3. On the subject of places in history, in monitoring the emails that fly back and forth as the introductory section of the paper is refined I learn that on the very first page of *Gravity's Shadow* I made a historical mistake. I tread here on the dangerous ground of attributing credit, but I now believe that the following was incorrect. I said that the claim that gravitational waves do exist "was recognized by the awarding of the 1993 Nobel Prize in physics to two astronomers, Russel Hulse and Joseph Taylor, for an indirect confirmation of the existence of gravitational waves," with a footnote explaining: "Hulse and Taylor observed the very slow degradation of the orbit of a pair of binary stars over many years and showed that it was consistent with the emission of gravitational radiation according to Einstein's theory." I now learn that the Nobel was awarded for the discovery of the binary system that was only *later* used for the gravitational wave finding. Hulse, though he was there when the discovery of the system was made, left astronomy before the analysis was done, and Taylor's partner, to whom the share of credit for the gravitational wave part of the discovery should go, was Joel Weisberg.

the time the scientists were writing the Big Dog proto-paper they did not officially know whether it was a historic event or a blind injection.[4] The Big Dog proto-paper is, perhaps because of this, quite different from the early November draft of The Event paper. Its abstract fits the rules for *PRL* papers in respect of length, whereas the later abstract flouts the rules in the expectation that the paper will receive special treatment because of its importance.[5] The Event's abstract is over 1,000 characters and the length of the early draft is ten pages. Big Dog's abstract begins quite differently—in the more standard way:

> We report the observation of a gravitational-wave signal in data from a joint science run of the LIGO, Virgo and GEO 600 detectors.

And the Big Dog introduction begins:

> General relativity predicts that two masses orbiting under their mutual gravitational attraction will spiral together as they radiate energy in the form of gravitational waves.

Compare this with what we find in a November 5 draft of The Event paper:

> One year after the final formulation of the field equations for General Relativity in 1915, Albert Einstein predicted the existence of gravitational waves.

4. Actually, many significant scientists did know, either because they had realized that small errors had been made in the process of injecting that "gave the game away," or because some cheating had taken place and some scientists had looked in the injection channels. Many others had good reason to believe that it was a blind injection even if those reasons cannot be shown to be sound; whether the reasons were actually sound or not doesn't matter so long as you believe they are sound. For example, I was sure it was a blind injection even though others tried to persuade me my reasons were not good ones; many of the scientists thought as I did.

5. The scientists cite the notorious BICEP2 claim as precedent for this. In that case *PRL* published a 25-page paper (the official limit is 4 pages) with an abstract of over 2,100 characters (the official limit is 600 characters).

Turning to the conclusions of the Big Dog proto-paper and the November 5 draft, the opening and closing sentences in each case are as follows:

Big Dog proto-paper

The event described here is remarkably well separated from the background in the compact binary coalescence search.

This event suggests that the next generation of gravitational wave detectors [12, 13] could detect significantly more binary black hole mergers than previously anticipated.

November 5 draft

As we demonstrated above, a gravitational wave signal has been observed by the LIGO interferometers with a very high degree of confidence. After a half-century of development, gravitational wave detection has now been accomplished. The first step has now been taken in the exploration of the universe via the gravitational spectrum.

Future prospects are indeed bright for the new field of gravitational wave astronomy.

More evidence for the unusually historical nature of The Event draft is provided by one the scientists, who emails:

Nov. 23, 11:56: I looked through a collection of major discovery papers (both J/psi papers, tau lepton, gluons/QCD, W boson, top quark, both accelerating universe papers, CDMS2, BICEP, etc) and all of them begin with the phrase "We report…" or "We present…" or "We have observed.…" Aside from following precedent, I like this construction, as it suggests we are impartially reporting the results of our experiment to the scientific community. I suggest rewording the start of the abstract to the following:

"We report on the observation of a strong coincident signal in the two interferometers of the Laser Interferometer Gravitational-wave Observatory (LIGO). The signal was detected on September 14, 2015 at 09:50:45 UTC, and was observed to sweep upward in frequency from 30 Hz to 250 Hz with a duration of 0.2 seconds and a peak gravitational wave strain of

> 1x10^{-21}. The signal is recovered by matched-filtering with a signal-to-noise ratio of 23.6 and is clearly visible in band-passed data from the detectors. A careful investigation of the state of the interferometers around the event time revealed no instrumental or environmental causes for the signal. We empirically estimate the significance of the signal using extended observations to be greater than 4.9 sigma, equivalent to a false alarm rate of less than 1 event per 100,000 years."

Though this sentiment does not turn out to win the day, symptoms of caution begin to grow. In terms of chronology, we now leap ahead so as to keep this section on the writing of the paper in one place. Thus we turn now to emails from December.

THE PAPER: PULLING IN THE HORNS

> **Dec. 15, 00:38:** We suggest avoiding the "Einstein history" in the final sentence of the first paragraph. Throughout the abstract, there are a number of historical asides that could be removed (e.g. "A century after the fundamental predictions of Einstein and Schwarzschild…").
>
> **Dec. 15, 00:01:** General Suggestions:
> To avoid the controversy associated with word "direct," and also to clarify its meaning, consider replacing, when appropriate: "direct detection of gravitational waves" [with] "direct detection of gravitational wave strain"

The term "gravitational wave strain" is technical. I don't really know what it means in comparison with "gravitational waves." Here the question turns on the nature of the paper and its intended audience—specialists of a wider scientific community. "Wave strain" won't work very well for them. This is something we will return to.

Another emailer wants a more nuanced claim in respect of the binary black hole observation.

> **Dec. 10, 15:41:** the first two paragraphs are somewhat misleading, and could be boiled down to something like: Gravitational waves produced

> during binary coalescence will sweep upward in frequency and amplitude as the orbit decays; then, following coalescence, the amplitude of the waves decay exponentially as the system settles to a stationary configuration. This morphology is observed in the data. The only known objects that are compact enough to form binaries whose orbital motion produces GWs in LIGO's band are black holes and neutron stars; however, a binary neutron star system would produce gravitational waves sweeping up to kHz frequencies before their coalescence—higher than those observed. Consequently, this system is most likely a black hole binary.

This emailer, then, prefers talk of "likelihoods" to talk of "observations."

TO BE OR NOT TO BE DIRECT

In the second week of December, a storm of emails—dozens of them—sweep through the collaboration arguing about whether the paper should present the discovery as the "first direct observation" of gravitational waves or even mention that this was a *direct* observation. This can be seen as an element in the overall "pulling in of the horns" but perhaps it is also encouraged by uncertainty about which of the ripples is being addressed by the paper: the second or the third? If we start with the third ripple—scientists in other fields and the scientifically educated component of the general public—it seems to me that they need to be told that this was not just an observation of gravitational waves but that there was something special about it that made it different from other observations or non-observations of gravitational waves that they had read about or heard about. The word "direct" captures this difference.

Unfortunately, or maybe it's not so unfortunate depending on how you view the job of the sociologist, it is impossible for me to simply report on how events unfolded as though I did not have a view of my own. My view is that, other things being equal, strong claims are the right kinds of claims to make in physics, because one does not want to be running away from the responsibility of making a claim. This argument will be set out at length toward the end of the book and encountered where we look at views about the Second Monday event, but the leaning that I have in this direction has already been documented in *Big Dog* so it cannot be hidden. In any case, it is already imprinted on the ether in other remarks that were commissioned by the University of Chicago Press for their blog. At the time of the BICEP2 claims, many people were asking if what had supposedly been found were the elusive gravitational waves for which scientists who began with Joe Weber had been searching since the 1960s, and many were asking me if my project was now over since the waves had been discovered. The blog post I wrote for Chicago explained that if BICEP2's results were confirmed (they were disconfirmed within a year), they would have found *primordial* gravitational waves—not what the post-Weber and interferometer scientists took to be the Holy Grail and not much to do with the science that I had been following. Here is an extract from that March 2014 blog:

> Matters get complicated because there are other ways to detect gravitational waves. … They have already been detected in this way by Hulse and Taylor—winners of the 1993 Nobel Prize in physics—who observed for a decade the slow decay of a widely separated binary system's orbit, and showed it was consistent with the energy emitted by gravitational waves [I should have mentioned Weisberg—see note 3]. … When (if) LIGO and the international network of interferometers start observing, they will be looking in different wavebands than did Hulse and Taylor, and they will be able to see many more of many different kinds of phenomena. The observation of a binary inspiral, or a supernova, or a neutron starquake will take seconds or less, not decades, and there should be many per year once full sensitivity is reached. The true justification for the interferometers is then gravitational astronomy—including our first look into the heart of colliding black holes—with the direct discovery of gravitational waves exciting but not so surprising as it once would have been.
>
> Now, if it is confirmed, BICEP has observed gravitational waves in another indirect way. The group has inferred their existence from the polarization patterns of electromagnetic waves (the microwave background). … Speaking now purely as my unprofessional self—a citizen with a schoolboy interest in science, but one who is perhaps biased by lengthy contact with these

> groups—I think building mind bogglingly fine gossamer webs that can capture exquisitely ephemeral waves is more exciting than inferring their existence from the movement of stars or from patterns in the much stronger electromagnetic spectrum. This is because it leads to more than new understanding: it demonstrates unprecedented control over nature and a heroic extension of our means to uncover its secrets.

The blog contained more discussion of what counted as "direct" and what as "indirect," anticipating some of what will be described below, but pretty well everything I said in that blog came up independently in the late-December emails. The most exact match to my conclusion, perhaps, is this email from the evening of December 13 with its immediate endorsement from another contributor:

> **Sun., Dec. 13, 12:47:46:** I think that the emphasis on "detection" is misguided. Our aim is not detection, per se. We need to emphasize that we are starting a whole new way of doing astrophysics, quantitatively. Fortunately, we are able to show that this method works and that we are already discovering the new physics that is discussed in the paper.
>
> **Dec. 13, 23:07:** Absolutely!

This is what is so extraordinary about this claim: not the discovery but the founding a complete new branch of astronomy and astrophysics. But it can only be achieving this kind of result because of what I called the demonstration of an unprecedented kind of control over nature, and this, it seems clear to me, is what is signified by that word "direct." If it had not been a "direct" detection, but one of the other kinds of detection, then no new branch of astronomy and astrophysics would be being founded.

The counterarguments—the arguments for not using the word "direct"—are (a) a commendable modesty; (b) a desire to avoid political unpleasantness among those who believe that other ways of seeing gravity waves are also "direct"; and (c) the view that every specialist will understand exactly what has been done and there is no need to spell it out when a lot of trouble and strife can be saved by leaving "direct" and similar words unsaid.

Let us start with (c). It is certainly true that every specialist will understand what has been done and how it differs from what has been done before, but here is where the ambiguity comes in about what the paper is meant to achieve. The paper is looking farther out among the social ripples to a much wider group of scientists and even toward the general public and the media, and they will certainly not understand how this discovery compares to previous discoveries of gravitational waves. Were that not the case, then lots of confused people would not have been asking me if BICEP2 meant the end of my project. To the extent that the paper looks outward, not only to the third ripple but also to what we might call a fourth ripple—history, with everyone agreeing that this paper is going to be a classic and read for years to come—it needs to distinguish itself from what came before as much as it can. This is the paper that will found gravitational wave astronomy and astrophysics, first "directly" seeing gravitational waves, first directly seeing a black hole (though the directness here is more dubious), and, surely, first directly seeing an inspiraling binary black hole system—which is close to unambiguous, barring Boson stars and the like. From this point of view, the words "first" and "direct" ought to be mentioned wherever possible, because the paper is addressing not the people who will understand without being told, it is addressing the people who will not understand and need to be told.

One emailer goes straight to the sociological heart of the matter regarding the audience being addressed, doing my work for me. He writes:

> **Dec. 12, 23:40:** Maybe we should look at the public perception of direct vs indirect detection of GWs? I googled "gravitational wave detection" news. Below are the first few results as returned by my browser. I think it's pretty clear what the perception is and what people expect to hear from us.
>
> http://www.economist.com/news/science-and-technology/21679433-novel-approach-observing-heavens-orbit-gravitys-rainbow
> "Physicists have compelling indirect evidence that they are real. (The 1993 Nobel prize for physics was awarded for observations of a pair of superdense stars whose orbits around each other are decaying in a way that can be accounted for only if gravitational waves are carrying away

some of their momentum.) But researchers have never seen a gravity wave directly."
http://www.bbc.com/news/science-environment-34815668
"If it's a success, one of Albert Einstein's greatest predictions will have been directly observed for the first time."

http://www.space.com/27510-gravitational-wave-detection-method.html"
Scientists have still not made direct observations of gravitational waves, although researchers continue to endeavor to detect them using experiments involving lasers on the ground and in space."

http://www.nature.com/news/freefall-space-cubes-are-test-for
-gravitational-wave-spotter-1.18806"
First predicted by Albert Einstein almost exactly 100 years ago as part of his general theory of relativity (see http://wwww.nature.com/relativity100), such gravitational waves have never been observed directly—let alone used to study the cosmos."

http://www.cbsnews.com/news/esa-spacecraft-to-test-gravity-wave
-detection/"
Gravitational waves, he said, are a 'direct prediction of the general theory of relativity, but they are yet to be directly detected.'"

http://www3.imperial.ac.uk/newsandeventspggrp/imperialcollege/
newssummary/news_1-12-2015-15-42-15"
Gravitational waves should pervade the universe, but they have not yet been directly detected because they are so small."

http://www.cbc.ca/news/technology/lisa-pathfinder-waves-1.3347724"
So far, gravitational waves have never been detected."

http://gizmodo.com/a-groundbreaking-physics-laboratory-is-about-to-blast
-i-1745273245"
Its goal is simple: using laser interferometers, the spacecraft will attempt to precisely measure the relative positions of two 1.8 inch gold-platinum cubes in free fall. Housed in separate electrode boxes a mere 15 inches apart, the test objects will be shielded from the solar wind and all other external forces, such that the tiny motions caused by gravitational waves can (hopefully) be detected." [This is a different experiment.—HMC]

http://nation.com.pk/blogs/02-Dec-2015/in-search-of-einstein-s-gravitational-waves"
One prediction of GR has been elusive however. That is, the direct proof for gravitational waves—ripples or fluctuations in the fabric of spacetime—caused by accelerating masses—which transport energy as gravitational radiation. While they have been observed indirectly before, there has never been a direct detection of these waves."

https://www.theguardian.com/commentisfree/2015/dec/07/einstein-universe-gravitational-waves-theory-relativity"
With Advanced Ligo up and running, and Advanced Virgo about to launch, it would be highly fitting if the first definitive detection of gravitational waves were to coincide with the centenary of general relativity. The universe is sure to spring surprises: this could yet prove to be Einstein's greatest gift."

etc etc....

So that is the popular vote: gravitational waves have not yet been directly detected but, according to some anyway, LIGO and the like could do it.

The discussion of direct versus indirect was anticipated in the case of Big Dog and can be found in the corresponding book on pages 197–200. Given BICEP's demise, the question turns on earlier observations of orbiting neutron stars made by Taylor, Hulse, Weisberg, and also Thibault Damour, who worked out the model and did the calculations that allowed Taylor and Weisberg to create a proof of the existence of gravitational waves by fitting their observations to a theoretical curve. Hulse has long left the field, and Weisberg does not seem to be an active participant in the debate; but Taylor and Damour believe that what they did should count as *direct* observation, even though few others do. Damour wrote to me in March 2011 (I am quoting from *Big Dog*):

> Evidently, there is here some sociology at work: people like Joe Taylor or me, have written in several papers such statements [implying directness]...while people who fought for the funding of LIGO etc tended to downplay the value of the pulsar-experiment/GR-theory agreement as a *direct* evidence, probably because they wished to insist on the novelty that will represent the

> first direct detection of GW's arriving on the Earth (in addition to the important scientific prospects opened by GW astronomy).

Taylor and Damour are very much liked among more senior members of the gravitational wave community and their work is hugely admired, so there are many who do not want to offend them.

Now let us look at extracts from some of the emails so as to get a flavor of the debate. Here I am trying to let the story emerge rather than present the emails in chronological order, though the entire storm lasted only a few days anyway. I'll start with cautious emails and work around to the opposing view from those who believe that the paper should state clearly that it is the first direct detection of gravitational waves. As explained, the modest emails are part of a general trend that can be spotted as the paper moves into and beyond its seventh draft—there is "a drawing in of horns." Those first drafts were full of confident joy, but now people are starting to worry about who they might upset if they appear too pleased with themselves and asking whether it would be better simply to state the facts and let the scientific community draw its own conclusions. We saw something similar with the Big Dog rehearsal.

> **Dec. 12, 00:58:** Tone: We have a great achievement, and I would hate to see us generate any resentment or animosity in the broader community. I think we should take a modest tone in this paper. This is where we are talking to our scientific colleagues and we should simply say what we have done clearly and without making claims of how important it is. I don't think we need to say (repeatedly) that this is the first detection of gravitational waves, or the first detection of a binary black hole. I suggest that we say this once, in the conclusion. Save the superlatives for the tweets and blogs that will be the primary way this result is communicated to the general public.
>
> I would also like to suggest that we avoid the phrases "direct observation" and "direct detection," at least in the title, abstract and introduction. While I know that we use those terms as a way of distinguishing what we do from other methods of GW detection... I have come to understand that there are people in our broader community who think that this terminology is meant to diminish the importance of their work. In keeping with the modest tone I

advocate, I see no benefit in using these particular terms. Indeed, without further definition, they do not convey any clear meaning ... We have a great result, and it will not be any less great without this characterization.

Dec. 12, 01:30: Personally, I have never felt motivated by the direct detection phrase and thought it was a red herring. We saw some masses, which happened to be mirrors, moving under the influence of gravitational waves. Hulse, Taylor and Taylor's later collaborators saw some masses, which happened to be neutron stars, moving under the influence of gravitational waves. What's the difference?

The difference is the Taylor crowd observed a distant GW transmitter and figured out how it worked! We figured out how to build a sufficiently sensitive GW receiver and *since we built it, we know exactly how it works*. If anybody misses the impact of those italicized words, check out BICEP-2's and Planck's experience to date. Those words in italics represent a huge advance for GW physics and astronomy.

A lot of the contributors are trying to resolve the problem of whether to call this direct or indirect observation of gravitational waves as though they were philosophers—the above email is an example. This applies equally to Taylor and Damour. Thus Taylor wrote to me (the quotation is also included in *Big Dog*):

> In the binary pulsar experiment, and also in a LIGO-like experiment, one infers the presence of gravitational radiation based on effects it induces in a "detector." If a ruler could be used to measure the displacement of LIGO's test masses, I would grant that detection to seem rather more "direct" than one based on timing measurements of an orbiting pulsar halfway across our Galaxy. However, LIGO can't use a ruler; instead they use servomechanisms, very sensitive electronics ... and finally long sequences of calculations to infer that a gravitational wave has passed by. Such a detection, like the binary pulsar timing experiment, is arguably many stages removed from being what most people would call "direct."

It is tempting to go down this road, which echoes the email of December 12, 1:30 (above)—if we want to know if this is direct, we can look at the detailed mechanism of the observation or experiment and try to work out what is meant by "direct." The trouble is that this would mean that in modern physics *nothing* amounts to direct observation, since every telescope of other observing instrument nowadays uses sophisticated sensors, banks of computers to handle the numbers, and complex statistical inference to work out what they mean. Even Galileo wasn't observing directly since the light that came into his eyes was mediated by a complex system of lenses. This kind of logical inquiry is not going to prove the matter in the way that mathematics proves things. Rather, the argument has to turn on regular usage of terms within the physics community—a kind of high-level physicists' common sense. Taylor, Weisberg, and Damour inferred the existence of gravitational waves to the point of proof through measurements of the orbit of a pair of stars. One might say if one were so inclined that they directly proved the existence of gravitational waves. But, as one emailer pointed out (I wish I had thought of it), if something had got in the way of the gravitational waves on their way from the orbiting pulsar to Earth, and diverted them off somewhere else so they did not impact on the Earth, it would have made no difference to that inference, so long as the radio waves indicating the way the stars were moving still got through. But in the case of LIGO, what is seen is the impact of the gravitational waves on an instrument built with the specific purpose of reacting to the waves and converting them into electrical signals so that they can be measured. And, as the emailer said, if something got in the way of those gravitational waves and diverted them off somewhere else, LIGO would not see them. So in commonsense terms—that is, the kind of commonsense that constitutes physics—only in the second case are gravitational waves impacting on the instrument: this kind of directness seems a lot more direct than the other kind.

Other emailers decide to do some more sociological work:

> **Dec. 12, 14:04:** concerning the phrase "direct detection of gravitational waves" there are two aspects that we should consider:
>
> —the sociological and political one: We know there are people who feel offended by us claiming this to be the first direct detection of gravitational

waves, as they think the observation of the energy loss of binary systems is a direct detection. Irrespective of who is right, our desire to avoid a fight over this term may outweigh our wish to call it a "direct detection."

As for the fight, the fear seems very real; this comes from a senior figure in the field:

> **Dec. 13, 08:45:** I want to convey an impression that I got last week, when I attended a meeting.... After I gave my talk (of course without mentioning THE EVENT), somebody in the audience asked the usual question:
>
> When do you expect the first DIRECT detection of gravitational waves?
>
> Before I could even answer, one of the most influential [European country] astronomers in the audience literally exploded, saying that these words were a blatant disregard for the work that had been done by others and that gravitational waves had already been discovered. In the coffee break afterwards a long discussion followed that for the first time made it clear to me that others have very strong feelings on this subject.
>
> I am now absolutely convinced that we have nothing to gain from using either the word direct or first in the paper and particularly in the title! By raising emotions in some colleagues it might even distract from the main message we want to bring across.
>
> Something like "Observation (or Detection) of gravitational waves from a binary black hole merger," is short, sweet, and describes the facts.

I wrote to this senior figure and he told me that the senior astronomer was generally very well-disposed to the gravitational wave detection project, so what I say now about the relationship between astronomers and physicists probably does not apply to this incident and is, in any case, speculative. The speculation is that some of the heat in this debate emerges from the context of the rivalry between physicists and astronomers. In the days when LIGO was fighting for funding, astronomers were fighting back. At that time the astronomers were furious about the "O" in LIGO which stood for "observatory," claiming that what was being built was a device that should be paid for by physicists, not an observatory that might attract funds from their own projects, and that, furthermore, LIGO had little chance of doing any observations for a very long time, and not much when it did finally go online, while telescopes were discovering new things about the heavens every day. And LIGO, indeed, made no positive observations for twenty-three years from the date at which funding was granted, so the astronomers were not being unreasonable. Astronomers might have been feeling twenty-three years of resentment at LIGO's promises to see gravitational waves, while all along, a couple of astronomers had done the job of confirming the gravitational wave predictions of general relativity using their existing telescopes and spending nothing but their salaries. It might even have been that since instrument-builders are no longer considered by astronomers to be persons of high status, astronomers might feel scornful of physicists building machines rather than watching the stars. And remember, in late 2015, when this discussion was going on, no one outside the select 1,000 knew that gravitational waves had just been detected because The Event was still secret, so astronomers would still be seeing LIGO as a set of optimistic and expensive promises with an ever receding delivery date.

Here are a couple more emails from the "not to be" camp:

> **Dec. 14, 09:12:** Rationale: Even if you feel these terms may or may not be justified, they are simply wholly unnecessary and at worst can only do harm. This paper will stand for its (huge) scientific merit alone, no need for adding any personal judgements or qualifiers that could be perceived as aggrandizing ours, or belittling others. History and the community will decide what was "first," "direct" etc, but such judgements should have no place in a paper like this.
>
> Let's stick to stating facts only!

> **Dec. 13, 15:25:** I also advocate we remove *direct* and *first* from any statement in the body of the text of the discovery paper and the companion papers. The facts should speak for themselves and we don't need to relay on some sort of anchoring for the readers to decide how important the results are.

Moving now toward the other end of the spectrum—those who want to distinguish between LIGO's achievement and those of the astronomers—we find the following:

> **Dec. 12:** We have built a machine, from scratch, that is a transducer of gravitational waves to electrical impulses. This clearly constitutes "direct" measurement according to typical usage of that word, and certainly anyone not entrenched in our field would agree with that.
>
> It's also a mind-boggling accomplishment that no one else can claim. I therefore assert that we can, with a clear conscience, claim the mantle of "first direct detection" of gravitational waves.

> **Dec. 13, 16:24:** if I put on my astronomer hat, I usually describe what LIGO as doing as "observing" as opposed to "detecting." The difference is that we have built a detector that a physical phenomena has deposited energy in—we made an observation of the phenomena by this mechanism. By contrast with Hulse-Taylor, the physical phenomena we are observing (what our telescopes are reporting on) is the decay of the orbit itself, not the GW. Again, in analogy with astronomy—LIGO is more like a telescope, observing a source directly. The "detection" of GW from Hulse-Taylor is like inferring the stars shine because we observe their mass evolving due to the burning of hydrogen—not the same thing as detecting the light of the stars themselves, but an incontrovertible finger pointing at the responsible phenomenon.

And an email responding to the indignant astronomer:

> **Dec. 13, 20:01:** It is all about politics. And in 20 years, people will look at our paper and then at the Hulse Taylor paper and no one will have ever claimed the first direct detection of gravitational waves. … I don't think politics should trump science and I have not seen a single scientific argument that supports their claim of a direct detection in this entire threa[d]. Does anybody know a refereed paper that even claims a direct detection of, not evidence for, gravitational waves? (Outside of Weber and others which are known to be wrong.) If there is a paper that claims a direct

detection, we can't claim anymore that we are the first unless we dispute the other paper. Little late for that.... I don't like to be bullied. Sorry.

Another emailer now finds the introduction to the award of the Nobel Prize to Hulse and Taylor:

> The good agreement between the observed value and the theoretically calculated value of the orbital path **can be seen as an indirect proof of the existence of gravitational waves. We will probably have to wait until next century for a direct demonstration of their existence.** http://www.nobelprize.org/nobel_prizes/physics/laureates/1993/press.html

Still more surreal, an emailer tracks down a reference to direct detection on a website on which LIGO tries to explain its purpose:

> The LIGO Scientific Collaboration (LSC) is a group of scientists seeking to make the first direct detection of gravitational waves, use them to explore the fundamental physics of gravity, and develop the emerging field of gravitational wave science as a tool of astronomical discovery. (http://www.ligo.org/about.php)

And another emailer finds a reference to directness in LIGO's own mission statement:

> As it is defined in the LIGO Laboratory Charter, LIGO's mission is to open the field of gravitational-wave astrophysics through the direct detection of gravitational waves. (https://www.ligo.caltech.edu/page/mission)

As another emailer puts it:

> **Dec. 14, 18:50:** First, just to weigh in, I think this is absolutely the first direct detection of GWs. I don't think the word "direct" is needed in the title (and I think "LIGO" is), but I don't think we should shy away from using the word "direct" throughout the paper. It seems silly to me to say for years that LIGO is working toward the first direct detection of GWs and then be afraid to actually claim that achievement when the time comes.

And another:

> **Dec. 15, 16:40:** ... People have given us money to build an instrument capable of _direct_ detection, we have been talking about _direct_ detection all along, everyone expects a _direct_ detection ... and when finally we get one, we don't want to claim it? Are we afraid of meeting expectations?

As explained, this is just a brief selection from dozens of emails covering the full range of opinions and expressing them in ways covering the full spectrum of forcefulness. What are the unfortunate paper drafters to do in the face of this storm? One cannot but feel sorry for them. Imagine trying to write a paper with 1,000 people entitled to tell you how to do it better and not at all shy of sending in pages and pages of comments on each and every draft. About 2,500 emails in total are sent to the paper-writing team, with nearly half the authors listed on the paper writing in.[6] A lot of those emails contain many, many suggestions. Many of the suggestions argue about the use of "direct." Nevertheless, the writing team's reaction comes as a surprise—at least to me and one or two others. The team conducts an opinion poll! Here are two out of the three questions, with the third question being somewhat more technical and, in a sense, even more of a surprise:

Poll for choosing the paper title; Rank the title from your favorite (1st rank) to your last choice (last rank)

A. Observation of Gravitational Waves from a Binary Black Hole Merger
B. Direct Observation of Gravitational Waves from a Binary Black Hole Merger
C. Detection of Gravitational Waves from a Binary Black Hole Merger
D. Direct Detection of Gravitational Waves from a Binary Black Hole Merger

6. Thanks to Peter Fritschel for these numbers.

E. LIGO Observation of Gravitational Waves from a Binary Black Hole Merger
F. LIGO Detection of Gravitational Waves from a Binary Black Hole Merger
G. Observation by LIGO of Gravitational Waves from a Binary Black Hole Merger
H. Direct Observation by LIGO of Gravitational Waves from a Binary Black Hole Merger

Use of "direct" (detection and/or observation) in the body of the text

A. No problem to use "direct" in the paper
B. Use "direct" once in the introduction and abstract only
C. Use "direct" once in the conclusion only
D. Don't use "direct"

Instructions: Use the buttons to add candidates to the ballot, and then drag to arrange them in order of preference with your favorite at the top and your least favorite at the bottom.

Poll conducted on December 16–17

At least one emailer is bemused:

> **Dec. 16, 13.25:** I am immensely grateful for the work you have done, and understand the paper committee's evident exasperation. Nevertheless I don't think a voluntary online poll, however well-crafted, offers an appropriate gauge of "consensus." I'd suggest we ask our Executive/Steering committees to take up these questions for reasoned discussion. Of course, they may always decide to conduct a poll…

Other emailers write in titles in addition to those on the list; one wants a "none-of-the-above" option, so the sense of commitment to the eventually chosen title could be gauged.

As of December 17, I am guessing, just from the tenor of the majority of the emails, that the chosen title will not have "direct" in it. I am guessing that direct will appear in the rest of the paper—I hope so—but I am still bemused that the word "direct" is not an automatic inclusion in the paper and the title given that we are now all reminded of the mission statements. As I wrote, on December 15, to one very senior member of the collaboration who wondered how I was enjoying this aspect of the debate:

> I think the [strangest thing] is the mission statement that your goal is to achieve a first direct detection [and] after fifty years and [a] billion dollars you achieve it, and then you think you shouldn't say so.

In a collaboration-wide teleconference (with about 290 nodes) held on December 17, however, the organizers of the poll, perhaps reacting to some negative reactions to the idea of reaching conclusion of this importance by polling, say: "We will use the poll results to guide us: the poll does not determine 'the winners.'"

At any rate, the poll results come in on December 20 and are clear, though the number of voters was only 288—a bit more than a quarter of the collaboration.

> Poll 1: No direct and no LIGO in the title. Preferences are Observation *or* Detection of GW from BBH merger. Poll 3: It is OK to use direct (detection/observation) in the body of the paper.

WHAT SORT OF THING IS THIS PAPER?

Because I know too few of the principal actors, I am not able to spot rivalries between groups with any degree of certainty. Nevertheless, as the debate about what should go into the paper unfolds, I gain the sense that some of the heat is being generated by individuals or groups who want to make sure that their particular contribution or style of analysis is represented in this, about to become famous, discovery paper. Others also tell me of the rivalries developing, and we've seen the warning to me about throwing more "chum" in the water.

Among some there seems to be a tendency to prefer an appearance of sophistication to a search for clarity and simplicity. But perhaps this point can be folded into the choice of what sort of thing the paper should be. I feel the early drafts are torn between different possible aims.

(1) The paper as proving beyond reasonable doubt that after fifty years of struggle and false claims, a real gravitational wave has finally been seen. This demands a long and technically complex paper aimed at the specialist community and covering all doubts. Its precedent could perhaps be the BICEP2 paper—twenty-one pages of argumentation, data, and figures—in the face of *Physical Review Letters*' normal limit of four pages. The BICEP2 team felt this kind of length was required to prove their case. Ironically, of course, BICEP2 turned out to be wrong only a few months later!

(2) The paper as a readable classic paper—the first direct detection of a gravitational wave (as well as a black hole and a binary black hole inspiral) even if those exact terms do not find their way into the title. This aim demands a short, clear, and simple paper with the detailed justification compressed or relegated to other papers, to be simultaneously published if it could be managed or to follow up a little later. This paper would be aimed at the community of physicists in general rather than the specialists. It should be short because nonspecialists' attention span would be short and they would be bored by endless detailed justification.

The scientists keep asking for changes that could be classed as referring to one or the other of these versions of the paper in the course of justifying the inclusion or exclusion of "this or that," but the models are mostly latent: the tension is rarely an explicit agenda item. Here are emails where the issue is made explicit:

Nov. 13, 5:44: First, I would like to say that I think that the paper is significantly longer than I would like. I am sure that we can get permission from PRL for this length, but the fact that we can get that permission is not a reason why we should. I want this paper to be read widely by the physics community, not just the people who already have an interest in GWs, but the broadest swath of physicist that we can get.

Nov. 16, 2:38: I want to weigh in, as others have, on the issue of the paper length and add my support to the "it's somewhat too long" view. I stepped back and tried to read the paper as someone outside the field, and came away thinking that, while everything I needed to understand the result (and believe it!) was in the paper, the level of detail in some portions of the text was more than would be needed by a general reader and could be handled by referencing the accompanying supporting papers. Others have offered some good suggestions as to where to trim.

Here, from a surprising and very senior source, is a demand to leave it to the reader to work out what is going on, though this same commentator was also one of the first to stress that this would become a classic paper. I would think that a classic paper must explain exactly what it is doing, without leaving the wider community to make its claim for it:

Nov. 16, 21:16: I tend to favor being conservative on our side and using "evidence for." I don't think it weakens the conclusions.

Here is another emailer who stresses the wider community:

Nov. 24, 10:25: I continue to worry about the readability of this paper, and hope that in the end it will be appealing and understandable to the broad spectrum of PRL readers.... There are my recommendations for drastic cuts in order to reduce the length of the paper. My goal, once again, is to ask you to think of the reader, who should be anyone who normally picks up PRL, regardless of what type of physics is their specialty.

Insofar as this is supposed to be the second kind of paper, I think I have something to offer and I try to put my point to some of those with whom I am corresponding. Everyone agrees from early on that the figures presented in the paper are going to be crucial in conveying a convincing message to a wider audience. As this emailer says:

> **Nov. 4, 18:54:** Figures where the result jumps out at the reader, have an emotional impact beyond the rational impact that prose rarely can achieve. We are trying to achieve trust and trust has significant components both rationally and emotionally.

I argue strongly that the figures should be reduced in number; they are reduced as drafts evolve so I can't be too far off the mark. On the other hand, I also argue that what are called "omega plots" are unreadable to those not skilled in seeing hundreds and hundreds of them over the years. Omega plots are present in the early drafts, disappear from later drafts, but come back in for still later drafts, so apparently I didn't get that one right. They can be seen at the bottom of figure 8.1, which is figure 1 from the eighth draft of the paper. I still don't understand why the scientists want the omega scans—it seems to me that they are a hostage to fortune, because when weaker signals appear the omega scans will show nothing and will actually reduce confidence in the associated claim.

I show the caption, as presented in the draft, below figure 8.1. What I consider to be a better version of this figure without the omega plots will very shortly be presented as figure 8.3, with my own explanation of what the plots mean.

I also argue that the figures representing the statistical significance of the event should depart from the convention used in the field. This is one of those conventions that have built up over the years, but it makes no sense to people like me—and I can argue that it is people like me for which the paper is being written, at least under one interpretation of what is going on. The convention is that the vertical scale of these figures—the scale that counts the number of events—coincident or noise—of each class is cumulative. This makes it extremely difficult to read what is going on—I notice

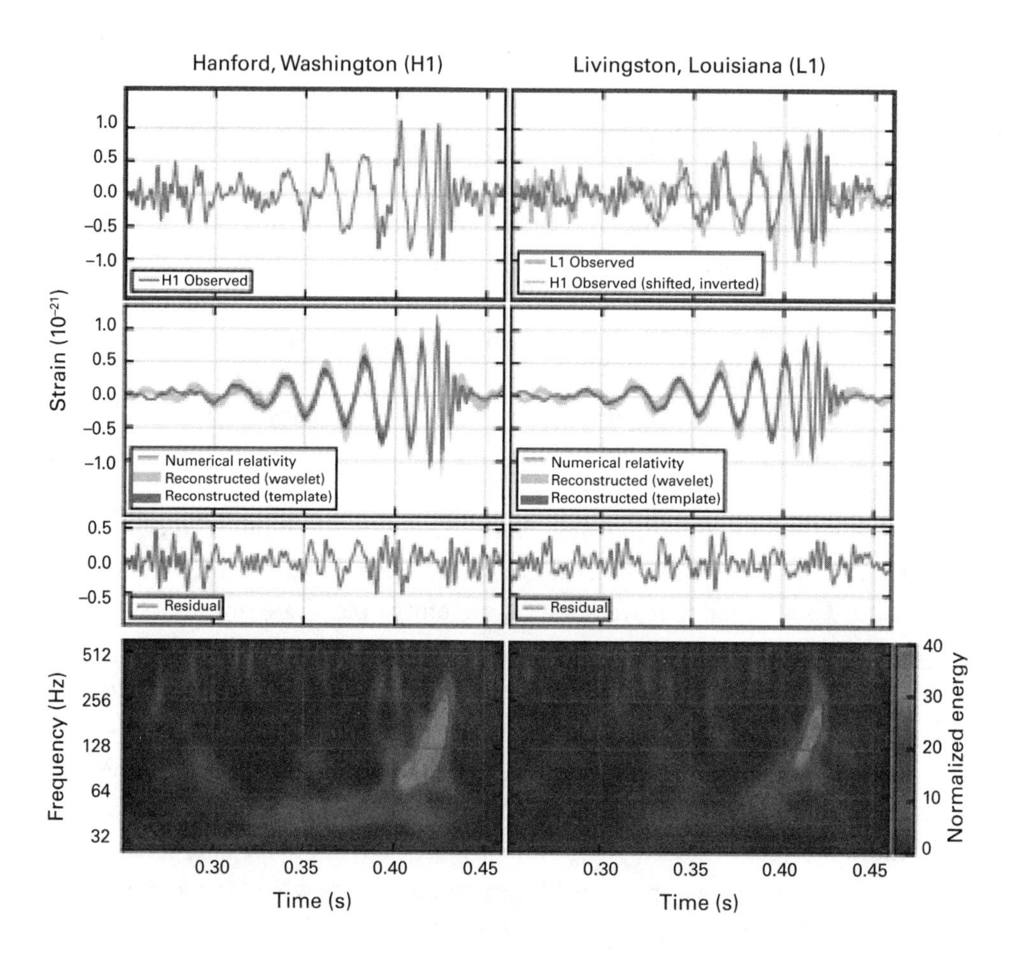

Figure 8.1

The final version of figure 1 taken from draft 8, the submitted draft: "FIG. 1. The gravitational-wave event GW150914 observed by the LIGO Hanford (H1, left column) and Livingston (L1, right column) detectors shown relative to September 14, 2015 at 09:50:45 UTC. Time series are filtered with a (i) a 35–350 Hz band-pass filter to suppress large fluctuations outside the detector's most sensitive frequency band, and (ii) band-reject filters to remove the strong instrumental spectral lines seen in the Fig. 3 spectra. Top row, left: H1 strain. Top row, right: L1 strain. GW150914 arrived first at L1 and ⊠ 7 ms later at H1; for a visual comparison the H1 data are also shown, shifted in time by this amount and inverted (to account for the detectors' relative orientations). Second row: Gravitational-wave strain incident on each detector (solid lines) in the 35–350 Hz band, calculated with numerical relativity [31], and 95% credible regions (shaded areas) for two waveform reconstructions: one that models the signal as a set of sine-Gaussian wavelets [32, 33] and one that models the signal using binary-black-hole template waveforms [34]. Third row: Residuals after subtracting the filtered numerical relativity waveform from the filtered detector time series. Bottom row: A time-frequency decomposition [35] of the signal power associated with GW150914. Both plots show a significant signal with frequency increasing over time."

some of the scientists having to be warned about this by others—and I argue that the scales should be noncumulative, which makes it easier to understand. The scientists do make this change—see figure 8.2a, which is figure 3b from the final paper (I have added the smiley faces—they will be explained in chapter 9—and the A and B labels, which will be explained shortly; figure 8.2b will be explained in the same place). The left-hand axis is now a simple count rather than the more complicated cumulative scale that was used in earlier iterations. So my thoughts about this aspect of the figures do end up matching those of the community.

LITTLE DOGS

We now have to depart from the main story of the figures to explain the intriguing story of "little dogs," because figure 8.2a cannot be understood without knowing what they are. Indeed, readers of *Big Dog* will be wondering why little dogs have so far made no significant appearance in this book.

Little dogs refer to a succulent paradox in gravitational wave data analysis that was central to the Big Dog episode. They take their name from the fact that the September 2010 blind injection became known as "Big Dog," but the name "little dog" has stuck firmly even though Big Dog is now just a memory. Everyone in this field knows what little dogs are, and it seems unlikely that, for the specialists, the name will change in the near future, though no one outside the field, except for readers of my books, will know what the term means. The term certainly will not appear in any of the published papers. A good test of whether someone is an insider in the field of gravitational wave physics would be to ask them what little dogs are, because the name probably won't ever enter the language of physics in general; it is likely to remain a term belonging to the "craft," at least until it completely disappears, which it probably will.

Little dogs arise because of the time-slide method of calculating how likely it is that a certain event could have arisen by chance. Remember, an event is traditionally thought of as a coincident signal from two detectors—this case H and L. A real signal has two parts—we'll call them H*S* and L*S*—one on each detector. To work out the background, one slides the

a.

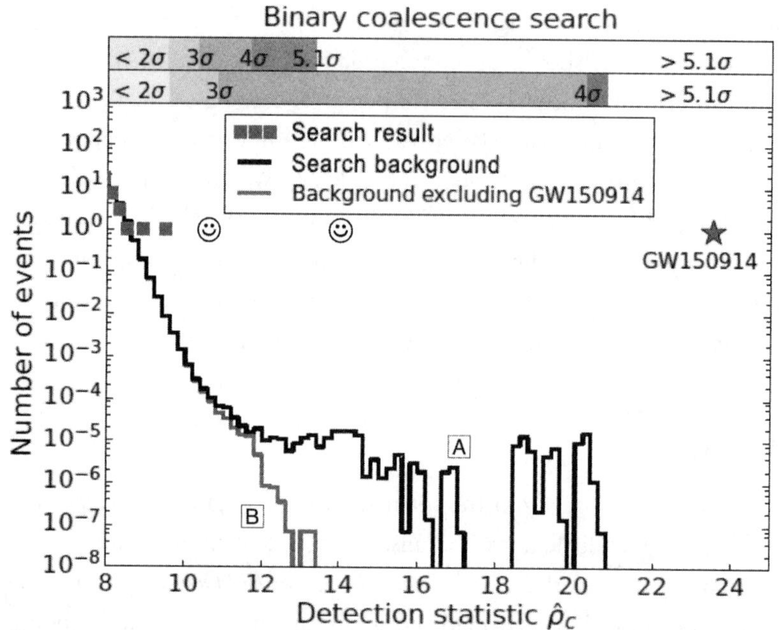

Figure 8.2a

The significance plot with some not strictly legitimate additions.

b.

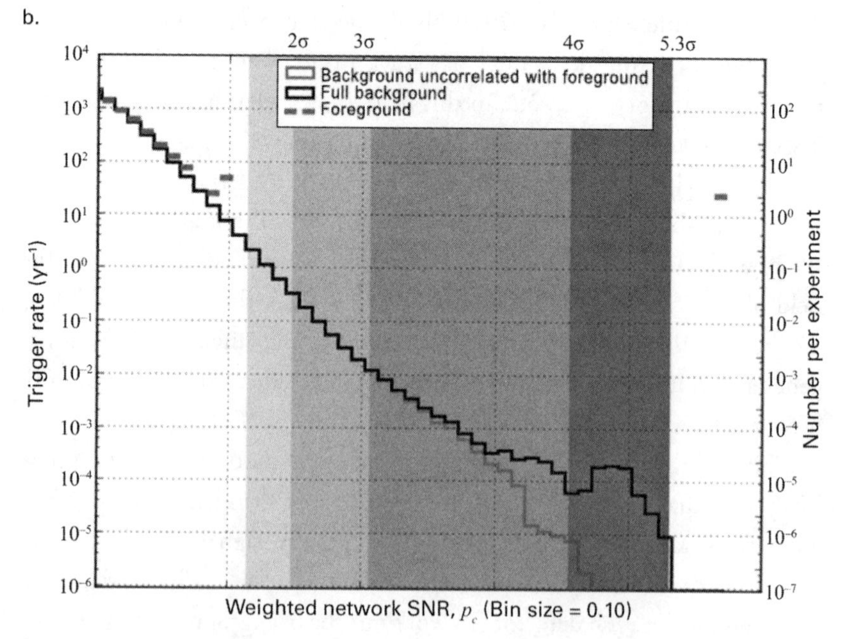

Figure 8.2b

The significance plot for the Boxing Day event, shown top right (see the next chapter).

trace from one detector along in time so that any apparent coincidences between the two traces—any pseudo-events—can only have been caused by random effects. Pseudo-events might be caused by a glitch, Ha, falling opposite a glitch, Lz, or Hb falling opposite Ly, and so on. One does lots of time slides to average out the chance background. Little dogs occur when one component of one of those chance events is part of the original putative genuine signal. A little dog could comprise H*S* along with Lc, or L*S* along with Hd, or some such. There can be a lot of little dogs. The problem is that if the initial H*S*+L*S* signal is real, then H*S* and L*S* should not figure in the background calculation, because they are not part of the noise; leaving them in exaggerates the background and makes the signal stand out less from the noise than it should. On the other hand, if the H*S*+L*S* signal is not real, then taking H*S* and L*S* out of the time-slide analysis understates the background and could lead to a false positive. Thus, one must know whether H*S*+L*S* is real before knowing whether to remove H*S* and L*S* from the background analysis, but one must know whether to remove H*S* and L*S* from the background analysis in order to work out whether H*S*+L*S* is real. As I said: a "succulent" paradox.

Big Dog contains a lot of mind-boggling discussion of this, but here things are much simpler. In the case of The Event, the equivalents of H*S* and L*S* are bigger than all of the noises in the interferometers. This means that something like an H*S*+Lz can never compete with H*S*+L*S* (which it could in the case of Big Dog); so, for The Event, the little dog problem does not arise. As Peter Saulson put it in one of our exchanges (Nov. 9, 14:27), "by a miraculous act of God, we didn't have to resolve the debate."

What about Second Monday, which is much smaller? Well, everyone agrees that signals associated with The Event should be removed when the background for Second Monday is estimated. There is now an agreed hierarchical procedure: one starts with the biggest event and calculates its background including the little dogs, then one removes its little dogs and moves on the next biggest event. The next biggest event will have its own event-specific little dogs, and these have to be included while calculating its statistical significance but then removed before moving on to establishing a background for the third biggest event (if there is one), and so on.

Now we can go back to figure 8.2a and make more sense of it. The black line labeled "A" is the background with little dogs included, while the gray line labeled "B" has little dogs excluded. The email of December 20, which can be found on page 115 above, explains why the little dogs are shown even though it makes for a very messy figure compared to what we would see if line "A" were removed.

A nagging doubt remains that was reflected in a brief argument at the beginning of December about whether it would be better to simplify things by removing line "A"—which came to be known as "the slaughter of the little dogs." The significance figures presented in draft 7 of the paper had little dogs removed, but by draft 8 they were back again.

I'll explain the question of little dogs as I see it. Even if one agrees that the little dogs should live in order to avoid an appearance of data manipulation, and even if one agrees that removing them cannot change the statistical significance of the event that created them, leaving them in after one has decided that an event is real gives a misleading appearance. The Event is either real or not, and once it has been decided it is real then it has been decided that its little dogs are not really part of the background. But in figure 8.2a, which is the modeled significance figure from the final draft, it appears as though they are part of the background, whereas really they are only part of the background used to decide that The Event was real.

The other feature of figure 8.2a, which we can now understand, is the two significance scales at the top. The lower one of these is the scale with little dogs present and is relevant to the calculation of the statistical significance of The Event—which is shown as a star. The upper scale of sigmas is the significance with the little dogs removed and is appropriate for calculating the significance of Second Monday, which is the rightmost square shape just short of 3 sigma. As can also be seen, The Event is well above 5.1 sigma with or without its little dogs, and that is why things are so simple this time around. There could, however, be mind-boggling arguments about little dogs pertaining to weaker events such as Second Monday.

Another change needs mentioning here. As can be seen, the statistical significance of The Event as per the pyCBC pipeline is shown in figure

8.2a as well above 5.1 sigma, but all drafts before draft 8 show it only as well above 4.9 sigma. We don't know what the true significance is; we can only set a lower limit, that limit being determined by the amount of available data in the time slides. Now, the community felt uncomfortable with presenting the result as "above 4.9 sigma" because the standard for statistical significance that allows the result to be presented as a "discovery" is 5 sigma, and they felt that to satisfy pedantic critics it would be better if the limit had reached 5 sigma. Unfortunately, when the "freeze" (or so-called freeze) on the detector state was relaxed after sixteen days of data-gathering, it would turn out that the resulting time slides, with their 0.2 second slide interval, came up with only 4.9 instead of the desired 5. So the community decided they should fix the problem and raise the statistical significance, all the while worrying that they might be seen as engaging in post hoc data manipulation.

There were two ways to raise the significance: do more time slides on the existing data by halving the time-slide interval and thus doing twice as many slides, or reaching into the O1 data further forward in time and adding it to the sixteen days. Some discussion established that a 0.1 interval would be technically acceptable. Note that nothing is at stake here in terms of belief in the event; everyone by this time knows it is an event, and no one's internal state in respect of their confidence in the discovery is going to be changed by raising the lower limit of the significance from 4.9 to 5.1. But this is the usual problem of setting a formalistic standard for something—in this case the nature of "discovery" rather than "evidence for." In reality, the distinction is less formal—it is essentially a decision with a justification based on communal understanding of what a "discovery" is. Here, the "image" part of the discovery was playing a huge but unspoken role and, in any case, it was obvious that though the *limit* was 4.9 sigma, the *actuality*, if only it could be known, would be well above 5. Nevertheless, what had to be achieved, it was felt, was a situation where one could say in public that the term "discovery" or "observation" was definitely justified by the 5 sigma criterion even if nothing "scientific" was being achieved by increasing the significance. As one of the emailers wrote (I can find only the report that was embedded in a thread):

—The 4.9 sigma statistical significance. I think the paper should state more clearly than it does now that the 4.9-sigma lower limit is just that (a lower limit) and is a limitation of how much data we have analyzed. However, I still worry that this fine point won't always be appreciated, and the event will be reported as a 4.9-sigma event. Why not avoid this and just include the next chunk of data in the analysis to get above 5-sigma?

Or, as the point was made to me somewhat more colorfully in a telephone conversation:

> Look, they formally established greater than 4.9—it's fucking a lot greater than 4.9, it's 5 [as you can see from the graph]. But somebody thought there would be some jerk who'd quote 4.9 and say it's not 5. So we came up with an answer that doesn't offend our standards very much, if at all, that lets us say 5.1 instead of 4.9 and we stopped some bullshit.

As regards reaching into O1 and adding more background data, Peter Saulson had said to me very early in the game, on September 22: "We will surely be using O1 data to obtain enough background for estimating the significance of GW150914" (see above, p. 32); but now that the decision to freeze had been made and to use just the sixteen days of data it was not so clear that dipping the bucket back into O1 was the right thing to do. As a senior member of the collaboration had written at the end of November:

> **Nov. 28, 3:58:** The Detection Committee agrees that we should be clearer that the 4.9 sigma significance is a lower limit, and that the real significance is significantly higher. It would be good to be able to improve on the significance limit if that is not too hard. Adding more data would be one way to do that. Increasing the number of time slides by going to a finer time step over the current dataset is another. We understand that some in the CBC group are trying time slides of multiples of 0.1 second which would increase the number of slides by a factor of 2. This result may be available soon. Adding more time is somewhat problematic. The data used in the current analysis were taken while the detectors and sites were in a weak freeze. This weak freeze was relaxed after the 16 days were completed,

> so some added work would need to go into Detector Characterization and review of the newly included data. Not impossible, but likely to lead to some small delay . . . [we] also discussed whether adding additional time or more timeslides would appear to be too much tuning of the data. The argument could be made that since we have only a limit on the significance, the additional effort to improve that limit is not aimed at changing the significance but only at refining our estimate of it. However, we need to be extremely careful. If we decide to try adding more time, or increasing the number of timeslides, we must be bound by whatever we find, even if we don't like the new number.

In the end, the 0.1 second time-slide interval is used and the lower limit of the significance is increased to 5.1 sigma. This does not seem like tuning, especially in light of the caveat expressed in the last sentence, but a ruthless critic might use it to make trouble.

A FIGURE THAT EXPLAINS STILL MORE CLEARLY WHY THE EVENT IS SO CONVINCING

Because I did not think all the figures were as clear as they could be for a reader like me, I asked one of the gravity wave scientists, Peter Shawhan, to make me a version of figure 1 especially for this book. I asked him because lots of people had been offering various versions of this figure and his was the nearest to what I wanted. He was kind enough to change his figure in various ways according to my specifications. This version is shown here as figure 8.3.[7]

The figure is the crucial evidence for a discovery and has six "features." Features 3 and 4 show the expected waveforms generated by the

7. Peter Shawhan writes: "Mostly what I did was to hack Matlab scripts written by Stefan Ballmer and Josh Smith (I believe) to add two plots and change the formatting around." He points out that Stefan and Josh should be acknowledged. Stefan Balmier also made some figures for me toward the beginning of the exercise. These were included in an earlier draft of my manuscript but are no longer needed—thanks, Stefan!

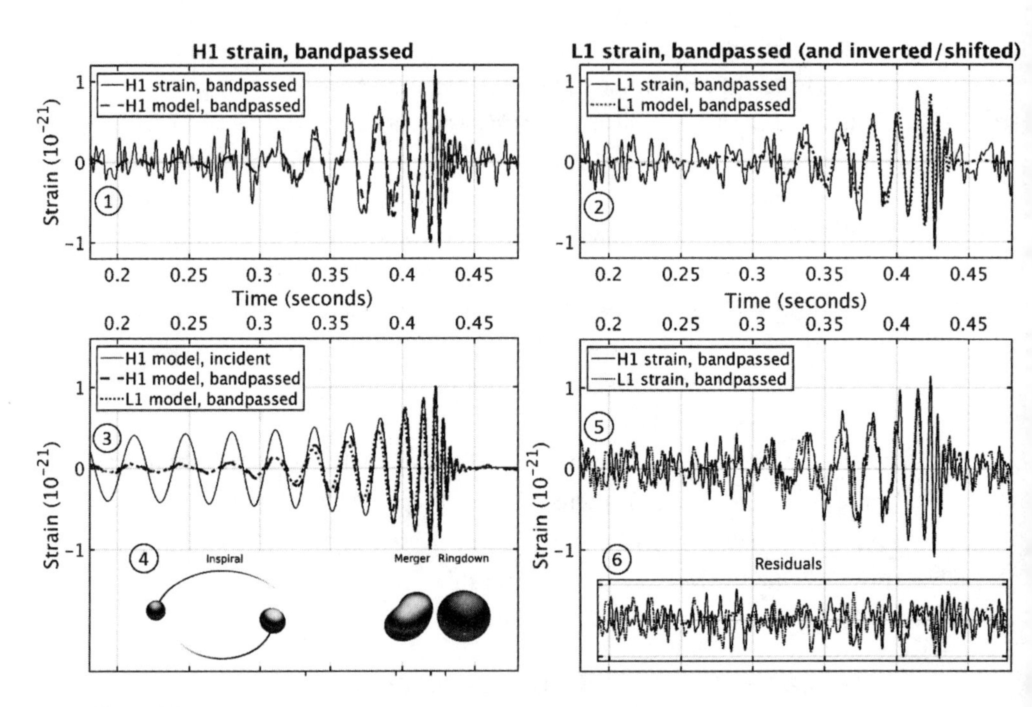

Figure 8.3

Peter Shawhan's version of the paper's figure 1, prepared for this book.

inspiraling, merger, and ringdown of the black holes. Feature 4 is what the scientists refer to as a "cartoon" depicting the stages of the inspiral and merger. The heavy line in feature 3 shows the theoretical waveform, while the two (dotted) lines, which nearly overlap, show what would be expected to be seen on the two detectors after their different orientations and time of arrival of signal have been taken into account. The detectors do not capture the entire waveform because, for example, their sensitivity is very much reduced at low frequency; that is why they would be expected to miss quite a lot of the left-hand part of the theoretical signal, where the black holes are still circling relatively slowly. That is why the dotted lines are different from the solid line in the left-hand part of feature 3. Expectations are also slightly different at each site because of minor differences in orientation and details in the way the instruments are built (something that turned out to be a positive feature when the possibility of malicious injections was being eliminated—see p. 18).

Features 1 and 2 show the measured waveform that impacted on the detectors superimposed on the expected waveforms shown in feature 3. As can be seen, the correspondence is close, meaning The Event was clearly seen at both H1 and L1.

Feature 5 shows the measured waveforms superimposed on each other (though the way they correspond can already be seen in features 1 and 2).

Feature 6 shows the residual noise when that part of the signal that corresponds to the expected waveform has been removed from feature 5.

The original plot is in color, which makes it easier to separate the various waveforms and thus see how closely they overlap. In the absence of color, figures 8.4a and 8.4b show zoomed-in versions of features 1, 2, 3, 5, and 6; the magnification is another way of seeing the convincing correspondences between lines or lack of correspondence where there should be none.

Incidentally, the final paper will have more than a thousand authors, of whom three are deceased, and they will come from 133 institutions. This list is to some extent a history of the detection of gravitational waves, but it is not a complete one. To pick out a few names, it does not include Joe Weber; Bob Forward, who built the first interferometer for gravitational wave detection; Guido Pizzella or Massimo Cerdonio, cryogenic bar champions; Brian Meers, who was the principle inventor of the crucial signal-recycling idea; Gerry Stapfer, who prepared the marshy Livingston site to take an interferometer; Bob Spero, leader of the Caltech 40-meter prototype group; Frank Schutz and Robbie Vogt, one-time LIGO project leaders; or Frans Pretorius, who first solved the recalcitrant problem of how to model the collision of black holes. At least one of these withdrew his name from the paper, and some, of course, appear in the reference list.

a.

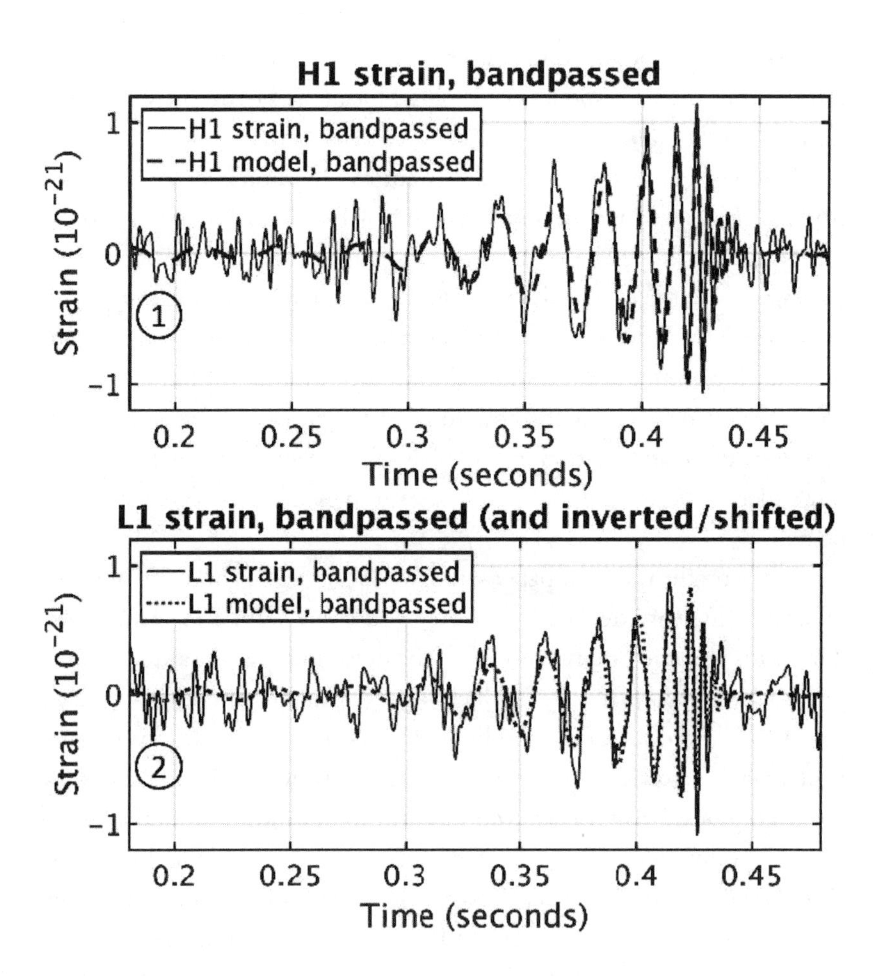

Figure 8.4a, b
Magnified features.

b.

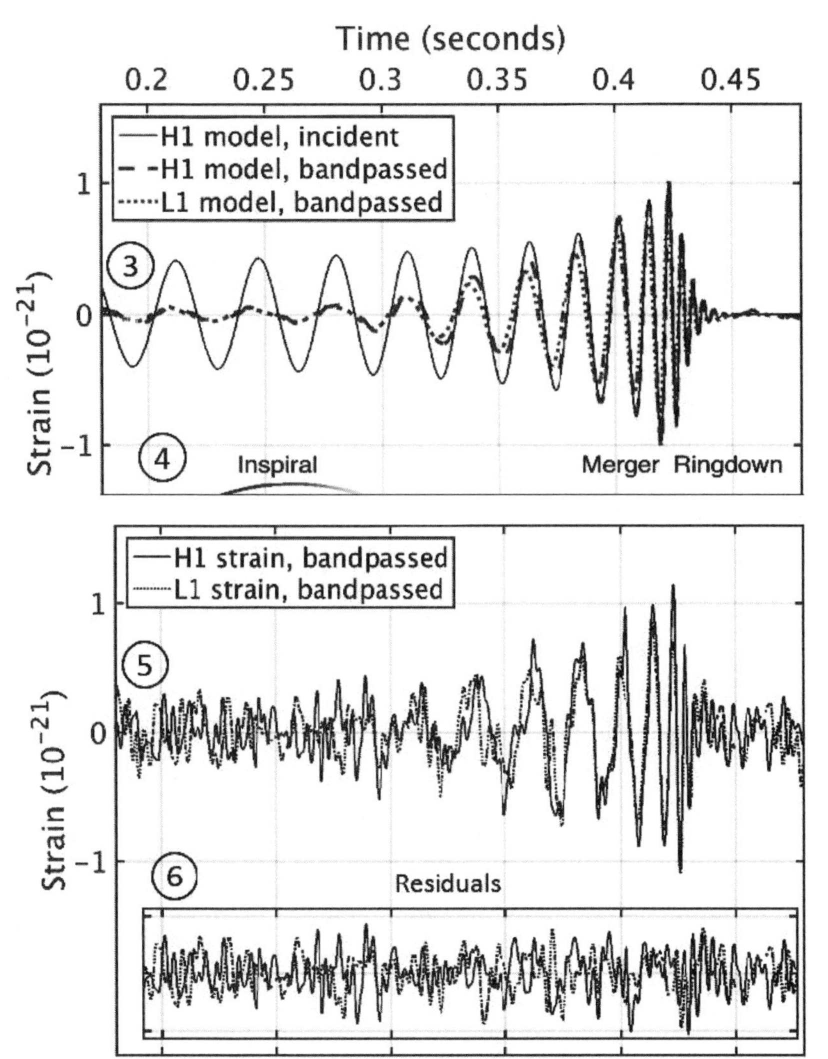
Time (seconds)
0.2
0.25
0.3
0.35
0.4
0.45
H1 model, incident
H1 model, bandpassed
L1 model, bandpassed
Strain (10^{-21})
1
0
−1
3
4
Inspiral
Merger
Ringdown
H1 strain, bandpassed
L1 strain, bandpassed
5
6
Residuals

9 DECEMBER, WEEKS 12–16

The Proof Regress, Relentless Professionalism, and the Third Event

December is an eventful month in several ways. We've already seen the discussion of "directness" presented in the last chapter. The other December events are symptoms of a change in the ambience of the community brought about by the grinding detection procedure including the process of drafting, which, it seems to me, is turning into a version of the proof regress—"How do we ever stop questioning?" (see below), and this seems as though it might prevent the paper ever being finished. Also, there is the appearance of a third event. This cheers everyone up again. And there is a new raft of rumors about the achievement, which depresses everyone.

CHANGING AMBIENCE AND "RELENTLESS PROFESSIONALISM"

The ambience of the collaboration is by now somewhat different; it has probably been changing for a while, but now it becomes obvious. From being characterized by warm good manners, a proportion of the emails now take on a niggly, "flame-like" quality. Perhaps bitterness is creeping into the email exchanges because this whole thing has gone on too long. People have stopped doing the hard work needed to disagree with someone without falling out with them—it takes too much energy—and now they just disagree "in your face." People are tired of the long-drawn out process and the stress of keeping a secret—I know I am becoming frayed at the edges by not being able to tell people why I am neglecting things that normally I would not be neglecting. I am having to rely on my colleagues' good will and their willingness to trust me and pick up some of my responsibilities without my explaining what is going on when it obvious that I should be doing a better job myself. Also, people want to be able to speak freely about their magnificent achievement, and yet, here we are at the turn of the year, and the earliest date for the press conferences is February 11—another six weeks of tension. We know that The Event is a gravitational wave, but, officially, we do not know and cannot know until the full detection procedure has been exhausted; and we cannot tell anyone until the paper has been submitted and accepted and the news conferences have taken place. Also, it seems, competition is eating away at the community: each group and subgroup wants its piece of the action to be seen as important. People are grabbing for bits of fame like starving people grab for dollar bills when someone throws a bundle of them into the air; in the gravitational wave business, we've been waiting a long time for scientific largesse.

The trouble is that The Event has violated all expectations, being so strong as, astonishingly, to be believed in almost from day one. So, three and a half months on, it is not surprising that there is a sense of exhaustion and anticlimax. Going back to the way events unfolded, that it was a publicly announceable discovery was pretty well certain once the box had been opened on October 5, which is day 21 and close to incontrovertible by the time the possibility of malicious injections had been seen to have been eliminated; this was during the community wide teleconference of October 29, which is day 45. We can say that after a month and a half it had become almost impossible to believe that The Event was not a discovery. The last two months have been a matter of refining the parameter estimation and writing and reviewing the paper, in preparation for what?—another round of internal reviews!

One senior member of the collaboration, however, tells me that the process is reasonable. While he agrees that the elaborate detection procedure was designed for a much more marginal event, he says that going through the entire procedure is still worthwhile. He says that it might result in small changes in the statistics—perhaps a false alarm probability of 1 in 110,000 years instead of 1 in 100,000 years. When I protest that this makes no difference at all, especially as it is all based on arbitrary assumptions such as what counts as the detectors being in "the same" state, he replies that it is nevertheless important to the community to practice their craft to the best of their abilities.[1] We will need to understand this way of looking at things a little later, so we will give it a name. Let's call it "relentless professionalism." Sometimes relentless professionalism is pedantry.

THE "PROOF REGRESS" AND THE NEED TO STOP QUESTIONING

The method of writing this paper is in some ways a most remarkable innovation in the production of a literary artifact. Succeeding drafts of the paper are presented to the whole thousand-strong community who are invited to comment. And about half of them do comment. This is great insofar as it means, as one senior member of the collaboration points out to me, that the paper is self-correcting; so much energy has been put into examining every single claim and every single phrase that if anything goes wrong, it doesn't go wrong for long.

But there are downsides: as the same senior member points out, it creates the problem that lots of people are trying to make sure they get some mention of their own particular contribution. As a result, the paper is not as crisp as it could be: it is a literary work that, though it started with a dramatic flourish, is in danger of coming to look like a paper written by a huge committee. It appears that the community is too democratic and too bureaucratically organized to allow someone to take the lead and pull it all together. Instead, when matters such as whether to call this a direct observation reach an impasse, we get a vote: a vote on the wording of a historic literary product!

One correspondent tells me, however, that I haven't fully understood the paper-writing procedure, which is less democratic than it appears. He uses a nice metaphor—the appearance of democracy is like the bullfighter's

1. The shift to 0.1 second time slides means that this figure should be 1 in 200,000 years. (It is 1 in 600,000 years before allowing for the trials factor consequent on searching three mass bins—see the discussion of opening the box on the Boxing Day event in chapter 10.)

cape. The democracy-cape attracts the attention of the community and reassures them, but behind the scenes the writers and their close advisors are making sure the paper makes sense. Given the excellent way the paper eventually turns out, this seems a credible estimate of what is going on.

Philosophically more interesting is that the mechanism, which allows anything to be questioned by any member of the community, tends toward everything being questioned. We know from the Duhem–Quine thesis or the experimenter's regress (see below, p. 263, and *Changing Order*), or Kuhn's "essential tension," or Planck's dictum that "science advances funeral by funeral," or *Big Dog*'s chapter 14 which shows that the Big Dog claim rested on twenty-five "philosophical" agreements, that anything can be questioned, and the limit on questioning things is the social cohesion of the scientific community. This, in a strong sense, is the central premise of the sociology of scientific knowledge; it is the engine that drives sociological inquiries about how scientists reach their conclusions, and it is going to make its presence felt over and over again throughout this book. Indeed, we have already seen it at work when the claim was questioned that The Event represents inspiraling black holes—they could be Boson stars, or anything else that physicists are clever enough to invent. At the time I remarked that I was tearing out my hair, but that was because I had allowed myself to slip too deeply into the role of a native. I was saying, "Come on, this is an inspiraling binary black hole according to the conventions that physicists use, so why question further?" If I had been preserving my estrangement a little more assiduously I would have been saying: "What a wonderful illustration of the fact that the limit on what can be questioned is not a matter of logic but is matter of social convention!" Because the presence of this sociological/philosophical driving force is going to show up over and over again I am going to invent a simple phrase to express this phenomenon. I am going to call it the "proof regress." Too much determination to get everything exactly right—too much relentless professionalism deployed where judgment is called for—would bring physics to a halt. Physicists have to circumvent the proof regress.

Thus, we know that The Event could have been the result of psychokinetic effects generated by the huge desire of a thousand people to see an inspiraling binary black hole and their subconsciously coordinating the collectively generated waveform via some kind of unknown communication channel, but we also know we had better not say so; that is beyond the bounds of social acceptability (see *Gravity's Shadow*, chapter 5). We know, however, that we are allowed to take seriously the possibility that someone maliciously inserted a fake signal; this is newly socially acceptable though still not the kind of possibility that we will publish. The invocation of the proof regress will be a reminder that anything can be questioned in logic; but not everything can be questioned in practice, or science would melt away.

THE ATTEMPT TO THAW THE FREEZE

I am now completely thrown by the questioning of another piece of taken-for-granted procedure, though this one has legs. It begins during a personal teleconference between me and a group of scientist that takes place on December 4. We are discussing the roles played by cWB and CBC in the discovery of The Event. cWB, remember, stands for "coherent Wave Burst," and it is a "pipeline" that scans the data for coincident events that look like gravitational waves while imposing little in the way of preconceived ideas of what they should look like. Before The Event, cWB used to be called an "unmodeled search," but the questioning process has led to it being pointed out that there are some assumptions built into even this pipeline. So calling it unmodeled is misleading and something less precise is now used in the description. CBC, remember, is definitely a "modeled" search, since it finds matches between proto-signals and the 250,000-strong template bank. Remember The Event was found by cWB because CBC's "online"—that is, real-time—detection algorithms were limited to low-mass binaries and The Event was a high-mass binary and therefore invisible to them. What I am now asking about in the teleconference is whether CBC would eventually have found The Event even if it did not spot it in real time. Everyone assures me that it would have found The Event because all the data is thoroughly scanned eventually, but it might have taken two weeks or maybe more.

But now a nice question arises. Remember that the senior management decided to freeze the state of the detectors a couple of days after The Event was spotted in order to build up enough background to complete enough time slides so that the signal would emerge with the standard of statistical significance demanded for such a "discovery"—5 standard deviations or "5 sigma." (As it happens the initial analysis with a 0.2 second time-slide interval took the significance to 4.9 sigma and, retrospectively, the interval was cut to 0.1 seconds so as to generate twice the number of slides taking the significance to 5.1 sigma—the result that was published—but that is another story.) To achieve the length of background thought necessary, the state of the detector was frozen for about five weeks, leading to the accumulation of 16 days of coincident lock with the devices in "science mode"—that is, fit to make observations. The "freeze" meant that an important repair that those responsible for running the machines wanted to complete was held back.

In the December 4 teleconference, I make a point about the freeze and its relationship to cWB, repeating what could already be found in emails such as:

> **Nov. 8, 00:27:** the most important added value of the cWB rapid alert on GW150914 is that consequent decisions and actions were taken in a very short time, in primis to freeze the operating point of LIGO detectors (to integrate homogeneous observation time for e.g. deep FAR investigations), and secondly to point LV attention to that specific time (including triggering and speeding up all DA/detchar studies: it simply became a must not to overlook it, and in fact we know that some bugs/technical inconvenients were rapidly corrected in other pipelines).

Amusingly, this emailer adds:

> The importance of the rapid alert was not just sociological: it did affect science,

thus drawing a contrast that renders my forty-three-year project otiose from the outset—but I am sure it is not meant that way.

Back to the teleconference: I put it to the scientists that even though CBC would have spotted The Event eventually, if it had not been for cWB spotting it in real time the detectors would not have been frozen, so that the length of background necessary for statistical significance would not have been generated and The Event could not have been announced to the world as a "discovery." Other emailers too have said the same thing:

> **Oct. 11:** Many people would say that we were very lucky that cWB was able to identify this event quickly. Remember, it was in the middle of the preparation for O1. It allowed us to respond swiftly to freeze the state of the instruments, which otherwise could have been altered in some unpredictable ways. Without knowing about the event, the offline analyses would produce results much later in the run (we may not even know about it now!) and quite possible these few days of data would not be seriously considered at all.

The answer that comes from some members of the group taking part in the teleconference takes me completely by surprise: they say that there was no such thing as "freezing the detectors" since the state of the devices is moving around all the time. I am so surprised by this that I think the scientists are trying to "pull the wool over my eyes," and I reply along the lines of "So you mean that when [very senior member of the collaboration] announced that we could not use any more background in the significance calculations than 16 days' worth because that was when the freeze was lifted, he was making a mistake!? And then someone smooths things over.

Immediately after the teleconference, I think over "what I should have said." This would have been along the lines of "I know that the state of the detectors is always moving about, and that is a problem for generating background; but that does not license making step-changes in the detector state and using data from the far side of the step-change in the same way as data from the near side of the step-change." In other words, it seems to me that it is well understood that the time-slide method is imperfect because of changes in the detector state, and it has long been recognized that indefinitely long stretches of data could not be used for generating background because data distant in time from an event were no longer representative of what was going on at the time (see *Big Dog*, 225, and the seventh philosophical agreement in *Big Dog*, chapter 14); but this is a problem to be recognized and handled with a compromise, not something that means you could use any data you liked. I will discover that the view I hold is also held by the experimentalists.

I think this is the end of it, but to my surprise I discover that at least some of the scientists are serious in saying that the notion of a freeze is untenable. They had already said so in an email I had missed before the December 4 teleconference:

> **Oct. 11:** I think you will find people who disagree that "freezing" the instruments (even discouraging maintenance or repair!) was necessary or desirable. In actual fact some things started malfunctioning and had to be fixed within less than a day of the alert. We can all have opinions about that, but we will never know what would have happened otherwise. CBC certainly planned to analyze and look at results from the whole of ER8, wherever calibrated data exist.

Furthermore, they are not ready to give up this view:

> **Dec. 22, 9:24:** btw, a number of people in CBC & DetChar think this "freeze the detector" argument is simply bogus. The detectors are moving about all the time and have to be pushed back towards their operating points if they stray too far, not to mention that one of them became seriously broken on Sep 15 ... they cannot be frozen even if we wanted to.

This argument turns nasty when it is suggested that the crucial importance of cWB's early sighting was to alert the operators at the sites that they must be especially assiduous in describing the machine's state—with the operators' replying that they were always assiduous and this remark was patronizing at best.

What I might have realized is that the notion of a freeze is vulnerable because whether the detectors are to be counted as being in the same state or not is one of my twenty-five "philosophical" judgments, and here it is being questioned. Permission to lift the freeze, by the way, was given in a note from the chair of the Detection Committee on October 9, but it was not actually lifted until October 20. The flavor of the decision and the way the freeze was lifted is nicely captured in the following email, which is a response to my follow-up question to Mike Landry in late January:

Jan. 30

Hi Harry,
I'm very sorry to have not replied to your first email. The crush of messages and tasks is really quite remarkable right now—not that I am complaining, given the circumstance.

There are freezes and there are freezes. Brian and I initially made a hard freeze, eliminating the Tuesday maintenance on Sep 15, and for a long while holding off even repairs of anything that were broken. However this is an untenable position in the long run. LN2 [a liquid nitrogen tank] needs filling, and things break so hard that you have to intervene.

The target for a 5-sigma detection was 16d of dual-coincidence data. This was an estimate, and it was not at all clear at that time how close or not we would be to a 5-sigma detection with that accumulation, not until the box was opened on the CBC analysis that included Sep 14. The 16d clock was set by the dual-coincidence data that were not vetoed by cat1 DQ flags; Laura Nuttall of the Detector Characterization group gave us a tally every few days, counting down when we were done.

Brian and I would watch Laura's tally of the background data, in order to assess when we were free to make any changes that we thought might risk the background estimate a little more, like changing the glitch rate (presumably for the better) in a significant way such that you could not use it any longer as GW150914 background.

Some changes to fix broken things went in place while background accumulation was underway, ranging from the-machine-is-down (Beckoff failure at LHO endstation, Sep 15), to there's a small peak in the spectrum (grounded air-handling unit external to LHO end station had to be re-isolated, Sep 29), to, we've left a beam diverter open and the end station EX, LHO Oct 6.

It is interesting to note, I don't think any change that modified the background glitch rate significantly was made in the entire run.... So the hard freeze, while completely understandable and conservative, was [not] necessary to preserve a background rate consistent with the natural fluctuations in the instrument.
—Mike

Others like Mike Landry and Brian O'Reilly, who understand the machine, nevertheless believe that the freeze was appropriate. Here is Brian's explanation:

Jan. 31

Harry,
There is another quite banal reason for taking a "hands off" approach while we collected background. There is always some element of risk when you touch the detector, that you will break something severely. We felt that even if the risks were low the best thing was to do as little as possible. Also, even minor changes can eat up many hours of Detector Characterization, Calibration and Analysis time later on, because you have to verify that nothing significant changed when you made some (allegedly) minimal change to the configuration.

As for the freeze extension: I looked at the Detection Committee minutes from Oct 9th. And as I recall it the consensus was to allow small changes to the instruments. As a practical matter though Mike and myself still retained full control of what could happen. We continued to be quite conservative in our approach, making sure that any change was reversible for example, and definitely not countenancing any major changes. I think this conservatism made the decision by the groups to analyze data up to Oct 20 an easy one, i.e. the instruments hadn't changed. The CBC box was opened on October 5th 2015 using 5 days of data. But the analysis groups were already thinking of folding in the data that had been collected after the GW150914 box was opened (that box covered 9/12–9/26).

"FOREGROUND" AS AN ESOTERIC TERM

Much less controversial, but interestingly indicative of the way The Event is focusing minds and of the positive side of the paper-writing mechanism, is another critical thread. Around the end of November the scientists notice that the term "foreground," which is just part of the ordinary vocabulary of gravitational wave detection, has no meaning, or a different meaning, in other parts of physics and astronomy. Worse, in other parts of astronomy it means another kind of confounding noise that prevents a clean signal being seen. The usage has been checked on Google and arXiv and it has been found that the only people who will understand it are gravitational wave physicists. That is no good for a paper that is expected to be read by a broad spectrum of physicists.

> **Nov. 26, 11:16:** For fun I performed a google search with the terms "foreground event" astrophysics and all but maybe one of the front page results were pointing to the GW community. This suggests to me that the GW community usage of the term is not more widespread in astrophysics and (unless described carefully) could lead to confusion.
>
> The search "foreground events" site:arxiv.org has a similar result. The term "foreground" seems to be used in SN searches (frequently without any definition) and appears to refer to the physical placement of the object as closer to the observer in an image—the "pedestrian" interpretation.
>
> My conclusion is that it is our jargon, and if we use it, we need to explain it.

I would guess that, in gravitational wave physics, the term "foreground" developed as a contrast to the more widely used term "background," meaning the noise from which the signal has to be extracted. As we've seen, in gravitational wave physics the background is generated empirically by counting the noise coincidences that are found in multiple time slides. So what have come to be called the foreground are the coincidences in real time with no time slides. Thus, in the captions to earlier drafts of the figure with the squares and the star that is used to indicate graphically the statistical significance of the result, real-time coincidences are referred to as foreground. But it is now becoming clear that this usage will either have to be carefully explained or abandoned. No one even thought about such things in the case of Big Dog, and it can only have been that there wasn't the same sense of writing for a broader audience because of the blind-injection possibility.

WHAT DOES "FAR" MEAN?

A second change that occurs concerns the meaning of FAR—false alarm rate—which is the basis of the statistical analysis. Here the intense consideration and reconsideration of every word in a paper that it is now known will be scrutinized by a skeptical community leads to a new way of thinking about the statistics that takes me by surprise. The only consolation is that nearly everyone else has failed to notice the problem alongside me until it is pointed out by one member of the community.

The statistics of a potential gravitational wave sighting, such as the Equinox Event (*Gravity's Ghost*), Big Dog (*Big Dog*), and The Event have, up until December 15 of this year, been explained in terms of both false alarm probability (FAP) and false alarm rate (FAR). In *Big Dog* I spend quite a few pages (starting around p. 220) trying to explain how they relate. But now an emailer is arguing that we should not mention FAR at all, or be very careful when we mention it, and it dawns on me that he is right. I notice that in *Big Dog* I have fallen into the exactly the trap he is warning about where I say (p. 220): "If the FAR is 1 per 40,000 years and the apparatus stayed on air for 40,000 years, it is likely that it would see one false alarm." I now realize that this is an incorrect and misleading thing to say.

It all goes back to the time-slide method for generating a noise background. Let us use the numbers pertaining to The Event. The analysis of the noise background is based on 16 days of running in a semifrozen state before the freeze was relaxed and some features of the interferometers were changed. Choosing a time-slide interval of 0.2 seconds, these 16 days' worth of data can be used to generate 100,000 years of background noise. If there are no other events that compare with The Event in that 100,000 years, one can see that the false alarm rate is better than 1 every 100,000 years. It has always struck me as a slightly discomfiting bargain that one could gain something like an idea of what would happen in 100,000 years from just 16 days of data, though I never had the brains or confidence to press that feeling of getting something for nothing to its conclusion. Instead I said the equivalent of: "This means that if the machine ran for 100,000 years we would expect fewer than one false alarm." But it is now pointed out that this is ridiculous—you could not know what would happen over 100,000 years from looking at 16 days. I write to the emailer:

> We have 16 days of data from which we construct 100,000 years of noise background. The puzzle you raise is how you get to say anything about 100K years given that you only have 16 days—seems like getting a rabbit out of a hat (of course, I understand the mechanism by which it is done but that is not your question as I understand it).
>
> Isn't this the answer: One is saying that if the machine stayed in exactly the state it was during the 16 days with the noise having the same characteristics, one could look for 100K years through such data without seeing such a signal in among the noise?
>
> This, of course, says nothing about a real machine running for 100K years, only an imaginary one.
>
> Is that anywhere near right? Answer should be real simple please.

The answer comes back:

> You are exactly right (simple enough?). If every Jan 1 there was a software glitch that looked like a coincident event there would be 100,000 of them in 100,000 years, but right now we would be ignorant of that.

So, after an embarrassingly large number of years misunderstanding FAR, I now realize that the large numbers of years of background we seem to be able to generate with time slides are not real years at all; they are just a number representing a result—the result of, in this case, something in the region of 7 million 16-day time-slides. They may add up to 100,000 years in all, but they are not 100,000 clock years, and they tell you very little about how the machine would behave after the 16 days are over. To determine that, you have to know how stable the machine is over the long term; and you cannot tell that from 16 days of data.

THE DETECTION PROCEDURE

So where are we at the end of December, three and a half months on from the day The Event was first spotted? I've already mentioned the document entitled "The Process for Making a First Detection" (appendix 1). The substance of this document was worked out years ago.

Where are we? We are coming to the end of Step 2. What is supposed to happen next is Step 3, which, as can be noted, is another layer of reviews. But hasn't this discovery already been reviewed to death—for weeks, a thousand people have been scrutinizing every line of the discovery claim. Going on seems pedantic at best. I think, and I am not the only one to think this, that the elaborate detection procedure was designed in the expectation that the first event would be a marginal one, so that if it was to be turned into a claim to be presented to the outside world, and given the history of gravitational wave detection that has created a culture of paranoia, one could not do too much checking. That is the problem I think we are facing at the end of December: too much success too early; this has meant that we are going on for far too long to prove something that is pretty well proved, and that is causing strain.

THE THIRD EVENT

On December 26, Boxing Day, a third event is detected. Once more it is an inspiraling binary black hole with component masses of around 20 and 6 suns. It seems to me be just another version of the second Monday event, but the community is much more excited by it than I expected.

> **Dec. 26:** My tree looked a little funny this morning, apparently, I forgot to open a present, and it was a gravitational-wave event. ;)

The community is treating it almost as though it is comparable to The Event itself; as Peter Saulson says to me, it is the "second shoe dropping" that we've all been hoping for.

> **Dec. 29 [Collins to Saulson]:** But you wouldn't be announcing this one if this was your first discovery.
>
> **[Saulson]:** We might, we might—we wouldn't be announcing it right away—we would be taking our three to six months to figure it out but I think this one is good enough that we might have written a first discovery paper. Ask me again in a couple of weeks.

I seem to be getting this wrong, and at first I think it is because I am much more enthusiastic about Second Monday than most of the others—that seemed to me to be the second shoe or, at least, a pretty heavy sock, and this seems like a slightly heavier sock to me—so this means less to me than it does to them.

But it turns out there is more to it. First, it is stronger than Second Monday, with a signal-to-noise ratio (SNR) of about 11.7 instead of 9.7. This seems to make a lot of difference. I am comparing it to The Event's SNR of 24, which makes both the others seem pretty similar to me. But it seems I am not technically equipped to understand that the difference between 12 and 10 is quite important—there is a gap in my interactional expertise.[2] It is important enough for the community to have sent out an alert to the EM partners, which they did not do for Second Monday.[3] Still puzzled, I ask if GW151226 would justify a first discovery claim—I know that Second Monday would not—and I get a strong "maybe" in response. Someone tells me that the statistical significance would be around 3.5-sigma or a bit higher—as opposed to about 2+ sigma for Second Monday—and

2. See Note XV in "Sociological and Philosophical Notes," 371.

3. Though the low latency of the Boxing Day event discovery was another aspect that encouraged them to alert the astronomers.

that would justify an "evidence for" claim. I don't quite believe it and think that what we are seeing is another indicator that the order of things has changed. I think the scientists are believing it because their mindsets have shifted; I think that a 3+ sigma event would not have caused so much excitement if The Event was not already out there. I understand and accept what one scientist tells me—that if nothing had been seen but the Boxing Day event they would have published a paper about it. It would have been interesting, but I don't think it would have convinced anyone, and that includes the community itself. I don't think that in itself it would have changed the order of things in the way that The Event certainly has.

Incidentally, the left-hand smiley face in figure 8.2a is roughly where the Boxing Day event would be located if it had been discovered within the 16 days of background used to place The Event and if it had been in the same noise "bin" (more on this below). It was not discovered in the 16 days so the background would look different it we were doing this properly, but the position of the smiley is still a good indicator of what this event means at this point in time. To me, represented in this way, it does not look all that different from Second Monday. The other problem with figure 8.2a is that the background plot with the little dogs is wrong for the Boxing Day event; figure 8.2b is a technically correct diagram showing how the Boxing Day Event would eventually be represented.

But there are also another couple of reasons for the excitement. First, there is something about the way it was discovered. As I have been slow to understand, most of the heat about the low-latency announcement of The Event has been generated by worries over what counts as retrospective design of analysis and modeling. Remember, the team is paranoid about freezing pipelines after the box is opened on the data; but this means that if the box is opened immediately by low-latency pipelines such as cWB, which then give an initial estimate of the parameters of the source—for example, that it is a binary black hole with certain component masses—then the offline analysts feel their hands have been tied before they can fully refine their techniques. One of the rows going on is about precisely this, with certain CBC advocates wishing that cWB had never done any analysis (and I speculate that this is why certain of these analysts want to say that the idea of a freeze on the state of the detector is bogus). That is, it is not so much intergroup rivalry as an argument over the meaning what it is to open the box and what it is to analyze retrospectively. It is another of these philosophical conundrums for which a pedantic solution is available but just does not work (see, once more, "the airplane event" as discussed in *Gravity's Ghost*).

I now learn that a characteristic of signals produced by massive systems, such as the roughly 40-solar-mass black holes comprising The Event, is that they are very short—the merger happens extremely fast. This means that it is very easy to confuse them with a "glitch"—a burst of noise in the interferometers—which also tend to be short. A longer signal will have plenty of time—a second or two—to reveal its non-glitchy character, whereas a short signal has only a few fractions of a second—about two-tenths of a second in the case of The Event—to reveal its nature. It was for this reason that the online template-matching pipelines were restricted by a low-mass cutoff point and therefore could not see The Event. CBC did not want to be hog-tied by too much information before they could do a thorough analysis of potential high-mass discoveries. But the perception had now changed and gstlal, a template-matching low-latency pipeline, has had its upper mass limit removed. It was gstlal that had seen the Boxing Day event. cWB had not seen it, and could not see it, because cWB is set to see only fairly powerful events—so cWB didn't see Second Monday either because it was too weak. Thus, the collaboration is very proud of its judgment, and this, it seems to me, has lent the Boxing Day event a certain glamour.

Still more important, the Boxing Day event, it seems, at least at first sight, is "precessing." That is to say, not only are the black holes circling round each other but, because they are spinning, the whole system is tipping eccentrically as it spirals-in, making the magnitude of the overall signal oscillate; there seem to have been a few such modulations in the duration of the signal. I ask one of the scientists who is an expert on modeling this kind of thing, Mark Hannam, what this precession would look like, and with extraordinary generosity he constructs a little movie, in color, for me, showing what might be going on in a system like this. Figure 9.1 is a sequence of stills from it.

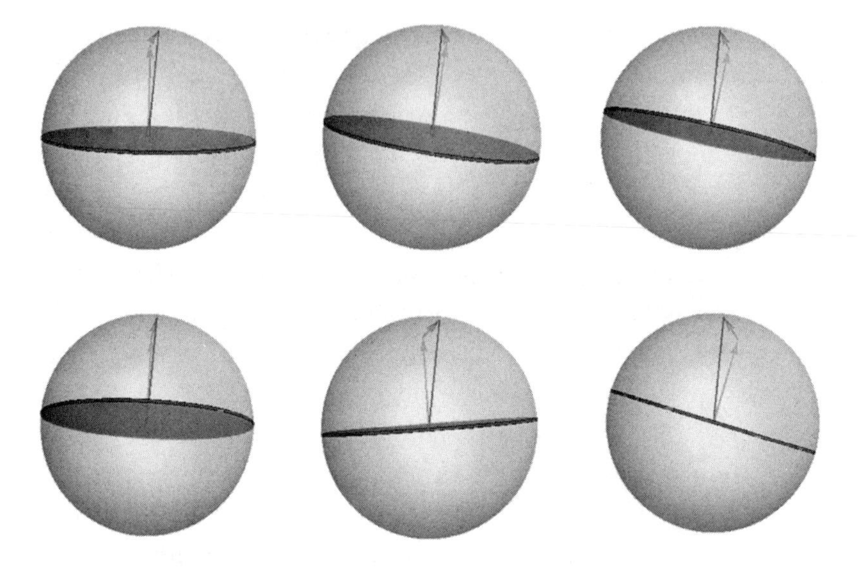

Figure 9.1
Stills from Mark Hannam's movie showing a precessing system.

The sequence reads from left to right. The first frame shows the plane of rotation of one black hole around the other with the axis of rotation being close to vertical. In the following frames the plane of rotation oscillates as indicated by the moving axis of rotation. The strength of the radiation emitted in the direction of Earth depends on the orientation of the plane of rotation, so the strength will change as the orientation changes.

If all this is confirmed it will mean that the Boxing Day event is a discovery in itself: while the existence of such systems had been theorized, Boxing Day would represent a first observation of this kind of black-hole binary system. This makes it still more glamorous, but it's not the sort of thing I would spot until it is pointed out to me because I don't have much understanding of astrophysics. My expertise, such as it is, relates to the interaction of gravitational waves and gravitational wave detectors. The interactions around the Boxing Day event make it clear to me that many of the community are plugged into astrophysical questions in a way that I am not. Here, for example, is an email from an astrophysicist member of the community, once more invoking the "N-word":

> **Dec. 27, 04:01:** this is just beautiful to look at . . . I have to pinch myself when realizing that nature is sending us these.

We may also speculate that the unexpected richness of the signal—the modulation on top of the signal itself—something that was not being looked for—would itself make the signal seem more real. Unfortunately, after some debate it seems that the group concludes that the SNR ratio is too low to enable them to be sure that the system really was precessing.

I write to a couple of the physicists and ask where things are going wrong—is it me or is it that the community is wearing rose-tinted spectacles, now that The Event has given it a license to see gravitational waves where it would not have accepted them before? One reply makes it clear that indeed, I am not grasping the difference in strength between Boxing Day and Second Monday, and I have to accept this. On the other hand, the other response, coming from a very senior, very experienced, and somewhat cautious physicist, agrees that this event goes a long way to relieving the worry about not seeing more than one event; but this physicist also says without prompting that Boxing Day is by no means strong enough to have justified a first discovery claim. So it looks as though I am not understanding properly, but there is also some degree of rosy tint in the spectacles; the order of things is changing.

DECEMBER RUMORS

We now hear of a fresh set of rumors. At the end of December an email comes around from the LIGO spokesperson with dire warnings.

> **Dec. 16, 19:37**
>
> Dear all,
> We have been made aware of new rumors about a GW detection by

LIGO, apparently started at recent conferences and other sources. Please remember to not share any results with non-LSC members.

You probably have seen the progress in the writing of a detection paper with GW150914; we will discuss details tomorrow—again, please make sure the drafts are not leaked to anybody outside LVC, this could lead to grave consequences (having a paper publishing a plot before we finish review, for example!).

If we learn about comments by an LSC member is the source of rumors, we'll take serious action. Be part of the solution, not the problem!

Please use the proposed answers below for guidance; you can also refer people to Dave Reitze or myself.

Let us know if you have any questions, or are asked any uncomfortable questions, or heard any rumors.

Next day there is a follow-up:

Dec. 17, 02:24

Dear LSC Colleagues,
I want to emphasize the importance of Gaby's message to you earlier today. We've recently been made aware of many rumors, mostly circulating among our colleagues or acquaintances in the scientific community, but also a few from journalists and science writers. In one case we learned of today, the journalist had very specific information about the LIGO-Virgo discovery.

While no one should be communicating with even trusted scientific colleagues outside the LVC, it is information getting to journalists that could cause us serious problems. Once a story is published in a paper or magazine containing specific and credible information, we will have no choice but to call a hastily prepared press conference to announce our discovery. If we don't, the press will tell our story for us the way that they want to.

At this point, I don't believe we're ready to go public, so I cannot overstate how important this is. Following the advice of media relations experts that we've consulted, we want to modify the guidance about handling press

> inquiries. **If you are contacted by a journalist, please refer any queries to Gaby, Fulvio, or myself.** Thanks for your help with this.
>
> ... We won't have an article to submit for at least a few weeks, so **if you are asked about any news from O1, you should tell the truth: "we are still analyzing and reviewing results, we will share the news when ready."**
>
> ... **We have evidence that several people in the LVC (including senior members) have shared many details with "close friends" in the scientific community.** ... Those "close friends" are naturally telling other friends, and asking other LVC members, putting us all in very awkward situations. ...
>
> It is customary for journalists to print stories if they obtain confirmation from more than one source. They may tell you they have heard already news from other sources (they've read the rumors!), and are only asking for confirmation.
>
> If a story appears in print in the next few weeks saying "Several LVC members confirmed to us that there has been a BBH detection with LIGO with XX masses, XX spin, on XX date" we'll have to call a rushed press conference with spokespersons and others saying "we have nothing to confirm, we'll call you again when we have news." This will give a very bad image of our collaboration (what are they hiding? they know already, why don't they say it?), and of course there could be even worse consequences if the story is different than what we'll publish (they have detected two strong events in the first month, they have detected a short GRB in coincidence, they don't release news just due to bureaucracy—I have heard all of these already!). ...
>
> It will not be good if funding agencies who have invested so heavily in our field are robbed of the press about the good news—this may effect funding in our field. Please be part of the solution, not the problem!

There is nothing new on the web, but I manage to learn more about these rumours. Their source is a German group or groups—senior people telling other senior people in funding agencies and the like, whom they think they have to tell for the sake of the future of the field (I know some of this is going on), and these people letting things slip to other people. The worst thing that has happened so far is that someone has written a paper, posted on arXiv, the physics preprint server, predicting that the interferometers will soon discover events just like The Event. Another person has also made such a prediction public because it has a bearing on the future of certain kinds of big-money research. Since no one expected to see black hole binaries before neutron-star binaries were detected, and did not expect to see anything at all for a few years while aLIGO built up to design sensitivity, both of these postings seem to have been based on knowledge of The Event, even though this is not made explicit. But casual conversations among the astrophysical community in Germany also indicate that the detections have become accepted as a matter of course. The timing and degree of detail included in the supposed "prediction" paper indicate strongly that it is based on The Event but, truth being stranger than fiction, there are suggestions that the author, who, it has to be said, was working on these kinds of ideas before the discovery, might want credit for making this prediction! (We will hear more about these premature papers when we discuss the *Sky and Telescope* rumor below.) People are fearful and depressed about these things; they feel that they have been let down by their trusted colleagues.

10 JANUARY AND FEBRUARY
The LVC-Wide Meetings and the Submission

Jan. 3, 20:16: For these reasons, and after consulting with appropriate parties, **Fulvio and I decided to move to Step 3 of the detection procedure.** This step involves charging the detection committee to objectively examine the detection case as presented in LIGO-P150914.

What has happened is that pressure has built up so that even though Step 2 of the Detection Procedure has not been completed, the leaders have decided to move on to Step 3 on the assumption that nothing much is going to go wrong during the remainder of Step 2—which is about reviewing what has been done. Step 3 is the objective overall examination of the claims made in the detection paper by the Detection Committee. It is expected that this will be completed quickly, making it possible for the LVC-wide meetings that will make the decision on whether to submit or not to be held in the week of the January 18. The schedule has slipped from what was proposed on October 9 by about five weeks; it is a hard five weeks. At this point it feels as though the six weeks between now and the press conference will be mostly a matter of discharging formalities. Note, however, that this leaves fewer than four weeks between submission of the paper and the press conference: that seems like a tight schedule, but the ground has been smoothed with *Physical Review Letters*.

In the meantime, circulars are going around about the upcoming American Physical Society (APS) meeting, which will be held in Salt Lake City in April. If all goes well I will be there, watching to see how the claims are received. Naturally the community wants to present The Event at the

APS, but the deadline for submitted abstracts is January 8. This means that abstracts have to be sent without "giving the game away." Abstracts "with sufficient vagueness" begin to circulate. Here is an example:

> Advanced LIGO recently completed its first observing run, which collected gravitational wave data with unprecedented sensitivity between September 2015 and January 2016. One of Advanced LIGO's primary goals is to detect and characterise gravitational waves from transient sources such as the coalescence of binary compact object containing neutron stars and or stellar mass black holes. We will report on the characterization efforts applied to past and future Advanced LIGO candidates.

Another abstract writer is warned to remove the word "successfully" from "Advanced LIGO recently successfully completed its first observing run" because "successfully" sounds like something was observed. Thus is the circle squared—everyone knows they are going to be talking about The Event, but at this stage it cannot be said. All this is happening around January 4. These couple of indications turn out to be merely the tip of the iceberg; dozens of emails like this are soon sent with participants warning others that a certain term or phrase suggested too much and indicating how it might be modified. An expert on language or spying might have great fun analyzing these interchanges.

JANUARY RUMORS

Two things happen on January 11. First, a new tranche of rumors comes to my attention as a result of a phone call from the *Nature* reporter whom I spoke to around September 26 (see p. 60). He has heard some new rumors and wants my opinion of them. These rumors are pretty startling, including one that the signal happened during the engineering run before the blind injection program had begun so it could not be a blind injection. Clearly someone has been talking.

I say, "I have not heard those rumors." That is literally true—I have not heard the rumors, so I am not lying; but I am certainly engaging in deceit and I wish I did not have to. This phone call lasts around half an hour, and though much of our discussion covers the early history of the field and such things as Ron Drever's contribution to interferometry, I am essentially engaged in deceit for half an hour. This isn't right and I feel terrible. I direct the reporter to speak to Gaby Gonzalez, the LIGO spokesperson.

I write to Gaby, telling her what has happened and saying that I think the reporter may have guessed I was being deceitful because, given my position in the field, I should have been far more excited by the rumors if I had not already known of the facts behind them. I indicate that I doubt she can "hold the line." Her reply shows that she is determined to try to do so, pointing out that the results have not even been reviewed internally yet. I think something is wrong here and that the scientists have become more concerned with the theater of revealing their finding in the most dramatic fashion than in telling the truth—which is their fundamental role in society. It is true that the results have not been reviewed internally, but everyone knows that this is real: why not just say to the world: "We think we have found something, but we can't be sure until we have finished our internal reviewing process—remember BICEP2. We are optimistic but will not make the announcement that represents what we consider to be a secure judgment until February 11. Science is more than hunches and hopes, and it takes months to confirm a complex and novel finding such as this might be." Once more, I think a detection and announcement procedure designed for a marginal event is causing trouble for this miraculously striking discovery.

And then, just after I write the above sentence in this book, at 8:45 a.m. on January 12, I get a phone call from my son, who lives in London, telling me that BBC Radio 4 is reporting that Lawrence Krauss has said he is sure that gravitational waves have been discovered. My son sounds really excited and has felt impelled to phone me to let me know about the broadcast. He has no idea of what I have known for the last four months but, of course, he does know this is my project. Fortunately, he is on his way to work so we cannot talk about it for long because I am even less happy about deceiving him. I am expecting a follow-up call from him that evening that is going to be difficult, but fortunately he has other things to do.

The official LIGO line is: "We are analyzing O1 data and will share news when ready. I'd say that it is wisest to just be patient." But does this not imply that data are being analyzed? Do I no longer have to pretend that I am doing nothing out of the ordinary? Can I now say that I have been watching this process of analysis for several months and reading about a hundred emails a day? I don't think I can lie to my family when they have heard this on the radio. Of course, I will take the LIGO line and explain that it still has to be confirmed, but I don't think I can keep pretending that nothing out of the ordinary is going on. And yet I do.

Before that phone call from my son, an email had come around listing the rumors then current:

> **Jan. 12, 02:01:** For some reason, there has been a new thread of rumors starting yesterday about LIGO detection(s):
>
> https://wiki.ligo.org/EPO/GW150914RumorCollectionPage
>
> Dave, I and several others have been contacted by many journalists today (Tech Insider, NRC Handelsblad, New Scientist, Nature News, Business Insider, The Guardian, Science News, ...) asking for comments on the rumors. We've said we're still taking data, and that analyzing, interpreting and reviewing results takes several months—all of which is true, **we [have not] yet reviewed results in the GW1650914 paper.**
>
> This kind of pressure is very damaging to our activities, especially at the time when we are trying to get not just the detection paper results reviewed, but when our attention should be focused on results we want to present in companion papers. The questions asked indicate that journalists have information from people with close knowledge (engineering run, high mass back holes, no blind injection, etc). Please refrain from talking ... to any non-LVC members about preliminary results—we need not only our internal review, but peer review before sharing results, and we don't have either yet.
>
> Please note that L. Krauss, the first person to print rumors in twitter, says he has sources in "the project." Inappropriate sharing of results with non-LVC

members is not just a very serious lack of ethics, it's also cause for serious measures for breaking collaboration rules.

Again, be part of the solution, not the problem!!!

The rumor gaining most traction is Krauss's tweet of January 11:

My earlier rumor about LIGO has been confirmed by independent sources. Stay tuned! Gravitational waves may have been discovered!! Exciting

Later the same day, Krauss writes:

Re LIGO. Caveat earlier mentioned: they have engineering runs with blind signals inserted that mimic discoveries. Am told this isn't one.

Krauss's tweet has also been picked up and made the basis of a long report in the *Guardian*.

The *Nature* reporter, in thanking me for our long conversation, sends me one of his sources. This was an extraordinarily well-informed and scientifically literate blog of Czech origin. Among other things, it claims:

A commentator whose name is known to us has told us that there have been two events … detected by the LIGO. A new rumor I got yesterday says that LIGO has "heard" a merger of two black holes into one bigger black hole. (I don't know whether this event is one of the two events from the first sentence of this paragraph.) (… the stellar black holes' masses are believed to be between 3 and 50 Suns or so). (http://motls.blogspot.co.uk/2016/01/ligo-rumor-merger-of-2-black-holes-has.html)

That blog does not appear to have been picked up by the media in general, but it is reported on *Nature*'s website.

This new set of rumors explodes across the newspapers and the web but the collaboration hold the line and simply refuse to say anything except that they are still analyzing the data. Astonishingly, it nearly works, and the rumors nearly die away. Thus an account of the rumors can be found on the version of BBC Radio 4's *More or Less* program that is promulgated on the World Service but is pulled from the domestic broadcast, leaving it short.[1]

But by now a careful reading of the rumors will tell you nearly everything you need to know. It has been said that two events have been seen (this was before the Boxing Day event, so it was accurate) and that at least one of them was a binary black hole (the Czech rumor), and it has been said that the event had occurred during the engineering run before the observation run had begun so it could not have been a blind injection (Lawrence Krauss). For a summary of these rumors see the second article in *Nature*.[2] One rumor virtuoso, Stein Sigurdsson, a Pennsylvania State University physicist, quoted in *Sky and Telescope* of January 13, and seemingly missed by *Nature*, has used the techniques used by generals in war to gauge the intentions of the enemy (as discussed in chapter 2, p. 23)—observing the patterns of activity associated with the massing of troops and materiel:[3]

> Sigurdsson points to a flurry of papers that have appeared this week on the arXiv preprint server that were curiously specific. Astronomers, says Sigurdsson, "posted somewhat different scenarios for ways in which you could have black hole binaries form, all of which coincidentally predicted almost the exact same final configuration, and said 'Gosh our model predicted that this very specific sort of thing will be the most likely thing that LIGO sees.'"

Sky and Telescope also quotes the tweet of another astronomer: "this seems a rather specific GW [gravitational wave] scenario to pull out of thin air?" Sigurdsson worked out something even more exact by doing what he called the equivalent of watching for late-night pizza deliveries at the Pentagon, which tell you when big events are afoot.

1. See http://www.bbc.co.uk/programmes/p03fm6b8. Thanks to Graham Woan for pointing this out to me.

2. Davide Castelvecchi, "Gravitational-wave Rumours in Overdrive," *Nature*, January 12, 2016.

3. Quoted in Shannon Hall, "About the LIGO Gravitational-Wave Rumor…," *Sky and Telescope*, January 13, 2016.

> When a few collaborators—who all happen to be members of LIGO—duck out of a future conference due to new overlapping commitments, it doesn't go unnoticed.... Based on dates cancelled, Sigurdsson speculates that an announcement will come from the team on February 11th.

This, of course, is dead accurate and known a month in advance.

EM PARTNERS REVISITED

In January the problem of what to tell EM partners raises its head again. The Boxing Day event has to be sent out as an alert, but the normal way of doing this would be to point the EM partners to GraceDB. The trouble is that the Boxing Day event, like The Event before it, is too strong to reveal at this stage—it would look too much like a discovery. All an alert needs to say is that something has been found that exceeds the alert threshold—which is a false alarm probability of one per month—so at the moment the community has to fiddle with GraceDB so that it does not say too much. Here is one member's response:

> **Dec. 4, 14:57:** I think we must find a better solution to communicate to our astronomer partners than to modify the Gracedb entry.... The idea of faking a different number in our database is creepy—both for our internal understanding... and also for our scientific "profile" (how would a skeptical reviewer of our results find the notion that we have several lab books with different numbers?).
>
> So, I strongly urge us to find a different and better way to communicate partial information to outsiders which does not involve changing values in our internal bookkeeping or lead to different numbers with the same label in different places, public or private.

The lower limits reported to the EM partners are enough to legitimize a search of the heavens without constituting a "discovery." But it is also thought that the EM partners should have some idea what they are looking for, so sky positions are presented in as much detail as possible along with some intimation of what the source of the signal might be. Subsequently, some members of the collaboration come to believe that the source of the rumors, given that rumor-spreaders such as Lawrence Krauss have stated that their informants did not come from the collaboration, could only be the EM partners; it is suggested that the EM partners, not having such a strong stake in the discovery, might be more careless about what they say in public, indicating that LIGO had spotted a trigger that was consistent with black hole inspiral. Even if it is not said to be a discovery, this would be enough to give rise to most of the rumors that have been promulgated since early on.

The withholding of information from the EM partners now gives rise to another kind of anxiety. A personal email I receive on January 17 puts it this way:

> I'm actually less worried about the wall of silence from the lvc in general than the disingenuous and slow data release to our EM partners. These are world experts trying to do the same science as us, with MOUs [Memorandums of Understanding] like us, but who have been artificially nobbled (for reasons that escaped me) and not given as much information as some random lsc undergraduate doing a course project on ligo data, who gets the lvc-all emails and can see gracedb etc.

As the end of the affair approaches a still more substantive tension arises. The intention has been to publish an early paper detailing all the searches by the many EM partners that have been executed in tandem with the triggers. Toward the end of January a draft of this paper is prepared for circulation to the EM partners—who would be coauthors. This draft, and its covering letter, are scrutinized to make sure that nothing more is being given away—that there is nothing that would affirm that a discovery has been made, only that a big search has been conducted on the basis of certain inconclusive triggers. The first sentences of the abstract read:

> A possible gravitational-wave transient was identified in data recorded by the Advanced LIGO detectors on 2015 September 14. By prior

arrangement, preliminary estimates of the time, significance, and sky location of the event candidate were shared with 62 teams of observers covering radio, optical, near-infrared, X-ray, and gamma-ray wavelengths with ground- and space-based facilities.

But the explosion of rumors and the determination to hold the line until the press conferences lead to the circulation of this paper being aborted. One of the scientists on the committee that took the decision explains it this way in an email to the collaboration:

> **Jan. 21:** . . . I went into the discussion satisfied that you guys had done a great job of scrubbing the paper and letter so as to give away no more information than necessary, while conveying a good sense of how the final paper will be structured. If everyone receiving that draft acted in good faith, I would not be concerned at all. What changed my mind about releasing the draft was reflecting on the recent timeline of rumors. On Jan 11 at 5:20 am EST, the updated false alarm rate was sent out, and later that morning, Krass [*sic*] "confirmed" his infamous tweet with a new tweet, and ever since, a workday hasn't gone by that someone at [at my university] hasn't asked me about the rumors. (I don't know if Krauss's original September tweet was based on an EM observer's report, but it wouldn't surprise me, since he disavows talking to anyone in the LSC.)

Since good relations between the collaboration and the astronomy community are vital for the future of this field, this is delicate ground. Note, however, that since the notifications to EM partners must, perforce, mention the date, and since there is nothing secret about the date of the start of official observations, this could lead straight to the deduction that no blind injections were involved, which would reveal that this to be a suspected binary black hole. So suspicions that the leaks could have come from the EM partners are reasonable, even if there has been no deliberate attempt to give anything away. The EM partners could consider that they were not giving away anything particularly sensitive since no one has said this was real, only that it was a "trigger," and in the past all such triggers have turned out to be false alarms. Nevertheless, the contents of the tweets could still be "scraped" from innocent discussions. Let me add that the logs recording where telescopes are being pointed are public, and quite a lot can be deduced from these.

On the other hand, one emailer thinks that the source is elsewhere:

> **Jan. 25:** The more dramatic leaks are undoubtedly internal. I personally know of examples of senior LVC members sharing intimate details of our plans with non-LVC colleagues; this has put me in a very delicate situation.... In addition, apparently [an important funding agency] knows about the announcement plans; this has trickled down to the [important physics institution] director's office, and the deputy director has subsequently contacted our... EM partners for comment. This has NOTHING to do with our direct contacts with our EM partners—the "leak" is coming from the highest levels of the US funding agencies.

Probably the leaks have more than one source. There is no doubt that more and more people are being told about the discovery because more and more people need to know in advance —decisions and plans depend upon it. I know that heads of departments are being told at a number of institutions. I know that funding agencies are being told. Even I have had to tell one additional person under conditions of strict confidentiality because one of our joint projects is being compromised by the amount of time I have to put into this project. And, right now, because I cannot get any concrete responses from the officials responsible for the main UK press conference, I am in the position of telling my son, who lives in London, that *I might* be coming to stay with him and his wife in connection with a business trip on February 11 *or I might not*, but I can't tell him yet—mysterious! The next time he asks, further prevarication will be so strange—especially as he has heard the rumors on the radio—that I am going to have to tell him too under an oath of secrecy (he doesn't ask).

I know that more members of families and the like are being told; one cannot maintain an indefinite number of disturbances in social life without starting to seem eccentric or rude. As time goes on, then, the number of people in the know is increasing fast and the pressure grows like a boiler about to burst. Once more, the attempt to maintain the secrecy and dissimulate leads, inevitably, to trouble, in this case the most important possibility being a breakdown in relations between gravitational wave scientists and astronomers; here secrecy may jeopardize the science, not just the integrity of the community.

JANUARY 11: THE SECOND SHOE HAS DROPPED AFTER ALL

Also on January 11, as if this wasn't bad enough, Peter Saulson writes to me that the box has been opened on the Boxing Day event and, under this new analysis, it turns out to have a statistical significance of 5 sigma.

> **Jan. 11, 18:59**
>
> Dear Harry,
> We just had the latest box opening (for Analysis Period 8), and the Boxing Day event turns out to be another 5 sigma event!
>
> Here's the open box results page:
> https://sugar-jobs.phy.syr.edu/~bdlackey/o1/analysis8/analysis8-c01-rf45dq-v1.3.4/7._open_box_result/
>
> ... So, even though the signal wasn't as strong as [The Event], this one would rank as a discovery all on its own. A big deal.
>
> Now people are starting to worry/realize that we'll pass our 4-detection threshold pretty soon (sometime during O2), and thus will transition into the Open Data Era. Life is moving quickly.

The right-hand smiley in figure 8.2a shows, in a technically incorrect way, where the third event is now that its significance has increased in this way. It is still a long way from The Event, but, being above the magical 5 sigma level, it would justify an independent discovery paper if it were the only signal there was. This makes a big difference. It now looks like a genuine confirmation of The Event—we are now out of the regime of single events and can forget about the Omega-minus and, much more importantly, forget about the monopole. This is huge scientifically and also, as I will argue, sociologically. Figure 8.2b shows the Boxing Day event's true significance represented with its own little dogs in the background plot. As can be seen, it is up above 5.3 sigma.

Why do I say when introducing this change in significance, "if this wasn't bad enough"? What's bad about it? There are two things. First, the rise to 5 sigma leaves no one in doubt that they've really seen gravitational waves and this means the deceit is still more marked—it is a real strain to think that we won't be sure for another few weeks until all the reviews are in: that possibility has vanished in terms of what ordinary physicists know. And, since secrecy about the Boxing Day event will be preserved beyond the press conferences, those meetings will be deceitful too.

The second shameful thing refers back to the sentiment expressed on page 29: "I hope no other events do show up, because if they do it will make this discovery all too easy and therefore less yielding in terms of *sociological* interest." Now that this second event has grown in significance to an extent that would allow it to rank as a first discovery, my life as a sociologist seems to have become much harder. So I am feeling guilty again—guilty for not jumping for joy with my colleagues but looking on this transformation with a sour gaze. January 11 has been a bad day for feelings of guilt. The problem will be dealt with in chapter 13.

The "4-detection threshold," by the way, refers to the agreement made within the collaboration that once they have confirmed detections of four gravitational wave events they will cease to treat their observations as secret and instead release all their data to the astronomical and astrophysical worlds as these data are generated. This, of course, is another symptom of the changing order of things: after four sure sightings, gravitational waves are going to be ordinary enough not to need special kinds of extra cautious confirmation.

I find myself asking Peter some technical questions about the way the significance of this event has jumped from less than 4 to around 5 sigma. The original 4 sigma estimate is based on the assumption that the background is what is represented in figure 8.2a—the 16 days of background for The Event. As far as I can see, this significance will change only if the background against which the third event is set is different from that of the

16 days. Have the interferometers suddenly become much less noisy? Does this have an impact on the notion of a "freeze"? It would if it were true! The same question strikes Peter and he tells me—to my guilty pleasure—that he can't understand it either, is beginning to doubt it, and that he may have to reverse his joyful attitude. He looks into it.

The answer is interesting. The Boxing Day event is less massive than The Event, equivalent to 25 of our suns in total rather than the 65 or so of The Event. What this means is that it falls into a different "bin" when it comes to matching it with templates because it has a higher frequency; there are three bins in all. Higher-frequency signals are longer so there is less chance of a noise signal having the same appearance—that is, this higher-frequency region is much quieter. That is why Boxing Day can be so significant compared to the background even though it is less energetic with only half the SNR ratio. To repeat, there are markedly fewer purely noise coincidences that come anywhere near mimicking such a thing in the higher-frequency bin so the likelihood that it could be caused by noise is reduced. This difference does not emerge until the box is opened on the relevant time slides, and that is why the statistical significance of this event has suddenly leapt upward.

THE LAST MEETINGS: JANUARY 19 AND 21

January 19 has been set aside for the equivalent of those collaboration-wide meetings that came at the end of the two blind-injection exercises, the Equinox Event and Big Dog, where, the community had to decide, first, whether it was ready to submit a paper for publication and, second, whether it was ready for "the envelope" to be opened—the envelope that contained the blind injection secret. These were huge and joyful meetings, each gathering the community from across the world to a hotel in California. The sun shone, the palm trees palmed, and the champagne fizzed—albeit in plastic glasses (singularly appropriate for blind injections).

This time there is to be no central meeting. "TeamSpeak," the teleconference system, seems to be working well enough for there to be no need to warm the planet with extra jet-exhaust. Everyone pretty well knows everyone already, the whole business has been done for months by telecon, and it is decided that the crucial meeting could be dispersed across the planet. So even if I wanted to travel to the heart of things, I couldn't have located it. In the days before January 19 I am working out where and how to spend that afternoon.

In a face-to-face meeting it is obvious that only a limited number of people can speak, but in a telecon situation there is nothing to stop everyone trying to make their contribution. To reduce the potential discussion points to a manageable number, the leadership has asked that participants to assemble into local groups and decide beforehand what points their *group* wants to discuss. This, as it were, reintroduces some of the constraints of face-to-face interaction. I could have stayed in my study, but I decide to ask the Cardiff group of gravitational wave physicists if I could join them so as to sample the atmosphere. It turns out, however, that the group from Birmingham University has decided to set up a British meeting and the Cardiff group has agreed to join them. So I write to Birmingham and arrange to travel up on the train with the gravitational wave physics group from Cardiff. We are lucky—the sun shines—but it is freezing cold and there are no palm trees in Birmingham. Worse, on the morning of the meeting the leadership explains that there were so many outstanding queries that the decision on the paper cannot be made that day, that a further draft—draft 10—would be prepared and circulated the following day, and that the decision-making collaboration-wide meeting would be held on January 21. So we've traveled to Birmingham for what was known in advance to be a meeting that would result in an anticlimax. It was a nice trip but not the real thing.

Worse, the meeting once more exhibits the problem of secrecy. And remember, in the matter of deceit I am not a distanced observer; whether I like it or not, I am a full-blown participant. For me the crucial discussion concerns whether the Boxing Day event should be mentioned in the paper. The Detection Committee recommends that it should, something that I think is vital given that Boxing Day is playing such an important part in building the scientists' confidence: they now know that they are no longer dealing with a single event and that the monopole can be forgotten (this is accepting what seems to be their collective view: that Second Monday did not count for much). But there follows a big up-swelling of objections along the lines that the collaboration could not yet be sure that, as with The Event, Boxing Day would survive the months of analysis required to confirm it. This line had justified the secrecy over The Event (even though everyone knew it would be astonishing if it was not real) and it is now being used to justify the same kind of withholding of information over Boxing Day.

The range of opinions runs from some who want a brief mention of the Boxing Day event in the discovery paper, through those who think that even if it is not mentioned in the discovery paper it should be mentioned in the press conference and opened up to collegial discussion, through those who think it should be treated with as much subterfuge as The Event. For the latter group the order of things has not yet changed; they want to push the logic of deceit, which might just be understandable in the case of a major discovery (but see chapter 13), into the indefinite future. Here is an example of a statement representing the extreme view:

> I think it would be a bad idea to mention the Boxing Day event in this paper. We don't have strong statements we can make about it and putting weak statements in the paper would only weaken this very nice result and distract focus from the main event. We will be asked if we've seen any more events but we've been asked for the last 4 months continuously if we've seen any events and we've given the same line consistently—that the data's being analysed and we'll tell you when we are ready. We just need to continue to do that—I don't see any problem.

This kind of remark confirms for me that deception is now becoming natural to the community; the community is a Dickensian academy and we are the artful dodgers.

BACK OF THE ENVELOPE

And so the big day is now to be January 21. One discussion that takes place between the 19th and the 21st may give some flavor of how these things can go. One of the leading members of the community had contributed a short analysis to the paper that showed why it looked as though the event was a binary black hole. The full proof would be really long and complicated—a job for the parameter estimation team—and we have seen that it has always been vulnerable to objections such as "it might be boson stars." But this member has come up with a nice simple way of showing that it almost certainly was a binary black hole. It is simply a matter of calculating the separation of the two circling objects as they spiraled closer and closer, which could be deduced from the frequency of rotation, and showing that two objects that close and that heavy but still not touching had to be so compact that the only thing they could be was black holes. Any other pair of other things would be too large, and if even a single one was a neutron star then the mass of the other would have to be so huge as to make the model implausible. Here is the opening sentence of that argument as expressed in draft 9 of the paper:

> A back-of-the-envelope analysis of the basic features of GW150914 points to it being produced by the coalescence of two black holes—i.e., their orbital inspiral and merger, and subsequent final black hole ringdown.

But this is an argument about plausibility, not absolute proof, and, as we have seen in the above remark about the Boxing Day event, there is a subset among the collaboration who think that all remarks that are merely plausible must reduce the credibility of a scientific paper like this one: "putting weak statements in the paper would only weaken this very nice result and distract focus from the main event." This is another example of "relentless professionalism."

There follows an exchange of more than thirty emails about whether this whole argument should appear at all and, even if it should, whether it should be described as a "back-of-the-envelope calculation." Here are a few indicative contributions:

Jan. 20, 01:23: When I first read this I had a fairly strong negative reaction to the usage of the phrase "back-of-the-envelope" to describe any analysis done for this paper. It seems like it undercuts the credibility of what follows. Do people feel like this is an OK statement to make? It seems like it would be better to say something that doesn't imply that we didn't do a careful analysis, even if it's not a full GR/NR analysis. Maybe "A cursory analysis…"?

Jan. 20, 08:40: Those words are important, because otherwise many co-authors are misled into thinking that this is a precise calculation, and kept wanting to give precise values. These words seem to be the most reliable way to communicate that the argument is an approximate one.

Jan. 20, 08:50: the _notion_ is important, but I agree that the wording "…envelope…" is not quite the right style for the paper. Something should be said to indicate that it is a top-level argument on principles.

Jan. 20, 13:03: I personally don't like that entire paragraph. This calculation is very nice—don't get me wrong. But it doesn't belong in the paper. Pick your reason.…

* Back of the envelope calculations are dangerous! They can give the "right" answer even if terribly flawed in the logic.… We do the full calculation because the details matter!
* Putting forth this calculation so far in the front of the paper makes it look like this is how we infer the source of the signal and how we think about parameter estimation. It looks amateurish. This is not how we do parameter estimation. We have done extensive tests that verify this signal is a BBH as predicted by GR and nothing else. We are not amateurs!…
* The calculation assumes much of what it's trying to prove anyway: "The most plausible explanation for this evolution is the inspiral of two orbiting masses m1 and m2 due to gravitational-wave emission." What you really mean is, given that the signal "looks chirpy," it must be a binary inspiral, because we can't think of anything else that does that; and given that it is a binary inspiral, it must consist of two black holes, because we can estimate its mass from PN, and this mass is too large for it to contain a NS.

We could poke holes in the argument all day. It's not airtight, it's not intended to be airtight. But everything else in this paper is airtight, and intended to be so! Jan. 20, 16:30: Oddly, I feel that the current discussion about excluding NSBH systems argues *in favor* of the back-of-the-envelope terminology.

The whole point of this text is NOT to definitively establish that we have seen BHs. That's what the rest of the paper, supplemented by the bulk of the companion papers, is all about. This argument is to give the "average physicist" some intuition for what is going on.

Jan. 20, 17:13: That's what I'm struggling with. I thought that this part of the paper was intended to be a simple "proof" that it must be BBH—that's what it seems to say. It sounds like it should either just discuss/interpret the signal to the extent that one can, without trying to prove with basic arguments that it must be BBH; or else state at the appropriate point that a detailed analysis (not presented here) provides the key link that proves that.

In the end, the words "back of the envelope" are dropped but not the paragraph, the first sentence of which in the submitted draft reads:

The basic features of GW150914 point to it being produced by the coalescence of two black holes—i.e., their orbital inspiral and merger, and subsequent final black hole ringdown.

THE FINAL SIGN-OFF LVC MEETING: JANUARY 21, REAL TIME

So, a bit later this afternoon the real sign-off meeting will be held at last. It will decide—unless something goes drastically wrong—whether the paper is ready to submit to *PRL*. *PRL* is ready to accept a paper that is fifteen pages long and have it reviewed in a week so that there will be time to make any requested changes and get it ready for promulgation before the press conferences. Here, for the sake of historical completeness, is the substance of the letter sent to the collaboration today:

Many, many thanks to the Paper Coordinating Team who after a heroic effort have considered all comments by LVC members and taken into account recommendations from the Detection Committee about the presentation of the GW150914 detection case.

As announced on Tuesday, we will hold an LVC-wide meeting on Thursday Jan 21 8:00 am PST (~9.5 hrs from now), to consider any objections to submitting the paper https://dcc.ligo.org/P150914-v10 to Physical Review Letters. The co-chairs of the Paper Team will comment on the latest changes, and the co-chairs of the Detection Team will be asked if there are any objections to the presentation of the detection case in the paper.

We will meet in TeamSpeak, in the LVC Plenary channel, no password. Please mute your microphone in TeamSpeak before you join the channel, and only unmute when you need to talk. Please do not type comments or questions in the chat area, it is VERY distracting and a sign of disrespect to the person talking. If you have a question, please type "hand up about" and add 2-3 words about the topic of your question.

We will not consider any further suggestions on the text, please only raise your hand if you have objections to co-authoring this paper, or if you have specific questions (not suggestions!) to the Paper Team or Detection Committee.

We will ask from eligible authors in the Aug'15 LSC author list + Nov'15 Virgo list (https://dcc.ligo.org/LIGO-M1600003) the following question:

Do you approve the inclusion of your name as co-author in the submission of LIGO-P150914-v10 to Physical Review Letters?

We will submit the paper if we have a substantial (>2/3) majority of "yes" answers from the votes received, with the vote closing at the end of the meeting (no later than 9:30am PST). Note that we will not exclude names from non-voters from the paper, but members can "opt out" of the paper if desired at any time before publication.

Looking forward to the meeting you all later!

I will be attending the meeting from my study.

WHAT HAPPENED

The meeting was so strange that I could not write it up until now—the following morning. There were intimations of trouble as the day wore on, some people being upset that a change had been introduced to the paper between draft 9 and draft 10 without the appropriate approval either from the persons initially responsible for those sections or from the Detection Committee. There were two small sections that were causing trouble, but we will concentrate on one of them as it turned out to be the focus of a huge row. Both were illustrations of relentless professionalism.

The change we will concentrate on is the shortening and simplifying of a paragraph that developed an estimate of the rate of binary black hole inspirals in the universe based on the observations that had been made. This kind of estimate is the preserve of astrophysicists, and astrophysics deals with approximate numbers when it comes to working out how many things or various kinds there are in the heavens. There had been considerable discussion over the months about whether it was sensible to estimate a rate based on only a single observation; the astrophysicists were determined that it did make sense while others thought it was an odd thing to be doing. Furthermore, this kind of rates calculation was something that anyone could do once they knew the nature of the discovery—its distance, component masses, and so forth—and the rates group were furthest advanced in their preparation of a companion paper, intending to try to publish it simultaneously with the detection paper. Therefore, they had to be happy about what was in draft 10 but also had to reconcile what was being said there with what they were saying in their companion paper.

I think what happened over a few hours on that Thursday afternoon (British time) says quite a lot about physics and physicists, so we'll spend some time on it. First, here are the two passages—the longer one from draft 9 and the shorter one from draft 10.

> By combining our observational results with an estimate of the detection sensitivity for binary black hole mergers, we can constrain the rate of stellar-mass binary black hole mergers in the local universe. An optimally oriented,

> optimally located binary system otherwise like GW150914 will produce a SNR of 8 in a single detector with sensitivity like those shown in Fig. 3 at luminosity distance 2.4Gpc (z = 0.42). Assuming that all binary black holes in the universe have the same masses and spins as GW150914 [107] and adopting a false alarm rate threshold of 1 per 100 years, we then infer a 90% credible range for the rate of 2 – 53Gpc^{-3} yr^{-1} (in the comoving frame). When we incorporate the full set of binary coalescence search results, properly accounting for each event's probability of astrophysical or terrestrial origin [108], and make more reasonable assumptions about the mass distribution [109], we obtain a higher rate estimate ranging from 6 – 400Gpc^{-3} yr^{-1}. These estimates are consistent with the broad range of rate predictions as reviewed in [109],with only the low end (<1Gpc^{-3} yr^{-1}) of rate predictions being excluded.

> These observational results constrain the rate of stellar mass binary black hole mergers in the local universe [108].We obtain a rate estimate of 6–400Gpc^{-3} yr^{-1}, consistent with the broad range of rate predictions as reviewed in [109], with the lower event rates being excluded.

What should be noted is the nature and substance of the estimates on the table. The estimates are numbers of events per year within units of volume of space: "Gpc^{-3} yr^{-1}" means yearly rate per cubic gigaparsec, which is a cube with edges roughly 3 thousand million light years long.

If the estimate is based on the observation of The Event alone, it comes out as somewhere between 2 and 53 events per year per unit volume. Note the huge range of this estimate: 2 to 53. If, however, Second Monday is taken into account and some recognition is given to its uncertainty, the estimate comes out as between 6 and 400 per unit volume per year. Those two possibilities are what are explained in the longer passage, whereas in the shorter passage only the second possibility is offered. The paper writers opine that they changed it because the longer passage seemed to involve a lot of complicated calculation of an outcome that was greatly uncertain.

There were intimations that this change was going to cause trouble in the course of the day before the meeting, with around ten emails pointing out that the shortening of the passage had not followed proper procedures and that it was misleading as it stood. The procedures that had been sidestepped were checking with the parties responsible for writing the passage in the first place and confirmation by the Detection Committee. But all this had happened in one day and the paper writers were swamped with thousands of comments.

The meeting started quietly in an air of suppressed excitement, but this problem slowly grew in salience. The chair of the meeting made it clear that no changes to draft 10 would be countenanced. This was because the major purpose of the meeting was for the 1,004 authors of the paper to be able to vote on whether it was ready to be submitted for publication and that vote had already started, so some had already voted on the paper as it stood. Therefore, the choices were limited. One could either vote for submission or against it, but the chair made clear that those who voted against it were honor-bound to ask to have their names removed from the author list—though, to confuse matters, since the vote was anonymous this could not be enforced. An additional problem was that because of certain technicalities, a few of those on the author list had to write in their votes, so they would not be anonymous. As the drastic nature of the vote became clear—approve this paper or remove your name from it—and the chair remained resolute in being unwilling to alter a word of the paper, the debate became more and more heated and entrenched.

Eventually the mood swung toward accepting the reduced length and the wording of the passage, but the astrophysicists insisted that the paper as it stood was incorrect: the rate should not be 6–400 per unit volume per year—the number based on taking into account Second Monday as well as The Event—but 2–400. Requests were made that this number be changed, but the chair insisted no changes to draft 10 could be made because these were the conditions of the vote.

The heated and quite unpleasant debate lasted at least forty-five minutes, with more and more coming out in favor of a change and contributors attempting to move that the vote be canceled and a new vote instituted for later in the day, and with others rushing in to second the motion and the chair explaining that no such motions would be accepted. Meanwhile the fixed deadline for the end of the vote was drawing closer and closer.

The rule was that more than two-thirds of authors had to vote in favor of the paper if it was to be submitted.

With, literally, a couple of minutes to go before the clock automatically ticked down to close the vote, the chair finally made it clear that there would be a possibility that the numbers could be looked at again during the refereeing process. This allowed one of the most adamantly objecting astrophysicists to opine that in that case a vote for the paper would be in order. At this point I, at least, was very worried that the vote would fail and the paper would not be submitted and that this terrible period of tension and secrecy would be prolonged. I sensed I was not alone.

After sleeping on it, the whole episode has come to seem to me like a kind of madness. Let us forget about the procedural issues and just consider what was at stake. What was at stake was the lower bound on an estimate whose upper bound was 400. The question was whether that lower bound should be 2 or 6. Who cares? What a totally inconsequential thing in the context of this magnificent discovery. Furthermore, whichever number is chosen, we know it is wrong because it does not take into account the Boxing Day event. It cannot take it into account because the community is bound by its oath of omertà—the Boxing Day event cannot be spoken of and so cannot be used to adjust the calculation. Furthermore, we know that this number is going to change every couple of weeks when the next observation runs find more events.

On the one hand, you have the first direct detection of gravitational waves leading to a completely new branch of astronomy; on the other hand, you have a lower bound on an estimate with huge uncertainties that everyone knows is wrong. We know that over a few years the range of that estimate will narrow right down as more and more events are seen, so we know we are looking at something utterly ephemeral—whereas the discovery paper is the opposite of ephemeral, it is the creation of a new normal and a whole new way of life for the astronomical community. How can something so ephemeral be jeopardizing something so lasting?

The answer, of course, is the nature of the academic world, the culture of physics, and *relentless professionalism.* The difference between 2 and 6 is of total unimportance to the world but is important for the pride of this group of astrophysicists who need to defend the result to their professional peers. This seems to me to be a pathology of academic life. It is exaggerated here because of physics' mathematical foundations. To me it seems something close to, if not quite as extreme as, the argument over the airplane event (see *Gravity's Ghost*, 27–32). We'll come back to it in chapter 13.

Incidentally, a couple of weeks later (February 5), an emailer talking about what facts about the discovery should be put into an easily accessible form echoes the same sentiment, remarking:

> I'm not sure about including the coalescence rate, as this will become obsolete (soon), while the rest will remain the cherished picture of our first born.

The clock ticked over to the 5:30 deadline and a few minutes later the votes were counted. The result should have been a formality, but it did not feel like that; I think everyone was on tenterhooks. The result came through: 587 for and 5 against! There was cheering. Maybe it was me alone who noticed that 412 authors had abstained or not voted. Under different counting rules the vote would have been about 80 short of a two-thirds majority. Never mind—in an instant the gloomy mood changed. Here is what was found on the chat window of TeamSpeak:

> **<17:31:32>** [just before the result] "Sathya1": we need to cheer up, pleaseeeeeeeeee
> **<17:31:49>** [just after the result] "valeriu": Eurovision!
> **<17:31:50>** "vicky1": Sathya, we'll still cheer with champagne
> **<17:32:06>** "Alan Weinstein": j-pop
> **<17:32:08>** "Peter Shawhan": Wow!
> **<17:32:12>** "Ilya": HURRAY!
> **<17:32:15>** "Jo van den Brand": excellent
> **<17:32:19>** "gmendell": Yeah!
> **<17:32:21>** "DanHoak": congrats all!!!!
> **<17:32:26>** "Fulvio Ricci": Great!!!
> **<17:32:29>** "vicky1": let's go!!!!
> **<17:32:29>** "Daniel Holz": FANTASTIC! Congratulations everyone.

<17:32:31> "Keith Riles": Splendid!

<17:32:31> "stan whitcomb": wheew! That was close!

<17:32:33> "Sathya1": Greaaaaaaaaaaaaaat

<17:32:34> "Gianluca": hurra

<17:32:35> "Nergis": hurray. Congrats Gaby and everyone!

<17:32:36> "AndrzejKrolak": well done

<17:32:38> "pai_arch": great

<17:32:48> "Dorota Rosinska": great

<17:32:48> "LHO MPR": Hurrah!

<17:32:51> "arunava": congrats everybody

<17:32:52> "Federico Ferrini": Evvvaiiii

<17:32:52> "neilcornish": Bit early for champagne here

<17:32:54> "Lionel London": @Alan: j-pop, that's great

<17:32:56> "vicky1": Thank you for all patience and extra-hard work

<17:33:07> "AEI Potsdam": Champagne is opened here. Congratulations all from AEI!

<17:33:11> "EGO Seminar Room": Congratulations…!!!

<17:33:12> "Tom Carruthers": Congrats everyone!

<17:33:15> "sanghoon.oh": Congratulations to everyone!!!

<17:33:19> "GariLynn Billingsley": Well done!

<17:33:25> "Eotvos": fantastic, congratulations!

<17:33:26> "Collin Capano": Darth Vader apparently has joined the call

<17:33:31> "Dave Reitze": WOW!!

<17:33:31> "Claudio Casentini": congratulations!!!!

<17:33:34> "RaRa": Congratulations all!

<17:33:34> "John.Oh (吳廷根)": Congratulations!

<17:33:46> "Leo Singer": Hello Stockholm, here we come!

And so on.

And then, for the first time in four months the emails went quiet—so much so that I had to keep checking that my Internet connection had not gone down. It had not: everyone was out celebrating. The number of emails collected by me after the submission, summed over the three days, Friday, Saturday, and Sunday, was 71, whereas for the same three weekdays in the two previous weeks it was 273 and 271!

Again, sitting alone in my study, I missed the celebration both in fact and in concept. I thought that kind of emotion had been dissipated by the aborted meeting of a couple of days ago and the row over the final submission. In any case, in social science you never celebrate the submission of a paper—that's just the start of your troubles. And, sadly, you never have anything as big as this to celebrate anyway. We'll return to this difference between physics and social science in chapter 13.

REFEREEING

Just a week later (January 28) the referees' reports come in and are transmitted to the collaboration. They are full of praise.

> These results are obviously going to make history.

> This paper is a major breakthrough and a milestone in gravitational science.... Its suitability for publication in Physical Review Letters is beyond question.

> It is an honor to have the opportunity to review this paper. It would not be an exaggeration to say that it is the most enjoyable paper I've ever read. The authors have clearly described the experiment and detection process and presented the evidence needed to demonstrate both the statistical and historical significance of the detection. In addition, it is a beautifully written paper that will be accessible to a wide range of readers and serve to inspire a new generation of physicists and astronomers. I unreservedly recommend the paper for publication in Physical Review Letters. I expect that it will be among the most cited PRL papers ever.

There are then a few technical queries and comments that the team believe can be dealt with before the weekend is out. It has also been agreed that the issues that gave rise to the huge row a week ago will be dealt with. In the resubmitted draft the rate will have gone down from 6–400 to 2–400 per unit volume per year, and there will be an extra sentence or two about estimates of the spin on the component black holes. There is only one slightly awkward note: one of the referees wants the claim about direct detection to be qualified. The editors advise, however, that in this regard nothing should be changed. By the end of the weekend we should have the final paper: it will be draft 13 though one referee insists on a final reading. Unless that referee is still unsatisfied, the press conferences will be held on February 11.

THE DISCOVERY PAPER—WHAT HAS BEEN ACCOMPLISHED

So now we have the final draft and it is a beautiful thing. Somehow, in spite of this ridiculous method of writing, with a thousand coauthors, the writing team has pulled it off. Even the voting for different aspects somehow worked out right. Look at the title:

> Observation of Gravitational Waves from a Binary Black Hole Merger

I wanted "direct" in the title and also thought it would be more accurate if "LIGO" appeared there too. But I was wrong; the title is a thing of limpid simplicity. I don't know how it reads to others, but for me, who has spent forty-three years waiting for this moment, it makes me shiver.[4] And "direct" is also right there in the last sentence of the abstract, where it should be, and is repeated in the last line of the conclusion:

> This is the first direct detection of gravitational waves.

That LIGO is the device that made the observation is also clear in the first sentence of the abstract:

4. The *New York Times* reported: "Robert Garisto, the editor of *Physical Review Letters*, said he had gotten goose bumps while reading the LIGO paper" (Dennis Overbye, "Gravitational Waves Detected, Confirming Einstein's Theory," Feb. 11, 2016). I think to know if one understands the emotional significance of scientific discovery, one must ask if reading a scientific paper can induce this kind of reaction.

> On September 14, 2015 at 09:50:45 UTC the two detectors of the Laser Interferometer Gravitational wave Observatory (LIGO) simultaneously observed a transient gravitational-wave signal.

And the paper retains that historical tone of the first draft—not quite so fulsomely, but that this is a significant event in the history of humankind's understanding of the universe is still there. The history element has survived the attempts to flatten the story out into a mere matter of measurement and calculation; here people have done things, not just had things happen to them. The beginning of the following sentence is still in draft 10 in spite of its necessity being disputed right up to January 19, though, unfortunately, the second half has been deleted:

> **Einstein understood that gravitational-wave amplitudes would be remarkably small** ~~and expected that they would have no practical importance for physics.~~

Even so, the paper is an artifact, fashioned to be an icon representing the thing in itself; the paper is like a symbol of office—the jeweled mace that indicates the power of a king. The thing that has been accomplished, for which the paper is a glittering symbol, is still more thrilling. Let us remind ourselves of what it is.

A half-century of effort, never rewarded with any scientific success, has been maintained through sheer determination, managerial virtuosity, and confidence in the theories that predicted it would eventually succeed. The first devices were built on the basis of hope alone, when calculation indicated they could never detect the waves; more and more refinement has driven the sensitivity up even though each succeeding technological generation had no scientific right to believe they would see anything. Advanced LIGO is, at least from today's vantage point, the first device that ought to succeed rather than ought not to succeed, whereas when the preceding generations of devices were built, hope had to triumph over calculation. From Weber's room-temperature bars to Advanced LIGO, the sensitivity to the squeezings and stretchings produced by the passage of gravitational waves has increased by, perhaps, 100,000. These things have to be guessed; measurements were not that accurate in the early days and the targets of the succeeding generations were different kinds of heavenly objects. But, forgetting these complications, because of the way volume of space explored increases as the cube of sensitivity, the likelihood of seeing a cosmic event has increased a thousand-million-million-fold from Joe Weber's first efforts to now. With no scientific success—in marked contrast to the development of particle physics, where each generation could build on the success of the past—half a century of scientific enterprise has been sustained.

The result has been, one cannot repeat it too often, the "Observation of Gravitational Waves from a Binary Black Hole Merger." It has been accomplished with machines of almost inconceivable sensitivity built to interact directly with the waves. Measurable impacts of gravitational waves on the detectors produce effects equivalent to a change of a width of a proton in the diameter of the Earth. This has to be sensed by the change in the length of the arm of an interferometer only four kilometers long; that means sensing a changes equivalent to 1/10,000th of the diameter of a proton. This is like seeing the change in level of the square mile of Cardiff Bay produced by the addition of a 1,000th of a drop of water.[5]

The detection is the most demanding confirmation yet of Einstein's general theory of relativity in the year of its one-hundredth anniversary. It is the beginning of a radically new field of astronomy that works outside of the electromagnetic spectrum, with gravitational waves being the only way to "see" black holes. Impossibly, black holes can now be "seen"! And black holes *have* now been seen—a pair of black holes spiraling into each other, leaving only one. A new astrophysical fact has already emerged: pairs of black holes can form binary systems and emit enough energy to decay and merge within the lifetime of the universe. This was not known until now.

This thing that has been seen is hard to grasp. During the course of a fraction of a second, the mass of three suns has been entirely converted into energy and emitted in the form of gravitational waves. Compare this with

5. In *Gravity's Shadow* I say 100,000th of a drop, but this is a mistake.

the mass of matter converted into energy in the most powerful explosions made by humans—about a gram for the first atomic bomb and about five pounds for the most powerful hydrogen bomb ever built. For a fraction of a second, that inspiral was, in energy terms, brighter than all the stars in the universe and, though it was a billion-and-a-half light years distant (a light year is about 6 thousand thousand million miles), it would have been brighter than the full Moon if its energy had been emitted as light. If it had been in the position of our Sun when it merged and, again, if it had emitted its energy as light, the solar system would have evaporated. And yet, to indicate how hard it is to see gravitational waves, if you had been floating in a space suit at a distance from The Event equal to the distance of the Earth from the Sun, you would have sensed nothing but a bang as the gravitational waves rattled the bones in your ear.

Peter Berger, in his book *Invitation to Sociology*, published in 1963, explains that the sociologist must be able to "alternate" between the worldview of those being studied and the estranged worldview of the analyst. The above chapters show me not just absorbing the worldview of those I study, but glorying in it; what a wonderful, fifty-year adventure I have been privileged to share! Now, as the remaining chapters unfold, the analyst, like the slave whispering in the triumphant Roman general's ear, must remind us that all we have seen are a few numbers representing strains on some mirrors, and all that has been described above is a structure built on trust and a huge and intricate body of social agreements about what these must be taken to mean. The trick is to be able to see things both ways and not to think that one diminishes the other. The trick is to allow each to enrich the other and to learn what we can.

11 THE LAST RIPPLES

From the Press Conferences to the American Physical Society and the Rest of the World

On February 5, I hear from the *Nature* reporter who points me to another rumor. *Science* has published an article confirming the rumors and now the entire story is out. They publish the following image of a tweet:

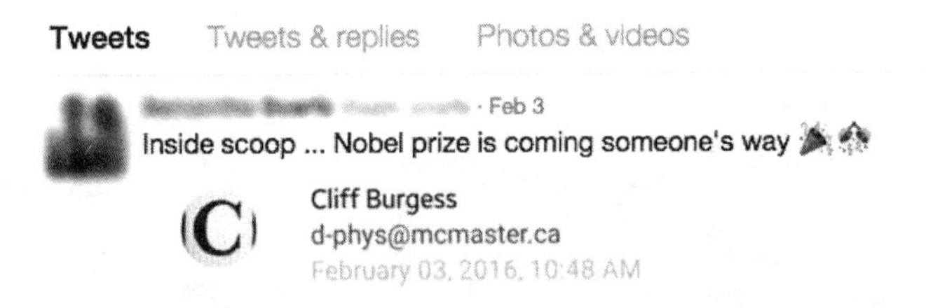

Hi all, the LIGO rumour seems real, and will apparently come out in Nature Feb 11 (no doubt with press release), so keep your eyes out for it.

Spies who have seen the paper say they have seen gravitational waves from a binary black hole merger. they claim that the two detectors detected it consistent with it moving at speed c given the distance between them, and quote an equivalent 5.1 sigma detection. the bh masses were 36 and 29 solar masses initially and 62 at the end. Apparently the signal is spectacular and they even see the ring-down to kerr at the end.

Woohoo! (I hope)

Figure 11.1

The paper finally published in *Physical Review Letters* is B. P. Abbot et al., "Observation of Gravitational Waves from a Binary Black Hole Merger" (2016a), http://journals.aps.org/prl/abstract/10.1103/PhysRevLett.116.061102.

The principal press conference is to be in Washington with satellite meetings all over the world—wherever gravitational wave work has been going on. There would be dozens of simultaneous meetings, many of them streaming the Washington conference as well as providing their own session. Figure 11.2 shows a collection of photographs from a few of the meetings:

I go to the London meeting, which turns out to be disappointingly low key with the organizers having chosen a small venue and seemingly trying to keep people away in case it was overcrowded; it certainly isn't—only a couple dozen people are there. But the very impressive Washington

Figure 11.2
A selection of press conferences on February 11 plus (top left) and a Congressional Hearing of February 24 (courtesy of LIGO).

meeting is streamed, and one can feel part of it and very proud when Dave Reitze makes his initial announcement: "We have detected gravitational waves; we did it!"

THE SECOND RIPPLE

On the platform in Washington, apart from Dave, are Rai Weiss and Kip Thorne, representing the founders of LIGO, and they are followed by Gaby Gonzalez. I am pleased to see that Joe Weber gets a mention as the founder of the field (with his wife, astronomer Virginia Trimble, prominent in the audience), and Ron Drever, now suffering from dementia, is described as one of the founders of LIGO. The Washington press release made mention of Thorne, Weiss, and Drever as LIGO founders.[1] I find a notable absentee from the press conference, both in terms of presence—he was giving a seminar at CERN in Geneva—but more so in terms of salience, is Barry Barish, who, with Gary Sanders, has rescued LIGO, and gravitational wave astronomy, from the brink of the grave.[2]

The other sour note as far as I am concerned is the fact that when questions are asked about other events in the data, the answer, once more, is deceitful. The Boxing Day event was a very important scientific fact making a significant contribution to the dissolution of what would otherwise have been the scientists' residual fear that The Event was another monopole—but the audience is not being told about it.

In the UK that evening, every news-broadcast is dominated by the discovery, mostly showing extracts and graphics from the Washington conference with interviews with British scientists. The coverage is hugely impressive, as impressive as for any scientific finding. I watch the 6:00 p.m. BBC news in a bar with a couple of colleagues from the physics department

1. See *Gravity's Shadow* for the whole extraordinary story of the founding of LIGO.

2. See *Gravity's Shadow* for a full account. Readers of *Gravity's Shadow* should note that Robbie Vogt's name does not appear among the 1,000 plus authors on the discovery paper. It was there on an early draft but he must have asked for it to be removed, I think possibly still angry at the way he lost his job as director.

in Cardiff who had been at the London press conference and we give each other high fives. Later, watching more news bulletins unfold with my son and his partner, it makes me very proud to be a part of what is going on.

But my job now is to pick up on the reception of the news. I listened carefully to the questions at the London press conference and at the Washington press conference, but no one is throwing doubt on what had been done, only asking for more information. A couple of days later I listen to a recording of the CERN seminar led by Barish, and there the scientists are prepared to make just a little more trouble.

Some are questioning the 5.1 sigma level of significance and how it was arrived at. Interestingly, Barish, when introducing the question session and making some suggestions for what might be asked, invites discussion of the significance level in the following way:

> I know people here love things like statistics and how to get enough sigmas to matter.

Barish is pointing implicitly to the way in which what counts as discovery varies from place to place and indicating the special way of thinking about it that is found among particle physicists. One questioner asks about how high the level of significance would go once the rest of the data from O1 has been analyzed and there is much more background available against which to set GW150914. This question is never quite answered except to say that the rest of the data has not yet been analyzed. The reason might be that no one on the platform wants to start a debate about what it means for the machine to be in the "same state" such that background from the rest of O1 would be comparable with background from the 16 days—the whole can of worms over whether the freeze made sense and whether the background was stable enough to justify the statistics. That might have introduced the thin end of a wedge into the soundness of the statistical calculation.

A related line of questioning refers to the possibility of further events; this seems to reflect the expected distrust of the single event, but it never surfaces. The compulsory subterfuge is deployed—it is said that there is a good chance that there could be another event in O1 but that the analysis has not yet been done. Once more, what no one here knows is that a good proportion of the physicists' confidence in The Event comes from the juicy Boxing Day signal. The Boxing Day event is part of this discovery even if it does not appear in the statistics. The fact that there is judgment behind the confidence is being obscured.

Then there are questions concerning how it is known that gravitational waves travel at the speed of light. There is no independent evidence yet that they do travel at the speed of light—there will be when some event is detected that emits both light and gravitational waves and the relative time of arrival on Earth can be compared, but that will require much more directional acuity and a lot of luck. Currently the question has to be answered in terms of the way all the evidence fits together—Einstein's theory says the speed of light is the speed of gravitational waves, and Einstein's theory also generates the waveform that was seen, and it is hard to see how it would all have worked out if the speed was something different—for example, the waveform would have been distorted and the signals from the two detectors, which fit over each other when one is shifted by less than the time it takes light to travel between the two detectors, would be fitting purely as a matter of coincidence between noise glitches and that is unlikely to the tune of 5.1 sigma. I think this question bears on the philosophy of the science; the discovery depends in many ways on the assumption that the gravitational waves travel at the speed of light—for example the whole coincidence and time-slide method would fall apart if the speed was something different—but the argument runs backward: we made this discovery that depends on gravitational waves traveling at the speed of light; if they didn't, we couldn't have made the discovery! It is pretty convincing but there is something curious about the logic. The scientists press on in a quasi-humorous way. But then one speaks up with a decisive remark (and remember, this is CERN):

> I think if those people who were asking the questions continue to ask the same kind of questions, it would be concluded soon that the Higgs had not been discovered.

It is being pointed out, once more, that one can always find reasons to disbelieve.

So far, then, it looks as though The Event is doing pretty well as far as the second ripple is concerned; particle physicists are just about ready to believe. Indeed, a few days later Barry writes to me:

> So far, the result is accepted without dissent in the scientific community. It will be amazing if there aren't some skeptics that raise their head after the publicity has subsided. But, I now doubt there will be much and once the Box[ing] Day event surfaces or other O1 candidates surface, there can be no doubt.
>
> I'm afraid after all your patience and years of following LIGO, the drama and subtleties of how the scientific community became convinced of the LIGO observation of gravitational waves will go down as having been just a "slam dunk." As a consequence, you will have to analyze how scientists became so quickly convinced!! For me as a scientist, that's all great. For you, as a sociologist, I suppose it is much less interesting than it might have been.

The editor of *PRL* was to write to the paper-writing team:

> The stat that really struck me was that in the first 24 hrs., not only was the page for your PRL abstract hit 380K [thousand] times, but the PDF of the paper was downloaded from that page 230K times. This is far more hits than any PRL ever, and the fraction of times that it resulted in a download was unusually high. Hundreds of thousands of people actually wanted to read the whole paper! That is just remarkable. [courtesy of LIGO]

So much for the reception by scientists well integrated into the mainstream community—very well-integrated in the case of CERN, since Barry Barish used to be a CERN particle physicist. I hope we'll find out more about mainstream physicists a little more remote from the core of physics when we get to the American Physical Society in April, but we can already see how things are going to go from the web: *Nature*, *Science*, the *New York Times*, announce the result on their websites, giving it wide coverage and vying for the best images and quotes: no doubts are expressed by the official sources.

Breaking a rule of mine, I send an email to the whole collaboration (I've done this mostly accidentally in the past), asking if anyone knows of any criticism of the finding. A paper by Stephen Crothers is brought to my attention (see below) but he is not part of the mainstream. Another paper has been sent around, claiming not that there was any problem with the actuality of the finding but that the sizes of the masses are wrong—they should be about half of what has been reported. This comes from a retired mainstream physicist, and a member of the collaboration tells me he would be surprised if it finds a publication outlet. Otherwise, there are simply no criticisms reported; one or two people write that they have been giving a number of talks and encountering absolutely nothing in the way of doubts.

An article written in 2013 by Louis Lyons, the doyen of the use of statistics in high-energy physics, argued that, because of trials factors and matters of prior expectations, gravitational wave detection would be insecure unless it was based on at least 7 sigmas, but no one is mentioning this now, not even Lyons. I write to Louis Lyons asking if he believes the detection of GW1509014 is insecure in light of his 2013 analysis. We have a long telephone conversation on March 21.[3] Lyons explains that he feels the coherence of the signals in the two detectors was hugely convincing. Another thing that becomes clear in this conversation is that scientists are hearing "5 sigma" but not really understanding that there is a difference between generating 5 sigma by collecting lots of weak signals until their aggregate statistical unlikelihood reaches the threshold and finding one signal and analyzing the background more and more intricately to show that this single event was very unlikely to have arisen by chance from its stretch of background (the 16 days). A determined critic could pick on this, but there are no determined critics in the mainstream or even in close fields such as particle physics.

In May, the award of two prizes is announced. The Physics Breakthrough Prize is a $3 million prize shared between "LIGO founders"

3. Lyons's article is "Discovering the Significance of 5 Sigma," arXiv:1310.1284 [physics.data-an]. In the telephone call it became clear that he was only intending these remarks as general guidelines and as a warning against erecting 5 sigma as a shibboleth.

Ronald W. P. Drever, Kip S. Thorne, Rainer Weiss, and "1,012 contributors to the discovery." The 1,012 are the authors of the paper and seven others including Frans Pretorius, Thibault Damour, and Rochus "Robbie" Vogt. $1 million is shared between the first three and £2 million between the remaining 1,012. In the same month, Drever, Thorne, and Weiss share the $0.5 Gruber Cosmology Prize. On the last day in May it is announced that Drever, Thorne, and Weiss are sharing the $1.2 million Shaw Prize in astronomy, and then at the beginning of June that the same three had won the $1 million Kavli Prize for astrophysics.

There seems to be something wrong with the system. Why do these people need so many prizes? They are already famous for helping in the detection of gravitational waves, and any prize other than a Nobel is not going to make them any more famous. As the economists would say, the marginal utility of these prizes is close to zero (except, perhaps, to the donors).

But there is also a problem given that the detection was a team effort. The Physics Breakthrough Prize makes a gesture to the team, but keeps the big three clearly prominent. All the other prizes echo the names of the Troika (see *Gravity's Shadow*) like the repetition of a familiar bedtime story for children. This is not how it was. *Gravity's Shadow* was not intended to be a history book, but its value as history is growing in the light of the pseudo-history being socially constructed by the prizes, a kind of still more schematic version of the notorious histories found in the introduction to science texts. Take any of the Troika away and there would probably have been no discovery; but the Troika could never have made the discovery alone. LIGO had to move from being a small science to being a big science. Drever could not even work within a big science environment; Thorne was helpless in the face of the disagreements between Drever and Weiss; Weiss was the only one who understood big science, but there is no evidence he could have managed a project of this size, and even if he could he would almost certainly have made the wrong technical choices, for example over the way light should travel in the arms. It was Vogt who first brought some kind of order into the science, and it was Barish, with his project manager Sanders, who actually made it work. The reason LIGO is receiving all the accolades rather than Virgo—a broadly similar device—is *because* Barish made things work. Others took it on after he left and did a tremendous job, but he had put the pieces in place through the application of brilliant technical judgment that, against all the odds, justified the continued funding of a dying project. And, of course, there are many others of whom you could say, "without X there would be no detection of gravitational waves"; the prizes turn science into a fairy tale.

Peter Saulson puts a more positive gloss on it. Without the fairytale there would be endless arguments about exactly who contributed more than who, who was dispensable and who was not, and so on. The fairy tale relieves everyone of that duty—that is its function.

Now we need to look in two other directions—to the general public—already represented by the amazing coverage on the news bulletins—and to scientists much more distanced from the mainstream.

BEYOND THE MAINSTREAM

I write to Davide Castelvecchi, the *Nature* journalist with whom I have been fencing, to apologize for all my deceit; he says it is just part of the game. But he gives me a lead to *Nature*'s huge and comprehensive coverage on its website. On *Nature*'s blog we find represented the scientists who don't believe. I have to do a triage because some are so strange that even someone whose mind is as professionally open as mine cannot use them. That done, some of the remarks of the survivors are useful for forcing us to ask why we do believe! What evidence do we have that these nonbelievers are wrong? The answer is *none*, beyond our readiness to accept a certain model of the way our society works. By stepping outside of this model, the critics define a boundary and force us to think about what exactly is inside the boundary. Here is the flavor:

> Pentcho Valev, Feb. 12, 2016, 1:12 p.m.
>
> Proof That Gravitational Waves Do Not Exist
> Here is an obviously valid argument: If gravitational time dilation does not

exist, then gravitational waves do not exist as well. The antecedent is true—there is no gravitational time dilation. Scientists measure the gravitational redshift but inform the gullible world that they have proved gravitational time dilation, a miraculous effect fabricated by Einstein in 1911: ... [several explanatory paragraphs with references follow].

"The gravitational red shift does not arise from changes in the intrinsic rates of clocks. It arises from what befalls light signals as they traverse space and time in the presence of gravitation."

Ja law, Feb. 13, 2016, 3:33 p.m.
I smell FRAUD. Here we have 1000 scientists who's livelihoods are on the line to find something with the millions of tax dollars spent over 15 years with nothing to show for it thus far. Now add to this the fact we are at the 100 year anniversary of einsteins prediction. Now add to this the prospect of Nobel Peace Prize money and fame and you have the perfect coctail for fraud and conspiracy to commit fraud. They want us to believe that they've found what we paid them to find yet do not address the occams razor here, That is 1) it's seismic vibrations they detected or 2) a blind injection meant to deceive. 3) anyone notice the supposed distance if this supposed collision of two black holes far exceeds the range the system can supposedly "see"? Yes, I smell fraud. Every supposed discovery this far to find GWs eventually collapsed under scrutiny. I don't see scrutiny here from the science community nor the Media. I see Brow wiping (whew-we still have jobs) backslapping, circle jerking applause and trips to the bank. Where are the real scientists? The ones dedicated to finding the truth rather than digging for renewed contracts and Nobel prizes?

Alone: bad. Friend: good!, Feb. 13, 2016, 6:02 a.m.
MICHELSON-MORLEY had a lot of fun with an interferometer way-back-when. These new guys are also using an interferometer but I don't notice any flagrant mistakes in their reasoning. But I just don't think things are actually working like they think and they don't really know what they are detecting. It doesn't matter though—it will be accepted and become another pillar of science even is it is wrong....

Chris Blake, Feb. 12, 2016, 10:16 p.m.

I would like to know how on earth these researchers can claim that this was caused by two black holes of precisely 36 and 29 solar masses when they can't even triangulate the signal and have no idea where these supposed waves are coming from?

Verner Hornung, Feb. 13, 2016, 2:38 a.m.

... To tell you the truth, I don't know how they distinguish the real shebang from inevitable vibrations and so on that cause the path lengths to fluctuate constantly: These can't be controlled to one part in 10^22, and only a single quarter-second event at 250Hz doesn't seem to provide enough to eliminate noise statistically as is done in particle accelerator experiments.

I write to a physicist, Reg Cahill, whom I have corresponded with for a while. He works in Australia at the physics department of Flinders University. He does not believe in relativity and has an alternative theory of light that rules out any possibility that LIGO could detect a signal. I ask him what he makes of the announcement that gravitational waves have been detected. He writes back (February 17):

> That "experiment" is a fraud.

He quotes the *New York Times*:

> The LIGO team includes a small group of people whose job is to create blind injections—bogus evidence of a gravitational wave—as a way of keeping the scientists on their toes. Although everyone knew who the four people in that group were, "we didn't know what, when, or whether." ...

Cahill is sure the supposed signals have been knowingly injected to support Einstein's theories.

> There is a long history of gravitational wave detections by various people using various techniques.... Most of these experiments were performed to detect light/EM speed anisotropy—the observation of which contradicts Einstein SR [special relativity]. In doing so the results showed non–Einstein

> Grav Waves. They also show that vacuum Michelson interferometers cannot detect light speed anisotropy or grav waves. One experimenter was Roland DeWitte using RF coaxial cables: he died of a "suicide." Stephan Marinov was investigating alternative theories of gravity: he also died of a "suicide" the day before he was to present his findings at a conference. There have been many more "gravity related deaths." There has also been an earlier fake experiment that stopped physicists investigating the gravity bore hole anomaly. The agenda is to convince academic physicists to not challenge Einstein theory of gravity, despite its numerous failings: bore hole anomaly, lab measurements of G, rotation curves of spiral galaxies, expansion rate of universe (here the red shift–supernova data shows a uniformly expanding universe. Einstein Gravity does not do that, so Dark Matter and Dark Energy fudges were introduced. In the future these unobserved stuffs would predict an as of now an unobserved accelerating expansion rate for the universe. A Nobel Prize was awarded for the "discovery" of this fudged acceleration. A Nobel Prize will be awarded for LIGO "discovery."

Cahill, and those sending commentaries to *Nature*, do not share the assumptions of the mainstream about whom you should trust and where you should stop questioning. They are prepared to question to the point at which the findings are no longer believable. That is one place where the second ripple spreads—we have to say the "second" ripple because some of these people are scientifically educated—they are sometimes highly accomplished scientists. Indeed, within a few days there are at least two full-scale articles, one of them Cahill's, on the preprint server "viXra" that purport to show that the discovery was not real. What is viXra? arXiv, the physics preprint server has already been mentioned a few times. viXra, which is 'arXiv' spelled backward, is a server developed in response to what is thought of as arXiv's overly restrictive promulgation policies; scientists whose postings have been rejected by arXiv can post on viXra, which has almost no restrictions on promulgation.[4] The first of these papers (viXra:1603.0127), which is widely circulated by its author, Stephen J. Crothers, appears on

4. See Note XVII in "Sociological and Philosophical Notes," 375.

the day of the press conferences; it is an update of an version posted a few days earlier. It is a long paper that argues that the finding cannot be correct because the underlying theory is incorrect. viXra:1603.0232, posted on March 15 by Cahill, is a densely mathematical paper with copious data arguing that the findings were something other than the scientists think; he has an entirely different theory of gravitational waves and claims that the signals must have lasted four seconds rather than a couple of tenths of a second. Here is part of the abstract:

> Experimentally it has been shown that such vacuum mode interferometers [like LIGO's] have zero sensitivity to gravitational waves, which have indeed been detected using other techniques over the last 100+ years. One such recently discovered technique uses quantum barrier electron tunnelling current fluctuations in reverse biased diodes. … These are Quantum Gravity Detectors (QGD). There happens to be an international network of such detectors, and the data from this network shows a significant event at the same time as the LIGO event, but extending over some 4sec duration. Previously in 2014 such Quantum Gravity Detectors detected gravitational waves generated by the resonant Earth vibrations, whose frequencies were known from seismology. It is suggested that the LIGO event may have been an Earth generated gravitational wave event that was detected by the electronics of the LIGO measuring and recording system, an effect previously discovered in 2014 using time-delayed correlated fluctuations in data recorded by oscilloscopes located in Australia and London.

So here we have an entirely different universe of gravitational waves and gravitational wave detectors coexisting in physics departments. Mainstream scientists simply ignore this kind of thing, as they must if science is not to dissipate; policy makers and social scientists have to find some way of handling it. It seems clear that policy makers have to base their decisions on the work of those supported by the funding agencies of the state and the outlets in which they publish. In this case, policy makers must take gravitational waves to be the phenomenon designed to be detected by the large interferometers, whose detection was announced on February 11 at the Washington, DC, press conference, the description of which can be found in the corresponding paper in *Physical Review Letters*. What other choice is there, given that, at best, only members of the core have any deep knowledge of the science and even they have to take most of what they know on trust? Social scientists have to find a way to describe the difference between the fringe and the mainstream that maintains the mainstream at the center of decision making even while illustrating the essentially social nature of the choices made by the gravitational wave physicists and the entirely social nature of the choices made by those outside of the immediate discovery network.[5]

THE THIRD RIPPLE

Meanwhile, gravitational waves are finding their way to the general public. There is massive television news coverage with never a doubt expressed; gravitational waves have simply "been detected"—exciting but no more dubious than, say, the Moon landing. What is happening is that gravitational waves are being "domesticated" in the same way as black holes or the Higgs have been domesticated. Everyone knows what a black hole is—it is a feature of everyone's day-to-day life embedded in a "semantic net" that includes "the cosmos," "the big bang," "Stephen Hawking," "brilliant scientists," "Einstein," "space," "alternative universes," "time travel," "wormholes," "astronomy," "rockets," and "being sucked into things"—and this is in spite of the fact that, before The Event, no black hole had been observed except by inference. As for the Higgs, everyone knows that it was found by the huge and brilliant team at CERN, but, familiar as it is, no one knows what it is. I know it is the last piece in the jigsaw puzzle of the particle "zoo" known as the standard model, but what I have is "beer-mat knowledge," good for answering questions in Trivial Pursuit but that's about it.[6] On the other hand, the fact that we can imagine encountering questions about black holes and the Higgs while playing Trivial Pursuit is one of the

5. See Note XVII in "Sociological and Philosophical Notes," 375.

6. See Note VIII in "Sociological and Philosophical Notes," 362.

things that makes them real: all this familiar knowledge makes stuff real. The Moon landing, note, is pretty real for everyone but, just as in the case of what is building in respect of The Event, there are conspiracy theories about that too; and, just as in this case, you have to stray from the mainstream to find them.

On Friday I gather a good selection of the UK print newspapers; they are big contributors to the domestication process. The *Guardian* is a broadsheet for the left-liberal middle classes and its news section is thirty-eight pages long. It gives the story the lead and whole of page 11. It had also given it the whole of page 3 on Wednesday, building the story on the rumors. On Saturday, the *Guardian*'s regular political cartoon (p. 31) features the Syrian peace talks represented as some kind of funny-looking celestial object with the caption: "Not gravitational waving but gravitational drowning." Thus do gravitational waves spread into the ordinary language.

The *Independent* has a similar readership to the *Guardian* but has a smaller tabloid format with seventy-two pages. It gives the story the entire page 1 and pages 6–8; it opines that this is "one of the greatest achievements in human history."

The *Telegraph* is another broadsheet, with thirty-eight pages in its news section. It is a right-wing, patriotic paper for the educated. It makes gravitational waves the second story on page 1, leading with:

> A British scientist who was pivotal in the project to detect gravitational waves could not celebrate the momentous discovery with colleagues because he is suffering from dementia.

This, of course, is Ron Drever. The paper also gives up pages 10 and 11 to the story.

Martin Rees, the Astronomer Royal, writes columns in the *Independent* and the *Telegraph*. He opines that this is of similar importance to the discovery of the Higgs; most other commentators say it is much more important than the Higgs, but Rees has long been said by gravitational wave physicists to be less than enthusiastic about the enterprise.

The *Daily Mail* is a "little Englander" tabloid with ninety-two pages serving those with strong right-wing opinions. It gives the story half of page

10, mistakenly claiming that Einstein predicted that colliding stars would generate gravitational waves that could be detected on Earth, whereas he actually thought they would remain completely undetectable.

The *Mirror* is a left-leaning tabloid with eighty pages. It gives the story most of page 21 but says that LIGO was invented by Thorne and Weiss, missing out on Drever.

The *Sun* is a tabloid with sixty pages that began its life by publishing photographs of topless models on the notorious "page 3" (now dropped). The only science I could find was on the bottom third of page 15, headlined: "Top Prof Dies in Rubber Suit with Dog Lead Round Neck." The "Top Prof" does not seem to have been one of the gravitational wave team.

The *Guardian* website of February 12 includes a hilarious cartoon—one of a series called "First Dog on the Moon," which anticipates one of my major sociological theses. The fourth panel of the cartoon opines:

> Obviously we can't see these waves—the only way we know they are real is by using another extremely sensitive device which detects scientists having feelings of excitement.

The excitement evoked in scientists by a gravitational wave is calibrated using the marginally smaller effect of a cheese salad sandwich as a standard candle.

Later I will discover that my major thesis about social construction, which turns on pointing out that no gravitational waves were seen but merely a few numbers that were *interpreted* as gravitational waves, has been thoroughly anticipated (albeit on a strange, flat-earther YouTube channel that appears to treat conspiracy theories as an art form): https://www.youtube.com/watch?v=7w05W0sOkEQ. It claims scientists have not seen gravitational waves, nor has their machine seen gravitational waves, but that the machine produces lots of glitchy noises out of which they have picked one and *interpreted* it as a gravitational wave.

A member of the LIGO team has put together a collection of newspaper front pages from around the world. And, as though to put an indelible stamp on the soon to be taken for granted nature of this exotic

Figure 11.3
Newspaper front pages from around the world (courtesy of LIGO).

phenomenon, in the United States the discovery is presented on *Saturday Night Live* and *The Tonight Show*, and, on Saturday, February 13, the humorous US radio show *A Prairie Home Companion* devotes about five minutes to gravitational waves. Gravitational waves have arrived!

The physicists continue to do my job for me by gathering more indications of the domestication of gravitational waves. On February 16, a French (presumably humorous) website normalizes the waves in contemporary fashion by calling for a ban on them and the distribution of protective helmets (http://www.tak.fr/pour-un-moratoire-sur-les-ondes-gravitationnelles/). This is a Google translation from the French with my minor edits:

> **For a Moratorium on Gravitational Waves**
> Bringing together hundreds of independent researchers, the "Collective for a moratorium on gravitational waves" (COMOG) sent us this platform. We publish it *verbatim* in our columns:
>
> In recent days, highlighting the "gravitational waves" continues to make the press headlines. Everyone welcomes this alleged "scientific breakthrough,"

which was published in the *Physical Review Letters,* a journal under orders of the nuclear lobby.

Now, our collective, consisting of independent researchers who wish to remain anonymous for their own reasons, is concerned about the apparent toxicity of gravitational waves.

To date, there is in fact no serious study establishing the actual safety of these waves. That is why we propose an action plan of four points.

1. We recall, first, that the oscillations of the curvature of space-time can present **health risks found**, especially on the neurological system of employees too long exposed to the gravitational waves. It is appropriate in this case to call for the government to strictly enforce the Labour Code, to limit the time of exposure to gravitational waves, and equip the wage earners with protective helmets.
2. To these health risks are added, as often environmental, of **deleterious economic effects**. The curvature of space-time is likely to cause untoward inconvenience, especially in the field of transport and travel. An example: if space-time is curved in the wrong direction when one performs a trip from Paris to Bordeaux, the journey can last more than twenty-five hours, according to our estimates. Gravitational waves expose the French economy to serious danger that cannot be underestimated.
3. It appears that the production of gravitational waves calls for **masses of matter and energy that are absolutely astounding**: black holes, neutron stars, washing machines, etc. We ask that the environmental and climate impact of gravitational waves be measured in France by an independent body and a carbon footprint be determined as quickly as possible.
4. As a result, we ask Ségolène Royal, Minister of Environment, Energy and the Sea, responsible for international relations on the climate, to apply the **constitutional principle of precaution,** and to take by decree related measures that are defined and recommended in principle 15 of the Rio Declaration. It seems to us urgent that France decide on a moratorium on gravitational waves.

If the government does not abide by these basic precautions, peaceful COMOG teams will be forced to resort to direct action. Within six months, we will proceed to the systematic dismantling of gravitational wave antennas. Our teams of volunteer harvesters shall uproot the plants of space-time curvatures. Finally, we will not hesitate to leave Paris to set up a **zone to defend** Proxima Centauri, even against the advice of the prefect.

We call upon our fellow citizens to join our fight. Gravitational waves, no thank you!

On February 18, the *Huddersfield Daily Examiner* (Huddersfield is a town in North England with a football club—"Huddersfield Town"—all about as provincially English as can be) carries a story about the "Huddersfield Town Supporters Association" (HTSA). It includes:

Our HTSA column last week touched upon the subject of regional supporters groups and their recent cosmic rise in popularity. The Laser Interferometer Gravitational-Wave Observatory (LIGO) can probably demonstrate whether this is due to colliding Black Holes over Bexleyheath.

And Barack Obama had tweeted on February 11:

Einstein was right! Congrats to @NSF and @LIGO on detecting gravitational waves—a huge breakthrough in how we understand the universe.

At the forthcoming American Physical Society (APS) meeting (see below), Dave Reitze, the director of LIGO, will present a slide showing a woman wearing a dress patterned with the waveform, an Australian competition swimmer with the waveform on his swim-cap, and a New York advertisement for apartments.

I attend two meetings in March: a general relativity 100th anniversary meeting at Caltech and the LIGO-Virgo collaboration meeting in Pasadena. Of course, the cat is now out of the bag so a big topic at the Caltech meeting is The Event. I follow Barry Barish onto the platform; he describes the technicalities and I talk about the way small science and big science had combined to create this possibility, with Barish bringing about the necessary transformation, and I talk about what it meant for me as a sociologist

Figure 11.4
More domestication of gravitational waves.

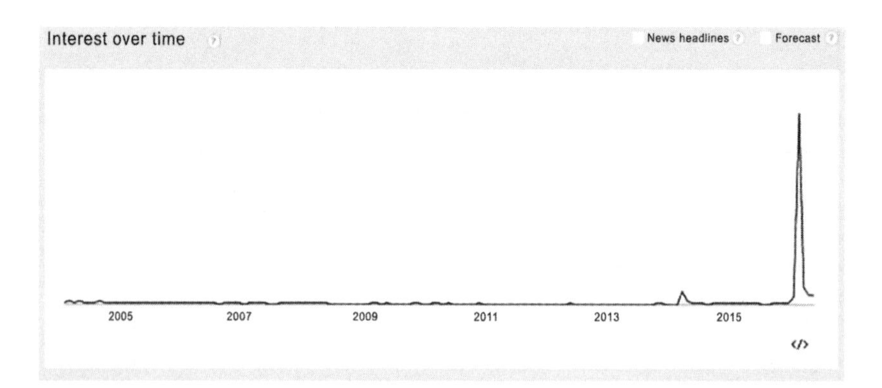

Figure 11.5
Enormous spike in interest in gravitational waves around February 11.

to be confronted by such a sudden and certain result. At neither meeting is there any criticism but the LVC is selling huge numbers of T-shirts and polo shirts with the waveform of The Event printed or embroidered on them: the waveform is becoming an icon! Many more such garments will be sold at the April APS meeting.

At the March meeting of the APS—a much larger meeting than the April meeting I am going to attend—a group of physicists who have nothing to do with LIGO or gravitational waves performed a song based on the Neil Diamond/Monkees' "I'm a Believer." The lyrics are as follows:

"I'm a LIGO Believer." Lyrics: Marian McKenzie. Tune: "I'm a Believer," by Neil Diamond (courtesy Marian McKenzie).

I thought waves of gravity were fairy tales—fine for dilettantes, but not for me.
What's the use of searching?
Noise is all you'll find.
I don't want to clutter up my mind—

[Chorus:] Then I saw the graph—Now I'm a believer!
You can laugh, and hold me in scorn.
I'm convinced, oooh, I'm a believer
In Weiss, Reitze, Drever, Gonzalez, and Thorne!

Einstein spoke of grav wave propagation.
Weber tried to find them on the moon.
BICEP2 announced them,
Then said "Never mind."
—Do you wonder I was disinclined?

[Sing chorus] [instrumental interlude] and repeat

The whole song can be seen and heard at:
https://www.youtube.com/watch?v=GN2sFasYCr0

What about social media? Google Trends (figure 11.5) shows the huge spike in interest in gravitational waves around the February 11 press conferences by tracking hits on Google. Unfortunately, we have only the normalized trend, the scale having a maximum of 100, not absolute numbers.

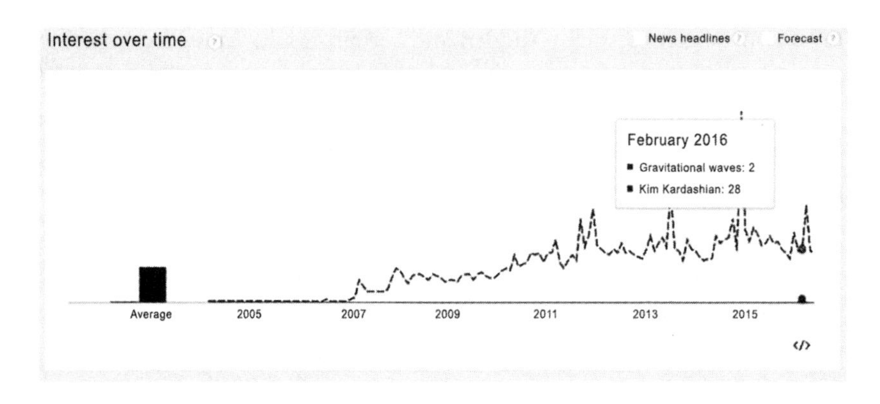

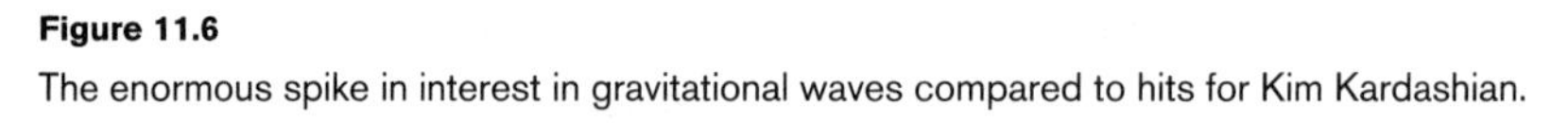

Figure 11.6
The enormous spike in interest in gravitational waves compared to hits for Kim Kardashian.

Gravitational waves are not, however, about to take over the popular imagination. Figure 11.6 shows the same spike in comparison to Google hits for Kim Kardashian, the reality TV star. Gravitational waves' *enormous* spike is a mere 2 percent as high as Kardashian's peak performance and only about 5 percent as high as her average day-to-day score. Aside from the spike, gravitational waves score zero when compared to Kardashian's average.

CONCEALING THE BOXING DAY EVENT

The April APS is fast approaching. At one time it was hoped to announce the Boxing Day event there, but this has now been ruled out; the decision was made at the Pasadena LVC meeting. I think this is wrong and damaging. What has led to the vow of silence is relentless professionalism. There are still arguments going on (I write this on March 30) about exactly how to calculate the significance of the event; the two different pipelines, gstlal and pyCBC give different results, and the parameter estimation is not completed as the team has decided to calibrate the apparatus more accurately before finalizing the calculations. Furthermore, the team wants to have a refereed paper in hand before announcing the result to the world.

In some ways this is laudable, but in other ways it is crazy. It is vitally important that another event has been seen, and no one is in doubt about its statistical significance even if they are not exactly sure about the number. That it has been seen makes a huge difference to the confidence of the community and to the way others outside the community would think about gravitational waves and act in respect of them. For example, though the community won't worry too much about this, it just might be that Reg Cahill could change his mind about gravitational waves, if he were to know of this event; the sociologist in me says he wouldn't, but scientists have to imagine it to be at least a possibility that there is at least a chance that he and his colleagues would abandon his alternative version of the gravitational wave universe if they knew there were a second event that they now had to explain. Though I am in correspondence with Reg Cahill, I cannot ask him the burning question: what does he make of the second event? I cannot ask him because there is not supposed to be any Boxing Day event according to the official account of things. More and more the decision to conceal is causing trouble. Here, six weeks after the press conferences, the true nature of the discovery is still unknown to anyone except the scientists who were involved. For three months, other scientists' actions have not been informed by what has been known since the end of December.

Now that the path-breaking announcement has been made, how can this continued concealment be justified? One understands the logic—the scientists can't have a peer-reviewed paper accepted until they agree on an exact level of statistical significance and finalize the parameters. But isn't this relentless professionalism getting out of hand? It is as though the results of the Trinity Test were kept secret from decision makers until the exact yield of the explosion could be calculated to the last decimal place. Here the exact level of significance does not matter any longer, and neither does the exact way the Boxing Day event is announced; what matters is that other scientists should be able to go forward in their decision making in the knowledge that what has been seen is not a unique event but has been repeated. The details can safely come later. All that needs to be said now is that there is a second event and the exact numbers will follow.

Incidentally, I have now seen the first draft of the Boxing Day event (GW151226) paper and I don't like it. I find it pretty unconvincing compared to The Event paper. The Event was convincing because of the overlap of the L1 and H1 waveforms, but there is nothing similar to be seen in this case; this event is more like what everyone expected to see first, and with an event like this it would have been much harder to convince the outside world that the impossible had happened. The reason is almost counter-commonsensical: it because the components of the inspiral are lighter, 14 and 8 solar masses instead of 36 and 29. This has some statistical advantages: this event lies within a lighter mass bin than The Event, with fewer large noise excursions, and it lasts for much longer—about 5 seconds instead of 0.2 seconds (though in the submitted draft of the corresponding paper this would turn out to be 1 second rather than 5, indicating how much less well-defined was this signal). For these reasons the statistics are as good as for GW150914 but the signal is far less violently energetic and does not give rise to startlingly visible waveforms. The first draft of the equivalent of the striking figure 1 in the original discovery paper, of which people say "you only have to look at it to know The Event is real," is shown in figure 11.7. It is figure 2 in the first draft of the Boxing Day event paper. As can be seen, there is nothing visible in the omega plots, and such apparent signals as have been extracted from them are some kind of reconstruction. The same seems to apply to the rather unconvincing waveforms.

Ironically, the physicists have been unlucky. As Peter Saulson writes to me on April 3:

> Now that we've had 150914, it is proving also to be a burden in the sense that we've all gotten used to (and, more importantly, gotten our "public" used to) the idea that you can see signals in a simple graphical presentation of the data. That just isn't going to be true in general and it is not true in this case.

In a long conversation I will have with Peter at the April APS (see below), I finally begin to understand a little more about why the findings are

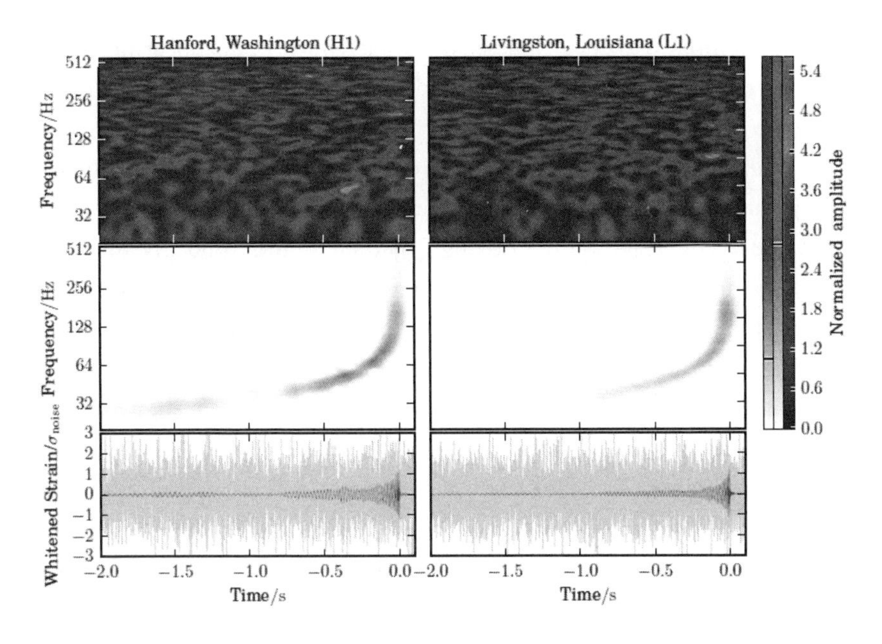

Figure 11.7
Figure 2 from the first draft of the Boxing Day event paper.

being presented as they are. I finally understand why the omega plots that I disdain are so loved by the physicists. The reason is, and it has only become clear to me after the promulgation of this draft of the Boxing Day event paper, that the signals that come from the sky are so different in duration profile and strength that one never knows in advance what kind of graphic will illustrate them best (I doubt that the scientists have fully realized this until now). Omega plots are used all the time by the physicists to tell whether a putative signal is really noise. If the excursion from regular noise is loud enough, it will show up on an omega scan and a quick glimpse can reveal that it is the wrong shape for a signal; that saves a lot of agonizing. A signal must have a characteristic banana shape, bending upward from bottom left to top right convex to the right—that is the "chirp"—and it appears on an omega scan with the most careful filtering and pre-processing that is needed to produce a waveform done automatically. So an omega plot is close to the first court of appeal for a loud potential signal. In the case of The Event, it does appear loud and clear on the plots, so the physicists are delighted. But weaker signals do not show up on omega plots, and that is what has happened with the Boxing Day event. I saw that coming. What I did not see coming was that something similar would apply to the waveform itself so that the only public evidence for the reality of the Boxing Day event—and, probably, for most of the events that are going to be seen in the future—is the statistics. Now the public, including the wider scientific community, has to be trained to believe in the statistics alone after being "spoiled" by the cornucopia of sweetmeats that was The Event.

Has the community been lucky with The Event? Yes, in that it is so convincing graphically that there is no argument and, as a result, the press conferences could be so dramatic. No, in that a whole new exercise in public education is now needed. If it had been the other way around and the Boxing Day event had come first, there would have been the usual arguments and doubts, but then the big fat Event would have come along and squashed them all triumphantly; that might have been better!

As it is, the Boxing Day paper is going to be surfing the wave of The Event. I argue that it would be better if the next paper should be a report on the entire O1 observation run rather than devoted to GW151226 alone. The next paper should be the one that ushers in gravitational wave astronomy, and its centerpiece should be something like figure 1 from the first draft of the paper we are looking at—see figure 11.8,. I say "something like" because I believe this figure should also include GW150914. What one can already see, and what one would see even more clearly in such a paper, is that there is a "line" of events all the way from the noise plots on the left to the statistically significant event on the right. In the figure one can clearly see two squares on the left, in the 2 sigma or less area, the rightmost of these representing Second Monday and the leftmost seeming to indicate a larger number of very weak events. Maybe this is another sign that I am not a real physicist, because I find that line really interesting even though the left-hand squares are not statistically significant when looked at one by one. But then I was always more excited by weaker events like Second Monday than most of the rest of the community.

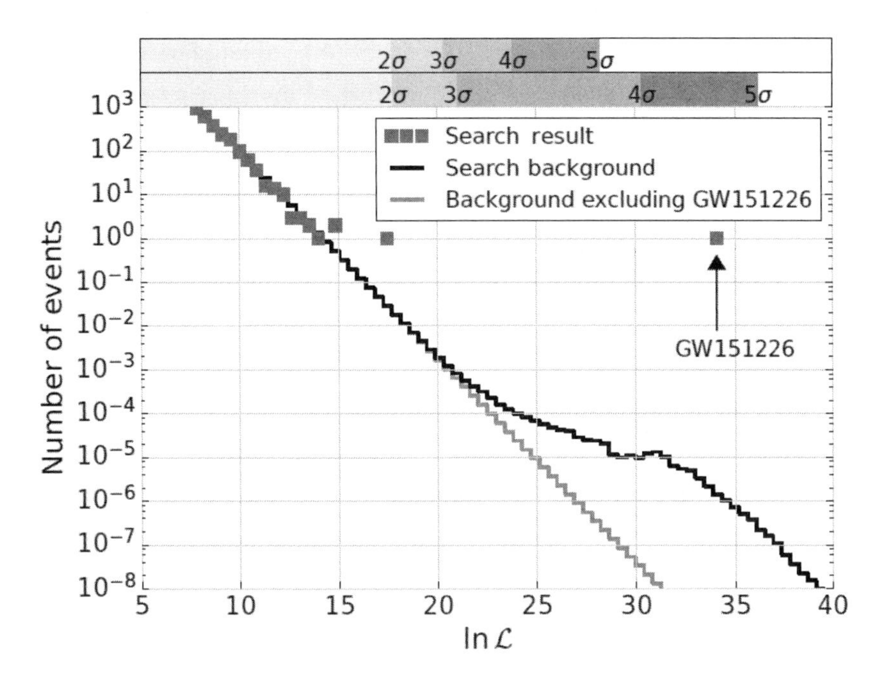

Figure 11.8
Figure 1 from the first draft of the Boxing Day event paper.

One influential emailer argues that it is pointless to waste time agonizing about the meaning of those left-hand, statistically insignificant symbols when, quite soon, we will have the many statistically significant results of O2 to investigate. But what I am seeing, especially in that very weak far left symbol, or symbols, as I will later learn, that represent, perhaps, eight events down near the noise, is Joe Weber's "zero-delay excess." If there are more coincident events than off-coincidence events, then they represent evidence for gravitational waves; they represent evidence for the large number of gravitational wave events that, in theory, must be hitting the detectors even though none of them is strong enough to be statistically significant. I am more excited by that than I am by the Boxing Day event. I think the question that people should be asking is how statistically significant is that *line*, not just how statistically significant is each of those events taken by itself. Or at least they ought to be asking how *interesting* is that line, and what could it signify even if each of its elements is not statistically self-supporting. Maybe this is me not being a proper physicist; or maybe it's the physicists being hog-tied by relentless professionalism and their desire to produce rabbits from hats rather than craft work that usually will, but sometimes may not, lead to places of interest.

THE AMERICAN PHYSICAL SOCIETY MEETING

In the hotel shuttle from Salt Lake City airport I find myself with four physicists going to the APS meeting; they are talking about LIGO. I ask them if they have any reason to doubt the discovery. One says that with a funding decision on LISA (the gravitational wave detector to be launched into space in the 2030s) coming up and it being the 100th anniversary of Einstein's theory of general relativity, it all seems a bit too good to be true. But he'd studied the paper and cannot see anything wrong. No one mentions a second or third event.

Over breakfast I talk to a particle physicist from Virginia who tells me he is here because of LIGOs finding—it has touched his soul. He remarks that there is no comparison with the Higgs—it would have been more interesting if the Higgs had not been found, whereas LIGO is astonishing; he feels emotionally moved by it. He, like everyone else, has no criticism of the result.

There will be many LIGO papers at the APS meeting and the organizers have arranged a last-minute plenary session on the discovery; it had to be last-minute because The Event was still secret at the time the conference was being organized. I talk to the scientific program organizer and he explains that this was so. He also tells me that there are few or no antirelativity people at the meeting this year and no special sessions arranged for people with unorthodox views; as a sociologist, I am sorry to hear it. He says the best I can do is listen to the question-and-answer periods at the gravitational wave sessions—which is what I had in mind. I go to the first session. There are no difficult questions but there is resolute concealment of the Boxing Day event—no one knows that the second shoe has dropped. The same applies to all the other sessions I attend.

The plenary is crowded. I guess there are seven hundred physicists there, and the program organizers confirm my guess. The physicists, of every type and level, tuck into pizza as the two slick presentations wow them. There are short question sessions but, as before, nothing serious.

At last, one of the gravitational wave scientists whom I chat with in the corridor remarks that he has come across a preprint that seems to fit what I was asking for—a serious criticism of the finding. It comes from Davor Palle, who, a Google search reveals, is a Fulbright scholar and has a few publications in physics journals, many postings on arXiv, and none on viXra or in the other recognized fringe journals. Palle claims that Einstein's theory does not prove that gravitational waves exist, while The Event could have been caused by the gravitational attraction of tides on the two detectors. The trouble is, as my informant explains to me, Palle has no explanation for the waveform, so this idea does not seem to have legs. But it is something. It does not seem to have been promulgated on arXiv.

This aside, for me, in terms of the resistance to The Event, the APS has been a washout; there has not been any resistance—not a whiff. The APS has confirmed what we were seeing all along: every physicist wants gravitational waves and every doubt has been handled in house. This is a huge success; but where is Robert Merton's famous "organized skepticism"—the willingness to criticize that is supposed to keep science on the straight and narrow? It's been organized, all right—organized into extinction by five months of self-imposed doubting.

12 CHANGING ORDER
The Long Aha!

In the late 1960s and early '70s Joe Weber claimed he was seeing gravitational radiation with his laboratory-sized, room-temperature resonant-bar detector. By 1975 his claims had lost credibility because he was seeing far too much radiation to make sense in terms of the theory and because it was agreed that his results were to be relegated to the not reproducible category. But his work at that time was still being considered by some—it was not seen as crazy by everyone. Indeed, so little was his work seen as crazy that Kip Thorne was still predicting the discovery of gravitational radiation just around the corner, and the National Science Foundation was funding a new generation of detectors, the cryogenic bars, which would be about a thousand times as sensitive. But if, as we now believe, Joe Weber should have been able to see an event roughly once every hundred, million, million years, then we have to say the world we lived in then was not the world we thought we lived in. More correctly, it was the world we thought we lived in *then*, but *now* that past world is a different place. That is the nature of the change.

The change has happened step by step and has involved scientists coming to treat new kinds of activity as routine. Joe Weber is the hero of this whole story because if he hadn't done something new—built his resonant bar—there might still be those arguing about whether gravitational waves could be detected in principle.[1] But Weber never managed to establish

1. See Dan Kennefick's book *Traveling at the Speed of Thought* for the theoretical argument; I am not sure of the extent to which Dan endorses my view that it was Weber's actions that brought this argument to closure.

room-temperature bars as the new normal—his was always a disputed science. When the funding switched to resonant bars running at near the temperature of absolute zero, then room-temperature bars and the associated activities faded away.

Later, after quite a fight, which seems quaint looking back but was a very serious matter at the time, the interferometers took over. Already we can see history being rewritten: now it said that it was never expected that the initial generations of LIGO would see anything while it was always expected that Advanced LIGO would. Neither of these things is true. One thing that will be hard to recapture is the opposition to the whole interferometer enterprise and the doubts that affected the science not only from without but also from within. I've exploited those doubts in my sociological work for four decades, mostly using them to show the "interpretative flexibility" of scientific claims: the same data can lead to different conclusions depending on how you argue. But my problem is that I now seem to have encountered an event without any interpretative flexibility—(nearly) everyone believed it from the outset, and that "nearly everyone" includes me!

To work for as long as I have in a field one must love it. I think that what was reported on February 11 is one of the greatest scientific discoveries of all time and is the culmination of one of the most astonishing passages of single-minded perseverance there has ever been in peacetime. I can give an insight into how things have changed by quoting from a couple of emails I received from David Mermin two weeks after the announcement. Mermin is a well-known Cornell solid-state physicist and deep thinker about, and elucidator of, quantum theory and relativity. He has long known my interest in gravitational waves but, without realizing just how involved I have been, writes to me on February 25:

> Dear Harry,
> Have you been enjoying the absolutely astonishing news from the newly upgraded LIGO? It's the biggest surprise in my scientific lifetime. I never thought it would work at all, much less discover so extraordinary an event so convincingly.

Mermin, with whom I have been enjoying friendly arguments for decades about the project and findings of sociology of scientific knowledge (see Labinger and Collins's *The One Culture*), had told me around twenty years ago that LIGO could never work because, among other things, it would be impossible to construct such a huge vacuum system; he thought the whole thing was a misdirection of scientific funds.

I reply to his first February 25 email explaining how deeply I have been involved in the discovery, and a little later he writes again:

> Dear Harry,
> I only learned [about the detection] when Dennis Overbye's front page article appeared in the NY Times. For me personally there are two ironies.
>
> You're one of them. When you told me you were about to start work on LIGO I complained that you had picked a project virtually designed to demonstrate that scientific discoveries remain forever (certainly for both of our lifetimes) shrouded in ambiguity.
>
> The other is my Cornell physics colleague Saul Teukolsky, who has devoted much of his career and the careers of many students to heroic calculations of the gravitational radiation emitted by two merging black holes. While the calculations were magnificent, I thought it was a pity (though I never told him) that they were about phenomena that would never be observed and were unlikely even to exist. Those calculations are now the basis for the identification of the 9–14–15 event, and the means for extracting the extraordinary amount of information that's been deduced from it.
>
> As I say in the last paragraph of my 2005 relativity book, "The process of discovering that one's former beliefs are wrong … is what makes the pursuit of science so engrossing."

So what we have to deal with here is how this remarkable thing has been accomplished in the face of the doubts of very deep thinkers like Mermin. I also have to deal with the teasing inquiry of my physicist friend Peter Saulson:

Nov. 23: So I guess that the one discussion that we need to have over our next beer is this: Back in the day, the take-away lesson that you wanted to teach the world (I thought) is that creation of new scientific knowledge is a difficult (and social) process. Clearly, sometimes it is. But here we've got a case for the textbooks where the transition from decades of struggle to settled knowledge is going to be settled by one glance at Figure 2 [now figure 1 in the discovery paper].

To repeat, after nearly half-century of exploiting the search for gravitational waves for examples of the endless interpretability and disputability of scientific findings, how do I cope with almost uniformly undisputed "discovery"—this wonderful "eureka moment" with its shocking marvelousness for me as much as the scientists? If there has been any inadvertent "philosophical massage"—deliberately choosing examples to suit my thesis—what do I do now that the masseur has delivered a knock-out blow?

THE NEW NORMAL

[Dec. 29, Peter Saulson, via telephone]: You know, by some scientific definitions, the date of a discovery is the date that the paper is submitted to the journal. So there is some part of our tradition that we love and believe in that says that since we haven't submitted the paper to the journal we won't be able to claim the discovery until the day we submit it. On the other hand, right, people started believing in this in the afternoon of [Sept. 14], OK? So you're pointing out that the d-word gradually condenses and there's a spectrum of when you can say that a discovery is made. People started believing in it within minutes to hours and will finish believing in it when we submit the paper to the journal.

In chapter 5, on page 80, at the beginning of the fourth week, I described what was going on as not so much an "aha moment" but something more like "A—ha—ha—ha—ha—HA!" At that point the "ha"s were the sighting of an unusually big event, the check that it was not a blind injection test, the graphic presentation of the coherence of the signal in the two detectors, the reduction in credibility of the malicious injection possibility, and the opening of the box. But that was only the start of things—we were still in step 1 of the Detection Procedure (see appendix 1); step 2 would not begin for another two weeks. And, of course, the three weeks' worth of work that had been described by that stage were nothing in the context of the fifty-year search for gravitational waves, which could be thought of as a hundred years if we go back to the first conception of gravitational waves with the invention of general theory of relativity.

In this chapter, I am going to look again at the process of scientific discovery making, with my 1985 book *Changing Order* very much in mind. The title signified that it was a study of change in science, treated as social change. The idea on which it was based was Wittgenstein's notion of the "form of life" as found in his 1953 *Philosophical Investigations*, and as interpreted via Peter Winch's little book, *The Idea of a Social Science*, which was published in 1958. Thomas Kuhn's *The Structure of Scientific Revolutions* was published in 1962, and it seemed to take the Winch–Wittgenstein notion of form of life and apply it to science, with the new label "paradigm." We can see what a form of life/paradigm is by considering just a few lines of Winch's book (around p. 120). Winch asks us to consider the difference between the discovery of a new germ and the discovery of the germ theory of disease. The discovery of a new germ is likely to be a significant scientific discovery, but the discovery of the germ theory of disease is much more than that—it is a change in the way we live our lives. For example, surgeons can no longer operate in blood-spattered waistcoats but must attend to hygiene. As Winch says, one cannot have the germ theory of disease along with dirty operating theaters, and one cannot have all that ritualistic scrubbing and hygienic robing before an operation without the germ theory of disease—it would not make sense. In other words, the new concept and a whole new set of ways of living are intimately bound up together: this is a form-of-life or a paradigm, when that term is understood properly.[2] Scientific discovery is not just thinking something through and

2. See Note IX in "Sociological and Philosophical Notes," 363.

not just measuring or observing; scientific discovery is creating new ways of acting and being. Scientific discovery is social change—it is changing the social order of things. That is why, though the general theory of relativity was, in one sense, "discovered" a hundred years ago, we are still discovering it now.

I am going to treat the change from not being able to detect gravitational waves directly to being able to detect them as a change in our form of life in the way that Winch treats the discovery of the germ theory of disease. It might be argued that The Event does not involve any major change of theory so that the comparison is ill drawn. But, first, as I can testify, it does comprise a major change in our "way of being in the world," and I will document all the changes that it is bringing about and try to predict a whole series of changes that it will bring about. The point is that Winch's contrast between the germ theory and a new germ is too sharply drawn. If you are a scientist working on stomach ulcers and suddenly discover the germs that cause them, this is not the discovery of the germ theory; but it is a revolution within the smaller scientific universe you inhabit. It is better to think of a kind of fractal model, with the entire germ theory at the top with many events of similar form writ small within it—the discovery of individual germs. The first detection of gravitational waves is one of those scientific changes—a change in our scientific form of life—that is just a little way below the top of the "physics of general relativity" fractal.

Wittgenstein had come to formulate the idea of a form of life because he wanted to understand how it was that words had meaning. He came to realize that dictionary definitions merely pushed the problem one stage back: for every definition one can ask how its constituent words have meaning. Wittgenstein concluded that the foundation of the meanings of words was their use—"Ask for the use, not the meaning." The meaning of the word "germ" was to be found in the way surgeons scrub their hands: we see no germs, we see only obsessive scrubbing—that scrubbing is telling us about germs. That scrubbing creates germs even as it destroys them, because it creates the order of things. But Wittgenstein did not discuss the way meanings change, only how they are maintained. My 1985 book was called *Changing Order* because I wanted to look at the way meanings changed, and that is why I looked at science—a perfect laboratory for investigating meaning change, because so much of the activity that constitutes change in science is located in well-bounded and therefore relatively easily observable spaces. Just imagine how much harder it is to investigate change in art, fashion, or politics.

Sociological research is a creature of its times. What needed doing in the 1970s and '80s was to show that scientific change was not the essentially "automatic" application of theory and experiment in the right mixture. At that time science had too much authority—it seemed to float above the ordinary world with its priest-like spokespersons in white coats willing to pontificate on a wide range of scientific topics. What had to be revealed was that scientific change was much more like regular social change than the idea one took away from presentation of science as a formula and scientists as brilliant computer-minds.[3] My key finding was that it was not possible to change order one way rather than another simply by looking to see which experimental results were replicable and which weren't. The reason was that experimental skill has a large tacit component and this meant that the competence of an experiment, or experimenter, could always be challenged should someone come up with a negative replication. That in turn meant that replication was not just doing experiments but also agreeing on who were the competent experimenters and who were not; and that made the process of replication look much more like social change. The argument over Joe Weber's early findings offered a perfect case study.[4]

3. Computer scientists at the time were trying to build programs that would do science, one very well-known one being known as "BACON." It was said to be able to deduce Kepler's laws of planetary motion—which it probably could so long as it was fed perfect data. But physical science is not just extracting equations from data; mostly it is extracting data from noise. See Note X in "Sociological and Philosophical Notes," 365.

4. I think my study was the first of its type—an empirical study of the way a scientific controversy was brought to a close—but it was soon followed by others that established this way of thinking. See Note XI in "Sociological and Philosophical Notes," 365.

In the three books solely about gravitational wave detection that I wrote before this one, the central theme is the same: how do the scientists reach the conclusions they reach and what is the nature of their agreement? Their agreement is always shown to have a large component of judgment and, because what counts as a credible judgment is a matter of social context, a large social component. Thus, in *Gravity's Shadow* I show how the claims of the resonant bar champions were defeated and how, in logic, it could have been otherwise. *Gravity's Ghost* and *Big Dog* are both painstaking examinations of how the scientists reached their conclusions about what the blind injections would have comprised if they had not been blind injections. Typical is chapter 14 of *Big Dog* where I show how much of what went into the decision about what Big Dog was, was based on twenty-five "philosophical," or social judgments; the contrast is with calculation and measurement. The same philosophical and social judgments happened with The Event too, but now something else is going on as well. The new thing is that within just the last few months we have seen a far more substantial change take place than ever happened in my previous studies.[5]

The dynamic of *Changing Order* is the contrast between the settled science of the TEA laser, which nobody questioned, and the controversial science of gravitational wave detection (along with bits of parapsychology). In the case of the TEA laser, scientists knew when they had successfully replicated the building of the device because it would do what it was expected to do—generate a beam of infrared radiation sufficiently powerful to make concrete smoke. No one doubted that this was what a proper TEA laser built in the early 1970s should do. The TEA laser was the equivalent of the discovery of a new germ within the existing germ theory of disease, and everyone already new how to act when a TEA laser failed to lase—try harder. Gravitational wave physics was radically different because no one knew what a proper gravitational wave detector should

5. Of course, my studies weren't the only ones, but I am trying to tell the tale of what I am doing. See Note XI in "Sociological and Philosophical Notes," 365.

do—should it detect gravitational waves at the high fluxes reported, or should it not see gravitational waves with that kind of energy? Because that question had not yet been settled, the argument about which devices had been competently constructed could not be resolved by observing their outputs. I called this the "Experimenter's Regress." The argument over the detection of gravitational waves was like an argument over the establishment of the germ theory of disease rather than the discovery of a new germ. In the last few months, with astonishing speed, gravitational wave detection has been transformed from a controversial science to a settled science—it is now TEA-laser-like. From now on we will be able to refer to what a gravitational wave detector detects in order to settle the question of whether it has been competently put together: if it doesn't detect the waveforms of inspiraling black holes, there is something wrong with it. That is the new kind of transformation.

Throughout the book I've mentioned "anticlimax." With this kind of social change, anticlimax goes along with the joy of discovery. On the one hand, there is the astonishing and beautiful discovery with its huge emotional charge and the "pinch-yourself" realization that it is real; on the other hand, what has been created is a new but ordinary world. Not that it won't be a wonderful world for the new science of gravitational wave astronomy, but, to repeat, all we are able to do now is "discover new germs"; we are no longer discovering the germ theory of disease. If I were writing *Changing Order* now, gravitational wave detection would fill the same role as the TEA laser. That so-much-more-solid transformation has to be explained along with the more immediate building of consensus over The Event. To explain it we will have to reach backward, before September 14, and we will have to remember what it felt like when the third event was found—when the second shoe dropped. In narrow scientific terms, that is the crucial happening even though it is nowhere to be seen in the discovery paper—and, as I write these words, it is still secret.

Or pretty secret! Rai Weiss has said something to the *New Yorker* that I thought was not supposed to have been said:

> Since the September 14th detection, LIGO has continued to observe candidate signals, although none are quite as dramatic as the first event. "The reason we are making all this fuss is because of the big guy," Weiss said. "But we're very happy that there are other, smaller ones, because it says this is not some unique, crazy, cuckoo effect."[6]

Even more strange, the *New York Times* of February 12 writes:

> According to Dr. Weiss, there were at least four detections during the first LIGO observing run, which ended in January.

But this appears to be based on a misunderstanding.

AN ALTERCATION IN THE MIDDLE OF DECEMBER

An altercation in the middle of December can act as a little illustration of the "long aha" because it shows how at that time no one quite knew whether gravitational waves had been discovered or not and not everyone agreed—this is a big group of people, and they are not all the same, nor will they all transform themselves into new kinds of people at the same rate. The topic of the altercation—a long-running one that has been discussed at length in chapter 10—is what to tell the EM partners, the astronomers and astrophysicists who search the sky for bursts of electromagnetic radiation or neutrinos. Though we were more than three months past The Event at this time, the partners may have stored data that they could search retrospectively if they knew they should be paying special attention. In the normal way, the standard for an alert sent to the partners is much lower than the standard for a detection; the logic is that a weak gravitational wave "trigger"—something that could never stand as a detection on its own—if combined with a strong source of another kind, could add up to something significant. The trouble is that The Event is not a weak gravitational wave trigger, and if the EM partners knew how strong it was they might react differently than they do when alerted about a weak trigger.

6. Nicola Twilley, "Gravitational Waves Exist: The Inside Story of How Scientists Finally Found Them," *New Yorker*, February 11, 2016.

Everyone had agreed that The Event must be kept secret until the press conferences. This gave rise to a problem regarding partners who might be specially well-equipped to see something associated with The Event. One physicist, A, suggested that a certain partner be sent an alert about a candidate event. Another physicist, B, violently opposed the idea:

> **Dec. 19:** I'd like to suggest that we share the draft paper with [a certain partner] concurrently with the detection announcement, not before.... If we give [the partner] a draft that refers to GW150914 as "a possible gravitational-wave transient candidate" and then two or three weeks later we give them a draft that presents the same data as a Nobel-prize-level event, they'll know that we were misleading them with the first draft. Even though we have valid reasons for being flexible with the truth, I don't think this is a good way to treat our colleagues.

A responded that it was just a matter of telling the truth, since The Event, at that stage, was nothing other than a candidate. A asked, "What is everyone working on if this is so obviously a real event rather than still at the candidate stage?"

> **Dec. 22:** You are advocating that the detection process is a done deal, and therefore we should not call the Event a "candidate." Not only is this argument misleading and dangerous, but it is also puzzling given our many colleagues working hard, even over the Holidays, to swiftly complete the analysis for all of us.

B responded:

> **Dec. 23:** Clearly the detection process isn't yet done, and there's still lots of work to do. However, the probability that we will publish is now so high that to present the event to our [partner] as a "possible candidate" is, at best, misleading.

And A answered:

Dec. 23: Clearly, either (1) the detection process is not yet finished and therefore the event is a candidate, or (2) the detection process is finished and therefore the event is not a candidate.

This thread became quite heated; but let us not worry about what was the right thing to do, let us worry only that in the last week of December it had not yet been established whether The Event was a mere "candidate" or a real discovery. In the last week of December, the order was still in the process of changing, and various options were still open about how one might talk and think about it. The argument between A and B shows that the disputed order of things had real consequences for whether certain courses of talk and action were more than just "choices" and bore upon whether certain actions were to be counted as having integrity.

Cognitive change is social change.[7] In late December, Every one of the thousand knew in their hearts that this is a discovery. But that does not mean they have all immediately become the different people that they will need to become, automatically acting in different ways as the discovery is established. We see scientist A asking what is all this work for if the discovery has already been established: scientist A is not yet a different person, even if scientist B is a different person. Maybe scientist A and his like-minded colleagues need more process and more time before they can change. Some are already living inside a "new normal"—a term I borrow from Peter Saulson, who wrote to me in November:

Nov. 10, 12:58: What is perhaps more significant is that we are seeing a "new normal" develop before we've even told the world about the discovery. We'll be blasé by the time we try to excite everyone else ;-) [Compare] how truly crazy we all were that first week or two, insisting on formal declarations that this wasn't a secret injection [because we couldn't believe it].

Likewise, in the second week of January, I asked Peter Saulson why my inbox is not flooded with emails celebrating the fact that the Boxing Day event has reached 5 sigma now that the box has been opened. He replied:

7. See Peter Saulson's email to me of November 23 (above, 258).

> As for why people aren't ecstatic: I don't know.... Maybe we've already expressed all of the joy we scientists know how to express over GW150914, and this is now the "new normal"? Sad if true.

But, in another sense, it would be far from sad. It would be exactly what should be happening—the establishment of the transformed ordinary. That would be a giant step forward from the first discovery, however anticlimactic.

That said, Peter felt that the significant change had happened before the Boxing Day event jumped to 5 sigma. On December 29, three days after it appeared, he said:

> Peter Saulson: I think it is the other shoe dropping that we've been waiting for... I have a good feeling about it and it will stop so much bullshit if it survives.... I'm going to bet a nickel that it will survive and then we will really be in a different [ball park?] when we've got a second event that we like.
>
> Harry Collins: Yeh, a different position but only in terms of your belief because almost nothing will change in terms of what's in the paper—right?
>
> Peter Saulson: That's right, either nothing at all or almost nothing but it will just change our attitude because that Second Monday event was so marginal that it was hard to get people to care about it but this one isn't. This one is kind of the statistical level, you know, we hoped, *maybe*, we would get something this loud for our first event: nobody dreamed we would get something as loud [as The Event].

Of course, the crucial social change is not within the community but will be in the wider world of science—the second and third sets of ripples. The wider world of science has not been privy to all the convincing things that we insiders have experienced; the wider world of science is going to see only the schematic bones of the thing. For them to believe it, they might need the endless reviews and discussions that the Detection Procedure demands even though the strength of this event is such as to render them otiose in terms of what the community understands; outsiders, without this "feel" for things that can come only with deep immersion in the process of the "core set," will need to reassured that everything has been checked a thousand times.[8] It was certainly potentially useful to be able to say that everything had been checked a thousand times when facing critics at the press conferences or the early meetings. The discovery is a process that will end only when gravitational waves are the new normal for everyone. Employing the metaphor from *Changing Order* and the 1975 paper before it, what we are watching is the strings being cut and the glue drying on the ship in the bottle of the direct terrestrial detection of gravitational waves. Very soon it will no longer seem possible that the ship was ever not in the bottle; and, for some of us, that moment is already here.

WHAT DOES THIS MEAN FOR THE SOCIOLOGY OF SCIENTIFIC KNOWLEDGE?

In *Gravity's Shadow* I say that scientists create their world out of interwoven relationships of trust—the things they take for granted and the people they listen to:

> Scientists know about things in the same way as we know about things: from hearsay. And even if you are one of the scientists I describe in these pages—one of the gravitational wave specialists—you know most of what you know about even gravitational waves from hearsay; that sounds odd, but think about it! Nearly all the science you know you learned from the printed page, the lecture theater, or other scientists' talk and actions. Even the results you know by so-called direct witnessing are tiny corks bobbing on a huge sea of trust—trust in the results of earlier experiments, trust in the colleagues who work with you, trust in the meters and the materials which make up your apparatus, and trust in the computers that analyze the experiment. (*Gravity's Shadow*, 5)

8. The "core set" is the inner group of scientists, usually very small but rather large in this case, who are actively involved in the processes of new science rather than watching from the outside; see *Changing Order*.

I pointed out above that we never see gravitational waves nor do we see black holes. All we see are a few numbers representing the shiverings of a ridiculously sensitive spider's web:

> Everything we know of stars and their cavortings we know only because of the ripples in social space-time [what people say to each other]. If there were no ripples in social space-time, there would be no cavorting stars in our universe (just as there are no cavorting stars in the universe of those who have no modern science, nor none of their gods or witches in ours). Thus one could argue that the causal sequence runs the other way, not from stars to human apprehension but from human agreement to the stars (as most of us believe it does for gods, witches, and the latest fashions). Everything you have just read about gravitational waves and their impact is … based on trust, hearsay, and socialization. (*Gravity's Shadow*, 14)

But now "trust" is no longer quite the right word. Nothing has changed in the philosophical logic of the situation, but everything is changing in the way we live our lives. It will soon no longer makes sense to say we "trust" the scientists who tell us about gravitational waves in the sense that it no longer makes sense to say that a child trusts its parents. To talk of trust made sense a little while ago, but now the nature of things has changed. "Trusting" is something we do in an active kind of way but the trust we have in our parents is like breathing, or better still, like extracting oxygen from the air. It is something that is a condition for living, not something we have to make sure we do in order to live. That is what it will be like to be on the other side of the scientific change we are witnessing: when I switch on the electric light, I am not trusting Michael Faraday and his successors; I am just doing the technological equivalent of breathing. If I reflect very hard I can discover that there is a logical sense in which something like trust is involved—I have to "trust" that there are no practical consequences to the problem of induction, and that the plastic or rubbery stuff that coats the copper wires will act as much as an insulator today as it did yesterday, and that switching on the light will not cause the bulb to explode or the entire universe to turn to ice, even though there is no decisive proof from past experience that such things could not happen.[9] But to say that I am *trusting* that these things won't happen is not quite right since I never think about them when I switch on the light any more than a child thinks about whether his or her parents are trustworthy. Switching on the light is just a feature of the world we live in. Gravitational waves are entering our world in that kind of way, just as black holes and the Higgs boson entered it not so long ago.

In *Gravity's Shadow*, I explained that "relativism" is a methodological approach essential to an assiduous study of the sociology of science but not a philosophical claim. On pages 756–758 of that book I described how I had built a model interferometer out of a laser-pointer and scraps of mirror and glass and it worked. I explained how delighted I was to have built such a thing—I went "whooping into the corridor" looking for someone to show. For me the interferometer was as real as switching on the light. I explained that this sense of the real was not fatal for the sociologist's project.

> There is no need to avoid the force of realism as it exhibits itself in the case of the interferometer. We can simply use the idea of the interferometer and interference fringes as the supporting framework for other kinds of arguments. On the whole, the exact choice about where to relativize and where not to relativize is not very carefully worked out; most of the time it does not need to be. Most of the time, some things are treated as scientific facts and some as facts-in-the-making depending on the dynamics of the story. Most of the time, the principle of methodological relativism, when it is applied to facts-in-the-making, needs be seen as no more than a version of a methodological guideline found in every science: concentrate on the explanatory variable. In this case, it implies that the science be "held constant," as it were. For facts-in-the-making, the science must not be taken to explain itself on pain of circularity and/or the dimming of the sociological gaze. (*Gravity's Shadow*, 758)

So, from now on, in my future writings about gravitational waves, they will be part of the framework, not the topic, so they won't need relativizing.

9. See *Changing Order*, chapter 1, for an explanation of this philosophical point.

But they *could* continue to be relativized, just as could switching on the light or the child's trust in its parents—that's just the philosophical logic of the thing. Go back in the book and we can find the philosophical logic; it appears whenever a calculation is based on a judgment, as calculations always are—for example, the twenty-five judgments found in chapter 14 of *Big Dog*. All the philosophical and sociological judgments found in that list are repeated here. Among those we have talked about extensively in this book is the judgment about whether the interferometers are sufficiently "the same" across the stretches of background noise used to set the statistical significance with time slides and how much manipulation of the machine is allowable in order to maintain that "sameness" without counting as post hoc "tuning" to a predetermined statistical target. We've seen the philosophical logic at work whenever scientists circumvent what we have called the "proof regress" by deciding that enough questioning is enough. We see it in the unquestioning acceptance by the community of the 5-sigma criterion and the 5.1 sigma in this particular case. We saw this lack of interest in criticism over the decision that this really was the observation of black holes even though one could invent other things to fit in the space; we saw it when the scientist in the audience at CERN warned against the equivalent of undiscovering the Higgs; we discussed the logic of experiment which rests on assumptions, or subhypotheses, as the philosophers say, that could be questioned if one were determined to prevent the order of things from ever being transformed. We thought about the same thing when we asked what is stopping us from saying that maybe these waves are caused by the immense force of mind-over-matter brought into existence by the intense longing to see them of a thousand scientists, or that the whole thing is a mass hallucination, or that it is a vast conspiracy by the scientists and me who have invented the entire story set out in these pages. And to make the point still clearer, we have looked at one or two scientists who do not share the mainstream judgment about whom to trust and when to stop questioning.

In a book about Uri Geller and spoon-bending children that I coauthored with Trevor Pinch in 1982—*Frames of Meaning*—we show how worlds change when assumptions change. We did observations in a tricky way on children who claimed they could bend spoons by paranormal means; we watched them from behind one-way mirrors. We watched the children cheat when they did not know they were being observed—all except one girl. We saw the spoon straight and then the spoon bent but we could not spot the cheating. We watched the video a half-a-dozen times until we found a moment when she could have bent the spoon by force, even though we didn't actually see the force applied. Then we stopped analyzing, satisfied that we had solved the problem. That is the normality we live in. But as we explained in the book, in a different normality our amazement would have been reserved for the children who appeared to bend the spoons by force rather than using their normal psychokinetic (in this reality we cannot call it "paranormal") abilities. Whoops! I've said too much. I feel the immense social force of the knitting of the brows. Yet, there is not a single reader of this book who can know "for certain" that The Event is not a paranormal effect, or a hallucination, or any of the rest of the possibilities. All this is old stuff, but it was the start of a journey that is now, in respect of the terrestrial detection of gravitational waves, coming to an end.[10]

Put on the relativistic spectacles, step back, and allow yourself to be estranged from the taken-for-granted, and see just how much everything depends on what you are prepared to accept on the basis of hearsay. What can even I be sure about—someone deeply embedded in the gravitational wave community for decades—nothing, if certainty is the direct witnessing of fact. I have not seen a thing directly; I've just heard a lot of stories from people I trust. To repeat the point, these are stories that from now on I will believe with as little effort as breathing.

The detection of gravitational waves is among the most solid and substantial pieces of our knowledge of the world. But consider what that solidity consists of—it consists of taken-for-granted reality engendered by our social existence. Live in a different society and gods and witches would have equal reality. We glimpse that possibility by reading the *Nature* comments and Reg Cahill's email. The tidy diagram that shows belief in

10. These questions about relativism were forged in the 1970s.

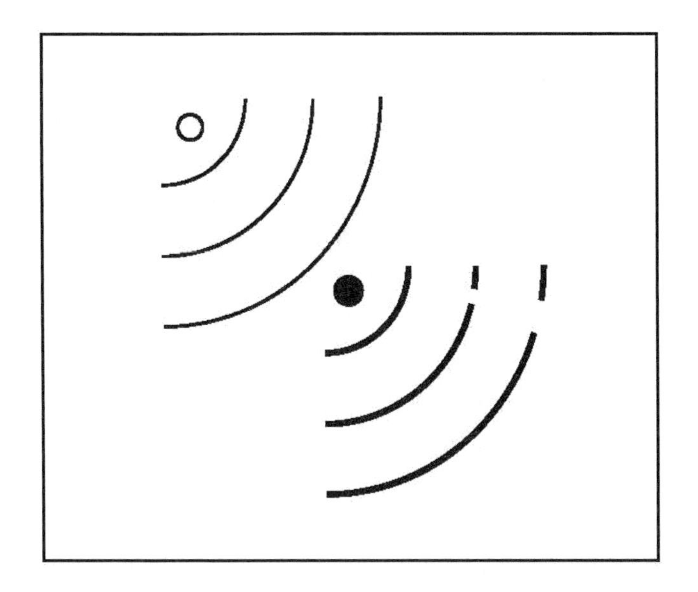

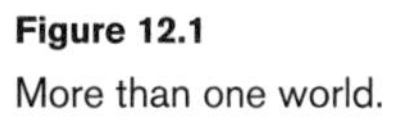

Figure 12.1
More than one world.

gravitational waves establishing itself as a series of ripples in spacetime and social spacetime is a little less tidy than drawn. It should look like figure 12.1, in which the last ripples are partitioned between mainstream science and mainstream society—the bulk of the ripples toward the left—and those who believe it all to be a conspiracy—the small sections toward the right. There is not one world, but many. Of course, somewhere further to right are still other worlds with their gods and witches and whatever else. A deliberately assumed relativism is the method that is best deployed to maintain this degree of sociological estrangement and, thus, to understand how the un-estranged world is stabilized. Is there really more than one world? Of course there is—it is a truism!

CHANGING SOCIETY

To repeat, we normally think of the arrow of causality as running from the stars to our views about the stars—let us say "from left to right" in the figures showing the sets of ripples, figures 7.2 and 12.1. Most of this book has been written in such a register—the left-to-right causality register. But if sociologists are to do their work properly, they must also be able to see and describe the world as though the causal arrow runs from right to left: from our social interactions to the stars. Look at the rightmost partition of the ripples and see that this indeed can be the way the arrow runs. No one has seen inspiraling binary black holes; all anyone has is reports of numbers emitted by impenetrable machines. A glance at figure 1 in the discovery paper for the rightmost partition of the second and third ripples spells conspiracy, not gravitational waves.

If one occupies the leftmost partition, however—as I do for most of this book—the big transformation between Joe Weber's time and now appears to score a goal for Peter Saulson when he asks me (see above) how I am going to account for the sudden acceptance of The Event. Advanced LIGO is the first generation of detectors that has a reasonable chance of seeing gravitational waves according to the theory that has, very roughly, been around since the 1970s. So that sounds simply like science has been done. But here we have to remember that it is only theory: all the experimentalists from Joe Weber onward were ready to say, "Maybe the theory is wrong, and if we seem to have seen something that conflicts with theory, then that is how science sometimes goes." After all, the detection of gravitational waves is proclaimed to be a proof of general relativity; so presumably the detection of much stronger waves, as the earlier experiments could have been interpreted as claiming, *could* have been a disproof—or a proof that general relativity needed fixing. And, of course, the earlier generations could legitimately claim to see things even without disputing the mighty general relativity; they had only to dispute our ideas about the distribution of sources in the heavens. Make the heavens extraordinarily lumpy, with a region of untypically strong gravitational wave emissions close to our solar system, and everything works out again (see *Gravity's Shadow*, chapter 5). A journalist from a popular physics journal contacted me as Advanced LIGO was about to go online and asked me what the consequences would be if things carried on as they always had and the new devices saw no events. I explained that general relativity was not at stake yet, and I did not think the theory of the detectors was seriously at risk yet; but if no events were detected, something would have to be adjusted in our understanding of astrophysics: there may be far fewer gravitational-wave-emitting events than was thought. So even the fact that a detection by Advanced LIGO fits the theory at last is itself a matter of social convention about how strong the theory is and what kind of theory it is.

And then, if we allow ourselves some crude Marxist-style determinism, think of the money and the effort! Think of how much has been spent and how many people have spent it, with all those people wanting to justify what they have done. Let me go back to my field notes written in 1997 (*Gravity's Shadow*, 540):

> Walking by myself around the site in the silent drizzle, I had a moment of estrangement from the project.... Suddenly I saw that this was madness on the grandest scale! All this money, all this effort, all this steel, all this concrete—for what? To try and see movements smaller than the nucleus of an atom!
>
> After the initial delight in the achievement, the physicists too felt a little humbled and frightened. One or two of them remarked to me that "this had better work," or some such, and they said it without a chuckle.

Of course, one could argue that the pressure applied just as much to Initial LIGO as the current model, but it seems the pressure is building up and up—excuses are running out.

And then, who is there to argue with this discovery? The bar proponents have long been defeated (*Gravity's Shadow*, Part III) and now the only people left who believe the interferometers are not the right technology, if there are any, inhabit the fringes of the scientific community. Almost everyone who is a viable expert on gravitational wave detection is on the author list of the discovery paper. Years ago, Barry Barish explained to me that there is a problem for big sciences because they absorb all the experts in the field and there are no outside reviewers left. In particle physics they solve the problem by setting up separate internal teams to compete with each other. Here, no one quite knows how *Physical Review Letters* will find suitable referees for the discovery paper, because all the obvious referees are already coauthors. There is, then, no credible opposition left. This explains why it makes sense for physicists to celebrate when they *submit* their paper—in big science anyway, that is the point at which they know that all the serious work has been done; they know there is no one out there who can mount a significant challenge. Is this not a social fact? Is this not at least part of the explanation of how this huge nexus of trust and assumption holds together?[11] As the ideas set out in *Changing Order* and related works have developed, it has become clear that the endless possibilities for argument about a scientific claim are realized only when there is determined opposition; here there is no opposition left outside of the fringe.

All this is, admittedly, a bit crude, especially in the face of the fact that while conditions have become so much more conducive to detection I know the scientists have still put enormous effort into trying to find out what might have been wrong with the signal. They do this because they are deeply honest people and because they don't want to be wrong and have someone else point out why. What are we left with? Our new beliefs, when you see them from the distanced and estranged perspective, are a consequence of the changed social order.

CHANGING ORDER

What other social changes will we see, what other changes in action and concept? What, in this field, is the equivalent of the surgeon's scrubbing, gloving, and robing? Here we are going to engage in that most disreputable and insecure of ventures: futurology. A lot of this might turn out to be wrong, but for the purposes of the exercise, all it has to be is *reasonable*.

Replication versus coincidence Let us start with figure 8.3 on page 168. Features 1 and 2 of this figure show the waveform as it impacted on the Hanford and Livingston detectors and show how well these measured strains

11. In social science one can never gather everyone together in this way. Indeed, the continuing existence of radically different "perspectives" is taken as a badge of honor.

fit the models. Feature 5 of the figure shows how well the measured strains fit each other. That is "the finding." But what is the nature of that finding? Is it a coincidence that stands out above the background, or is it an observation in one detector matched by an observation in the other detector? This is more than words; it does, or could, represent a historical change.

In the early days, resonant bar detectors recorded pulses of energy. A single detector could see nothing except energy jumps of unknown origin—the stars or noise. It was impossible to know what any one energy jump was. "The finding" was an excess of coincidences at zero delay compared to the number of coincidences in the time slides—what Joe Weber called a "zero-delay excess." Findings were coincidences and they were nothing else but coincidences. Gravitational waves had no identifying features except what we might call their "coincidenceness." But, as we noted, this changed with the interferometers because they are broadband instruments and can see the shape of pulses—their changing amplitude as the wave passes through the detectors. Gravitational waves now have identifying features beyond their "coincidenceness"; they have a waveform that can be matched to a model. It seems to me that this ability to match to a waveform has not been given sufficient salience because the scientists are still thinking in terms of coincidence as being the vital feature of gravitational wave detections. Adalberto Giazotto said to me that The Event was so strong that "even a single detector could make a discovery claim." This does not mean that, today, a single detector would be accepted by the wider scientific community as having made a discovery claim, but it does mean that we can see features 1, 2, and 5 of figure 8.3 as showing, not a coincidence, but a signal on one detector with a replication on the other detector; and "tomorrow," I am saying, this will be the new routine. Replication of a signal seen on one detector by another detector is a far more powerful claim and is really the reason the scientists believe in The Event. The "this is a coincidence with only one chance in 200,000 years of being seen by chance" (as established over 16 days) is the statistical way of looking at it, but "this is a recognizable signal in one detector which has also been observed by another detector 2,000 miles away" is the way of looking at it that is captured by that word "coherence." It is the coherence of the signal that first convinced everyone it was real. The statistics are mainly about convincing everyone else. The shift to thinking of replication as opposed to coincidence, and the possibility of seeing an event with only one detector because it can be identified by template matching, is a change in the order of things that I think we can see happening even though it may not yet have been as fully recognized as it should be.

The change that is going to happen is that single-detector measurements are going to be more and more accepted because less and less proof will be needed to confirm the observation of a gravitational wave under the conditions of the new normal. Gravitational wave signals are going to become more ordinary than noise in the detectors. One may put this in terms of Bayesian statistics—"everyone's priors will have changed"—or in terms of a change of our way of being in the world—like taking paranormal spoon-bending to be more ordinary than metal distortion with physical force!

Low latency and relaxation about criteria for detection A feature of the new normal, then, is that gravitational waves will be easier to detect:

> **Oct. 1, 21:26:** It has always been the goal to move toward doing searches in low-latency. This follows a general trend from early initial LIGO (~year latency) to enhanced LIGO (~week latency) to advanced LIGO (~seconds latency). At some point, detectors will just be running continuously, and divisions into observations runs, or periods of observation runs, are becoming increasingly arbitrary. We ultimately want to produce alerts / notices / whatever in real time…
>
> Given an iron-clad detection, I think whatever risk there once was of a low-latency search affecting other searches is now diminished, and once there is more than one detection then I expect that all things truly will be equal. (Well… of course the search with full data quality and better calibration will be better, but it won't be adversely-impacted by having done a low-latency search.)…
>
> In the long-term, then, I think we will be wanting to run in low-latency over the whole parameter space. Which means, the sooner that we try to get

> things working, the better, knowing that experience will help us improve the search . . .
>
> We want to further gravitational wave science, and this is best done if we can make discoveries public as soon as possible. . . .
>
> It is time to stop pretending to blind ourselves—it is no longer productive.

This email says it all—everything will run at the lowest possible latency as the conditions for seeing a signal become relaxed. The pipelines that can detect signals in real time will become the norm, with the offline analyses becoming backups used mostly in cases of unusually faint but, for some reason, especially interesting signals.

It follows that all the stress over blind injections will be relaxed: there will be no more rows over airplane events and the like, and the idea of opening the box will disappear for regular signals with well-defined waveforms such as inspirals. For such signals, everyone will forget that there ever were little dogs, or even time slides, because waveforms will take over from statistical analysis as the chief identifier of signals except for very marginal and unusual events.

It follows, of course, that things like Second Monday will be *events*, strong enough to be added up with all the other such events to produce event rates, not things that are whispered behind the hand. We see the change happening already. An email of November 25 included the sentiment, "Many people have said that the [Second Monday] event 'looks good' and that there's a feeling it may well be real." That this could be said about such a weak event is a sign of the changing priors or the changing order of things. We will see more of this in the postscript. We can see what is happening as a change in "evidential culture" (see *Gravity's Shadow*, chapter 22), as gravitational wave physics turns into gravitational wave astronomy. Astronomers and astrophysicists have to work out the rates of different kinds of events in the heavens, and they cannot base their estimates on the strongest events only or the rates will not be as realistic as they could be. Compared to physicists, astronomers and astrophysicists live in a world of speculation; in terms of the three dimensions of evidential cultures (*Gravity's Shadow*, 397–398) they have a higher tolerance for evidential collectivism, their evidential thresholds are much lower, and they are willing to embrace higher levels of evidential significance for weak data.

Soon, then, there will be no hiding of things like the third event because everything will be out in the open—at least when it comes to familiar phenomena like inspirals. Already there is an agreement that after four events have been seen, data will become freely available—we are nearly there (though I also hear that the community may be looking for ways of going back on that promise, such as claiming that it applies only after the detection of four inspiraling binary neutron stars; but I also hear that some are proposing to relax things and start releasing all the data now).

Secrecy Secrecy will not go away entirely, but what will be secret will be types of signal that have not yet been seen. The first sighting of continuous gravitational waves from a pulsar will be greeted with excitement—not as great as this, but much greater than the sighting of another kind of inspiral. The first paper on the stochastic background will be the same, and even more so for the first detection of the cosmic background of gravitational waves (remember the fuss over BICEP2). Maybe we will see the gravitational wave signatures of starquakes on neutron stars or cosmic strings.

The language of gravitational wave physics The language of gravitational wave detection developed a life of its own with specialist terms like "little dogs" and "foreground"—where "foreground" means signal, not noise, noise being what it means in every other astronomical specialty. But it looks like these terms are going to disappear as the domain prepares its face for the outside world—as the second and third ripple come into focus. As we have seen, the term "foreground" has been stripped from later drafts of the detection paper and the term "little dogs" never appears even though the dogs themselves did—instead they are described by some complex circumlocution. In the caption of what is now figure 4 of the discovery paper we find:

The tail in the black-line background of the binary coalescence search is due to random coincidences of GW150914 in one detector with noise in the other detector.

In a few years it may be that no one will remember what little dogs are or the whole semantic net into which they fit—"slaughtering the little dogs," "time slides," "blind injections." The origin of the term will be of historical and philological interest only—it arose because, by chance, a "blind injection" appeared to originate in the direction of the constellation *Canis Major*! Currently, this terminology is indicative of membership in a specialist community—no physicist outside this group will know what little dogs are—soon it won't be, because *no one at all* will remember what little dogs are!

How discoveries are announced Something else that I predict will happen sooner or later is that the objects that are discovered or seen will no longer be gravitational waves, because gravitational waves will simply be the medium utilized by gravitational wave observatories. We don't title papers "Observation of Light from Supernova"; we say "Supernova Observed" because the medium is no longer worthy of remark. Soon, gravitational waves will no longer be a medium worthy of remark. In a few years the title of a paper reporting an event similar to The Event will not be "Observation of Gravitational Waves from a Binary Black Hole Merger" but simply "Observation of a Binary Black Hole Merger." Thus will the "O" in "LIGO" fulfill its potential.

Internal changes in status and salience What is going to happen is that the different groups within the collaboration are going to shuffle around in terms of status. One of these shuffles has already been alluded to. Pretty well throughout the domain of the interferometers, the salient detector group has been those looking for inspirals. It is inspirals that have been in the cross-hairs because they are the heavenly event that it was thought would be first detected, and this was inscribed in the metric for the range of interferometers—the distance at which an inspiraling binary neutron star system could be detected. Now this has happened, albeit it is a black hole rather than neutron star inspiral that has been seen; and since it seems that these are going to be seen even at the current level of sensitivity about once every thirty observing days, and since the level of sensitivity, all being well, is going to go up by three and the rate to one per day as Advanced LIGO shakes down, inspiraling black holes are going to become very ordinary as far as gravitational wave *detection* is concerned. There is plenty of exciting science to come for these objects—the detectors are not yet sensitive enough to explore the characteristics of the ringdown of the merged black hole—which would enable us to say that for the first time we had seen a single black hole directly—and the steady accumulation of numbers and sky positions of the multiplicity of black hole inspirals will enable maps to be constructed that may uncover heterogeneities that reveal the evolution of the Universe, and so on. But the next exciting *detection* is going to be of a neutron star inspiral, about which different kinds of things will be learned.

After this, however, the cross-hairs will shift to the continuous wave group and the other groups who have been waiting in the wings for their moment on center stage. I am guessing that their status will shift with the shifting of the cross-hairs.

A much more poignant and serious shift in status and prestige is already happening. It reflects what happens in other big science projects. Great machines are built that corral new features of the universe; but once they are built it is the data analysts and theorists who take the stage, and those who built the machines fall back into the wings. In astronomy, the names of telescope builders do not appear on papers; in particle physics, the names of accelerator builders do not appear on papers (though *detector* builders may hold onto their status for longer).[12] When it comes to the direct detection of gravitational waves, the achievement is to have built machines of this extraordinary, unbelievable sensitivity in the face of fifty

12. Thanks to Barry Barish for a conversation on this matter. See Note XII in "Sociological and Philosophical Notes," 367.

years of scorn; it is to have shifted the science from small to big and to have held together a huge collaboration to do it. But now the excitement is not on lasers or mirrors or vacuums but on black holes—on what the numbers mean, not whether or how to generate them. Careers and hearts have been broken in conceiving and building the astonishing machines that can measure, remember, a change equal to the width of a proton in the diameter of the Earth; and yet all this has merely set the scene for what one of my respondents referred to as "the surging over-ambitions of the theorists."

We've see the change even as the discovery paper was being written, with one set of respondents complaining to me about the way descriptions of the instrument were being written out of succeeding drafts. Who will be remembered in a few years? Not those who were remembered a few years back—unless they win the Nobel.

I am someone who feels closer to the instrument than the theory and closer to the experimentalists than the theorists, so I experience this change as a kind of inevitable Greek tragedy—the physics community destroying its parents. But Barry Barish, who has seen it all, tells me that it won't happen quite so fast as I think; the instruments are still a long way from the kind of sensitivity they need to see the details of these events they are now just detecting and a long way from the sensitivity needed to see all the other fascinating things that can be seen only with gravitational waves. Indeed, there are plans for an "Einstein Telescope"—a 40 km interferometer, ten times as long as the LIGO machines, or a 10 km instrument, the sensitivity of which is enhanced by being constructed in deep tunnels with the application of cryogenics. With either design, much greater sensitivity can be achieved at all frequencies. So Barry tells me that the age of instrument builders is not yet over in terrestrial gravitational wave detection. We shall see.

Funding patterns In today's world one would be a fool to try to predict precisely what will happen to funding for gravitational wave detection. But something will surely happen. What would make sense, if the world were in a less chaotic state, would be that the building of a detector in India, long teetering on the brink, would now be confirmed.[13] We need sensitive detectors spread across the globe if sources of gravitational waves are to be pinpointed allowing multimessenger astronomy to come into being—astronomy that correlates gravitational wave sources with electromagnetic or neutrino events in the heavens. (Black hole inspirals should not be a source of other kinds of signal, but inspiraling binary neutron stars should be.) Maybe funds will be found for a detector in Australia—the talent is there, and at one time it looked as though a big device would be built near Perth; many Australians are on the author list of the detection paper. Maybe more money will go into the British German GEO detector, but it is not in a particularly useful location with Virgo located near Pisa—but perhaps there are different kinds of things it could do to complement the bigger instruments. Maybe new funds will be found to help Virgo forward so that it can play a more significant part in the search. One perspective says that LISA, the nascent interferometric detector in space, scheduled for launch in 2032, will be brought forward, while another perspective suggests it might be canceled now that much less expensive ground-based detectors have shown they can see everything except the most massive of black holes where ultralow frequencies are needed. I cannot imagine that the Einstein Telescope project—the building of one or more ground-based interferometers with ten times the sensitivity of Advanced LIGO—will not be given accelerated consideration. More sensitivity is needed not only for the understanding of the population and distribution of the sources of gravitational wave events but also for the measurement of the ringdown of black holes that have just merged. More sensitivity is needed to see more kinds of gravitational wave events. More accuracy, and therefore more sensitivity, is needed for multimessenger astronomy because it won't really work unless we can pinpoint sources. Giazotto's call for confirmation of single gravitational wave events by electromagnetic radiation may be misplaced because there is too much going on in the heavens and, today, we don't have

13. The building of the Indian detector was confirmed by India's Prime Minister on February 17, six days after the press conferences and a few days after I wrote the sentence.

the ability to pinpoint sources of gravitational radiation. Just before the announcement there was talk of publication of a paper that would link The Event with an unexpected burst of gamma rays, but, as one of the emailers put it (February 7):

> The fact that there is a quasi-temporally coincident gamma burst seen in one part of the huge range of possible positions in the sky of the BBH is not enough evidence to say "would"—it _must_ say "could be an unexpected EM counterpart." ... Exciting if true, but I have serious doubts that one can claim more than coincidence in time of two transient events. Lots of things were going on in the universe at that moment in time.

Our view of the heavens It puzzles me, but because the gravitational waves from inspirals fall into a frequency range of human hearing, scientists keep insisting that what we are now doing is *listening* to the skies instead of looking at them. It puzzles me because you cannot hear gravitational waves unless you are right up close, as in the imagined scenario discussed at the end of chapter 10 where we would be the distance of the Sun from The Event. Maybe scientists are reaching toward the analogy of the ear because the mirrors vibrate in response to a gravitational wave (or they would if they were not held still), and maybe it is because the phenomenon falls into this frequency range that scientists are tempted to transmute the signals into sound so that you *can* listen to them. Two black holes merge 1.3 billion years ago and we are asked to listen to the sound. It is the most unimpressive sound it is possible to imagine and I would have thought the physicists would want to forget about it.[14] The only comparable incongruity I can think of is from chapter 31 of Douglas Adams's *Hitchhiker's Guide to the Galaxy*:

> And so the two opposing battle fleets [joined up] to launch a joint attack on our own Galaxy. ... For thousands more years the mighty ships tore across the empty wastes of space and finally dived screaming on to the first planet they came across—which happened to be the Earth—where due to a terrible

14. You can listen to it here: http://www.popsci.com/listen-to-sound-gravitational-waves.

miscalculation of scale the entire battle fleet was accidentally swallowed by a small dog.

That, it seems to me, is how listening to gravitational wave chirps compares to black hole inspirals.

But, there *is* a relevant difference between our ears and our eyes that maps onto the interferometers (but not the resonant bars): ears are multi-directional—we hear from all around us—whereas eyes have to be pointed. I suggest that a combination of this feature of our ears combined with the desire of physicists to turn gravitational signals into sounds of corresponding waveforms—something that is going to be useful in the control rooms of the interferometers as more and more signals are detected given increasing sensitivity of the devices—is going to change humankind's vision of the heavens.

Our model of the heavens is a place of peace and stillness, the fixedness of the stars being a necessary standard for navigation and bringing out the special quality of meteors, comets, and the rare visible supernova. EM astronomers already know that the heavens are in turmoil; everywhere they look they see explosions and other such happenings (see the email above: "Lots of things were going on in the universe at that moment in time"). But the ordinary public does not "get" the vision of the sky that the EM astronomers have, because you have to look through a telescope to see what is going on and you see only a tiny fragment of the sky with each viewing. But in the near future we will be unable to avoid hearing the heavens popping, crackling, and boiling: that source of eternal stillness in our lives will be lost. Poetry will change: pity!

We are now, of course, way out on the third ripple, but these are the kinds of changes in the order of things that will make gravitational waves a normal part of our lives. Thus will the order of things be transformed.

RELATIVISM REVISITED

To repeat something that deserves repeating: Is the first part of this chapter meant to prove that the world is socially constructed and that nothing "real" happened? No; here relativism is used as a method—a way of concentrating on the way belief changes and, in this case, the way social order changes. To do this properly the analyst cannot allow him- or herself the luxury of cutting inquiry wherever he or she chooses by citing what is "rational" or "irrational" or falling back on what scientists come to count as "the truth of the matter." What I cannot afford to do is allow myself to accept that The Event causes scientists to believe in it simply because it stood out so far and so clearly from the noise. When Steve Fairhurst says to me on March 16 that it would have been much more trouble to convince the scientific community if the biggest signal had been the Boxing Day Event, I accept that what he says is correct; but I have to think about why anyone believes any of it anyway. The hope is that this kind of approach will enrich the understanding of how science works.

Should scientists adopt a relativist perspective? Definitely not; the world of the scientist had better be real. Where else would scientists find the energy to do their work and the willingness to ruin their social lives for the sake of 1/10,000th the diameter of a proton? The scientists should continue to thank their lucky stars—literally—that The Event was so big and so sharp. On the other hand, if they, like the social scientist, could learn to alternate between the realist and the relativist viewpoint, they might find it interesting, so long as they could firmly switch off the relativist perspective most of the time—just as I have to switch off the realist perspective, if not most of the time, at least when I am writing a chapter like this one. All one needs to approach the world in this dualistic way is to be a philosophical agnostic about the truth of the matter—that is, not to believe that the relativist (or the realist) perspective is so ridiculous from the outset that there is nothing to be gained by adopting it from time to time.

13 ON THE NATURE OF SCIENCE

TOO TRUE TO BE GOOD

Some of the questions and controversies that have arisen throughout this book have to do with the very nature of what is being accomplished—the very nature of physics. In this chapter I want to explore and explain what physics is or, more accurately, what I believe it should be. The question turns on what kind of truth physics should be trying to deliver.

Mathematics is an enterprise where a published paper is like a house of cards: a single mistake in a proof and the whole thing collapses. The logical aspect of philosophy is similar. Sociology is rather different—it is far more robust in the sense that a paper in sociology does not collapse if it contains the odd mistake; it is not the kind of argument that can be referred to as a proof. Just look at what I am arguing here—it is not proof-like; it is more or less "persuasive." If I get a sentence or two wrong, or make a mistake in one of dozens of discrete arguments presented in the book, it does not bring the whole book down. Physics has both kinds of argument within it. It has the mathematical kind—the calculative kind—and it has the persuasive, plausible, kind. We have already seen this: an example of the calculative kind is the estimation of the false alarm rate from the time slides; an example of the plausible kind is the question of whether the "freeze" made any sense and whether the instrument could be said to remain the same over the time from which the time slides were generated or if a longer time could be taken. An example of the calculative kind is that the estimate of rates of black hole coalescence coming out of the O1 observations was

2–400 per unit volume per year and not 6–400 per unit volume per year; and example of the persuasive plausible kind is whether this calculation was worth setting out in a long paragraph or a short paragraph in the discovery paper and whether the difference between 2 and 6 was worth arguing about.

A problem that physics has is that one kind of argument is sometimes confused for the other kind, and that can be damaging to good decision making. Physicists, I suspect, have a tendency to believe that decisions based on quantitative arguments are always best—but they aren't, not always. Physicists know this to the extent that one of the things they are very good at is approximating: they know when a result is presented with too many significant figures, leading to a misleading and spurious appearance of accuracy; getting the right number of significant figures is one of the arts of physics (note that you cannot have too many significant figures in mathematics).

Sometimes physicists (or mathematicians) do not notice that it is time to switch off the calculation and just try to get "the feel of things," something that human beings are very good at and that can often produce a better result than calculation. We know from *Gravity's Shadow* (561) that Ron Drever was very good at feeling his way through a technical problem and could turn out to be right in the face of others' mathematical proofs that seemed to show he was wrong. We learn to recognize people in the collaboration who are almost never wrong when it comes to a calculation but may not be so good at making a judgment, and we learn to recognize those who are good at judgments and not so good at calculations; I can recognize these types and I could name the people. The collaboration works because there is a division of labor between all these types and their capabilities. The collaboration fails to work when one party or another insists that their way of thinking is the only right one.

All this is closely related to the shutting down of the proof regress at the right time; physicists have to know when to stop arguing, criticizing, and calculating. We saw the point made in respect of when to stop arguing that it might not be black holes in the inspiral, and in a few other places. We can go back a bit further and relate the problem to what, in *Big*

Dog, was called "the epistemological sentence" (e.g., 231). Remarking on whether Big Dog appeared to be real (assuming it was not a blind injection), one of the scientists wanted to add to the end of *that* proto-discovery paper a sentence: "However, we cannot completely rule out other possible causes, and look forward to more sensitive detectors being brought online to observe ... numerous gravitational-wave events." That sentencc is logically immaculate in its caution, but that does not make it science; science has to be ready to be wrong.[1] Always asking for more proof or, what amounts to the same thing, always inserting reservations to cover one's back, is, at best, asking physics to shape its findings to a form more suitable to mathematics or logic, where things are absolutely provable.[2]

The anthropic principle—the view that the fundamental constants must be what they are because if they were different we would not be here to ask about them—is not science, because it is represents the wrong kind truth. It is true but it could not be anything other than true. It is a bit like the kind of truth that mathematics or philosophy aspires to. It is *analytic* truth rather than *synthetic* truth. Science must aspire to synthetic truth, and synthetic truth can never be quite secure. Nowadays there is some question of just where string theory lies in respect of this distinction: if it bears on nothing observable, is it really physics or is it just mathematics?[3] The truth to which at least some versions of religion aspire—revelation—may not be analytic but it falls into the same category as far as the aspiration is concerned because it is intended as absolute truth: "God is good," not "God is almost certainly good," or "We think god is good but there is a possibility that further evidence will show that we were wrong."[4]

1. This should not be confused with Popper's idea of falsifiability, which refers to well-defined empirical claims; we are discussing the overall kind of truth that science seeks. See Note XIII in "Sociological and Philosophical Notes," 368.

2. See Note XIV in "Sociological and Philosophical Notes," 370.

3. For questions about string theory, see, for example, Smolin's *The Trouble with Physics*..

4. I am not arguing that one cannot be both scientific and religious, only that they are different ways of being in the world. Of course, there are conflicts over substantive beliefs, like

That the components of The Event are black holes is not a truth like "God is good" or "The square on the hypotenuse equals the sum of the squares on the other two sides"; it is a truth that says "We have done our best and done it with integrity." It means, "Given that this is the only kind of truth there is in respect of this matter, you should choose to *act upon it* even though it does not cover the unforeseeable." It means "We sincerely believe, following our best efforts, that the world has changed and a new kind of order is now in place." So that is how a physicist should decide on whether to include a sentiment that covers all possibilities just in case they are wrong or whether, in this case, they should say something along the lines of "We think we have seen black holes but there may yet unobserved objects that could masquerade as black holes." The physicist should decide *not* to say such things because physics has to take the responsibility of being wrong from time to time yet still tell the rest of us what to think in respect of the physical world. To say these kind of things would be to try to be, as we might say, "too true to be good"—too true to be good by the standard of physics.

Blas Cabrera's monopole claim was not bad physics, it was great physics. The thirty-three-year old monopole still hovers over us, but Cabrera is said to have got away with it for what I think are the wrong reasons. As a respondent wrote to me:

> **Nov. 2, 16:31:** He was very careful in the paper to not claim a discovery in the paper (probably saving his career).

But Cabrera should not have had to save his career by covering his back; it should be recognized that physics will be wrong from time to time. Likewise with Joe Weber, who is now accounted as shamefully wrong, but without whom we would not be where we are today. And likewise the other half

---

the age of the Earth (because the facts produced by the institutions are incompatible) or intelligent design (because the methods are incompatible in respect of such a claim), but there are many things a religious person can believe while still being a scientist.

dozen or so positive claims that have been made over the fifty-year history of gravitational wave detection; they're just part of science.[5]

One emailer, defending a proof that the component masses of The Event were black holes, wrote something similar:

> **Nov. 11, 13:54:** All discoveries hinge on certain key assumptions. So this is not a "proof" in the mathematical sense of the word. . . .
>
> Nevertheless, the basic physical argument here is an important one, and in my opinion it *should* be presented. Other important discoveries in astronomy, astrophysics and relativity use similar logic. For example, the discovery of the Cosmic Microwave Background radiation was "proof" that the universe was once very small and hot. The discovery of rapid regular radio pulsars was "proof" that there were compact rotating stars with dimensions measured in the tens to hundreds of km. The discovery of the Higgs particle is "proof" that masses are generated via electroweak symmetry breaking. And so on.

That argument is right for physics. Physics is a fallible thing, and it should be presented as a fallible thing.

"OH WHAT A TANGLED WEB WE WEAVE, WHEN FIRST WE PRACTICE TO DECEIVE": GRAVITATIONAL WAVE DETECTION AS A BEACON OF VALUES

In this section, I act as "a self-righteously moralistic person who behaves as if he or she is superior to others"—that is, "a prig." It's not because I feel superior to others, it's because I want the scientists I describe to be superior to others and I am going to whine at them about it. The superiority of scientists is the topic of another book that, coincidentally, I am completing with a coauthor at the same time as I am writing this one. It is called *Why Democracies Need Science.*

5. I am not so sure that this applies to BICEP2 since they may have "jumped the gun" in terms of the regular standards of physics—but I have not researched the case myself.

The argument of that book is that in contemporary Western societies there are fewer and fewer institutions one can trust—the banking and financial systems, once synonyms for honesty, are now examples of ruthlessness; businesses of all sorts draw customers in with cheap deals and then rely on customer lethargy and inattention to raise their charges year by year; lawyers maximize the extraction of money from insurance companies for trivial injuries for which we all pay through premiums, while the law's adversarial system makes winning more important than truth, with the winners generally being the rich; tax-avoidance accountants rob the poor to benefit the rich; more and more sports, including athletics in the latest twist, turn out to be corrupt, with doping and match fixing and the bribing of officials and administrators becoming the norm. Nowadays, as Mrs. Thatcher said, "greed is good": therein lies dystopia.

Science is one of the few institutions left that still has integrity built into it. Of course, not all sciences direct themselves primarily at the truth, but, it seems to me after forty-three years living with it, gravitational wave physics does. We need to preserve gravitational physics, and other sciences like it, to be a beacon for democracy so that our grandchildren may still know what a democracy is like.[6]

It seems to me that, in a small way, toward the end of the story of The Event, gravitational wave physics lost its balance. It didn't lose its balance as far as *seeking* the truth is concerned; it lost it in terms of *telling* the truth.

INSTITUTIONS THAT NEED LIES

It is said that "the first casualty of war is truth" and so it should be. In war the aim is to win, and deceiving the enemy is a noble action; keeping up the morale of civilians by concealing defeats is another act that attracts no moral sanction. But war is not peace—we know that democratic values are shed in wartime and so there is nothing for democracy to learn from war.

6. For a more complete list of the ills of contemporary science see "Envoi: Science in the Twenty-First Century" in *Gravity's Ghost*.

Lying and deceit are integral to the profession of stage magic, but no one decries stage magicians for it. The audience is knowingly deceived—that is what they pay for.[7] We can call this "participatory deceit." When, in the 1970s, Uri Geller fooled some scientists with his tricks, stage magicians were treated as heroes for revealing them. Subsequently, *Nature* invited a stage magician to unmask the claims of a French champion of homeopathy—Jacques Benveniste—and published the magician's findings. In these episodes there was a lot of confusion about the moral basis of the scientific enterprise.[8] Science has nothing of moral significance to learn from stage magic.

Under certain circumstances, deceit is even integral to science itself. The supposed "gold standard" of medical science—the double-blind control trial—turns on deceit. Neither experimenters nor patients are allowed to know whether they are receiving an active drug or a placebo; secrecy and subterfuge are vital. But everyone knows what is going on, and participants volunteer to be deceived; this, once more, is participatory deceit.

Now compare the secrecy and deception in gravitational wave physics. Let us start with the most innocuous—the blind-injection trial. We have two instances of this, the Equinox Event and Big Dog. There was never a question that either of these events involved any moral impropriety even though they involved secrecy—for months the scientists didn't know whether what they were dealing with was real or a fake (or they would not have, had everything been done exactly right); this was participatory

7. The plot of an interesting novel—*The Prestige*, by Christopher Priest—has a magician pay Tesla to build a device that can physically transport him from place to place. To use this as a stage act, however, he has to engage in further subterfuge so that the audience believes it is a trick, not the far more marvelous actual physical transportation

8. See my early book coauthored with Trevor Pinch, *Frames of Meaning*, for a detailed examination of the Geller incidents and the associated interactions with magicians. David Mermin writes, "This process of discovering that one's former beliefs are wrong, and the painstaking search to identify the old errors, enabling one to construct better founded beliefs to replace them, is what makes the pursuit of science so engrossing. The world would be a far better place for all of us if this joy in exposing one's own misconceptions were more common in other areas of human endeavor" (*It's about Time*, 186).

deception. Indeed, if there was moral impropriety anywhere it lay with those scientists who looked into the injection channels so as to decide for themselves how much time they were going to waste on an analysis of Big Dog. That aside, there is no doubt that these blind-injection challenges were a great success, an extraordinarily useful rehearsal for what was to come.

But the context and the extent of participation changed subtly over time. By the time Advanced LIGO was going online, I was no longer a willing participant to this kind of deceit. I could not waste my time on another blind injection. And I am sure many more of the physicists were of my opinion, and many would have looked in the blind injection channels had the regime of blind injections gone forward. So deceit would have built upon deceit. In my case it certainly would have done—let me explain.

It was going to be impossible for me to spend another three or four months of my life on another blind injection challenge, having written up two such events already.[9] During the meeting in Budapest that preceded The Event by a couple of weeks I had sat down with Dave Reitze, the director of LIGO, and explained the dilemma: if I was to put the proper effort into analyzing the first discovery, which we hoped would occur sometime in the upcoming three years running of aLIGO, I had to know if I was looking at a blind injection or the real thing. If I did not know I would not be able to analyze until after the press conference, and that would not let me finish the job properly. Though he did not say much, I understood him to be indicating that my dilemma would be solved. The trouble was, as we both agreed, that scientists might be watching me, and if I did not take the many transatlantic flights (which at that time we thought would be needed) and show up at the meetings, they would guess that I knew something that they did not. So during the Hungary meeting (September 2) I wrote to Dave and explained:

9. I can say now, at the end of January 2015, writing this book has been exhausting and I would not have had the will to do it had I thought that this might turn out to be a fake.

> Also, thinking more about procedure, I will be explaining to anyone who asks that I have to make decisions that take into account the balance of my health, its current state, and my burden of work and so I will be picking and choosing my meetings and will be going to fewer than I would like to attend if I was fully fit. So I will be betting on getting myself to the most productive events and just hoping it works out as I cannot cover all possibilities but will try to fill the gaps with study of the email interchanges.

At the time it seemed that any discovery (or blind injection) would involve a flurry of meetings and flights, and here I was setting myself up to lie and deceive—I was working out how I was going to deceive even my good friend and faithful informant, Peter Saulson. It was going to be ghastly. "Oh what a tangled web we weave, when first we practice to deceive."

Setting my special situation aside, it seemed to me that the time for blind injections had passed; they had done their job. Now they were going to be more trouble than they were worth, at least if we are thinking of scientific worth. I argued this with anyone who would listen. But the trouble was that the leadership wanted blind injections for nonscientific reasons: they wanted them so that they could keep any proto-discovery secret by telling journalists and rumour-mongers that anything they had heard about could be a blind injection. I think this became the chief motivator for the blind injection challenges. The chief motivator, in other words, became not scientific gain but deceit.

Fortunately, The Event took place in ER8 so every member of the community knew, once the initial paranoia had run its course, it could not be a blind injection. Fortune lifted me out of my moral dilemma in respect of Peter and removed a moral dilemma for those scientists who were going to cheat by looking in the injection channels. But since the majority probably would not have cheated, if The Event had come a few days later during what could have been a blind injection regime, it would have been a disaster for the collaboration. We can see this by comparing the unfolding of Big Dog with the unfolding of The Event—a unique and fortuitous natural experiment.

Big Dog and its predecessor did act as really useful training sessions, reducing the time taken to analyze a signal from eighteen months in the case of the Equinox Event to six months in the case of Big Dog and five months (which was expected to be three months) in the case of The Event. Big Dog might have been a real event as far as most the scientists were concerned, and this meant there was a lot of excitement and a lot of fun. I still remember the animated atmosphere of that meeting in Krakow on September 21, 2010, when the existence of an energetic signal was first revealed. From the stage they announced: "This is a detection and we are going to start writing a detection paper," and my section heading became "The Bombshell."[10] But there was no really deep emotion. In the terms used at the time of The Event by one of those responsible for drafting the paper's abstract: there was no "choking up" because of the "palpable sense of history." And this difference, as we have seen, is clear in the style of the corresponding papers. Only in the case of The Event would someone write:

> **Nov. 2, 16:52:** The title, and the subsequent discussion in the net [should] reflect the historical experimental and technological feat of making a FIRST, DIRECT, COINCIDENT observation of a gravitational wave. … After an effort over more than five decades, [this is] the very first detection, and it should be heralded as the success of the painstaking work in many institutions to make this historical observation possible.

As I've already explained, part of the difference arose because a proportion of the scientists were convinced Big Dog was a blind injection before the corresponding proto-paper was finalized, but, still more, it was because *all* of the scientists knew it *could have been* a blind injection. While we can become deeply emotionally involved with a film or a novel or a poem—we can suspend our disbelief for an hour or two—we cannot fully suspend our disbelief for months. And this means that if what we are analyzing and writing up *might be* unreal, the emotions are diffused; one cannot

10. *Big Dog*, 184.

become emotional or think portentous thoughts if one is aware that one might be taking part in a charade. What the comparison of Big Dog and The Event reveals is that blind injection challenges can never be the same as the real thing even if it were to turn out, when the envelope is opened, that the challenge was the real thing all along. Doing the real thing and doing something that *might not be* the real thing over an extended time is not the same. And as the whole history of gravitational wave physics shows, the best science needs the kind of effort that is driven by emotional commitment.

Thus, as I write, a dozen companion papers are being polished. One has already been submitted for publication. But there were no companion papers written in the case of Big Dog—it just didn't feel the same! And, as we have seen, the tone of the Big Dog paper was that of a regular science paper—there was no historical resonance. In the case of Big Dog there was no attempt to ask *Physical Review Letters* to change its rules for this extra-special paper and allow a much longer abstract and sixteen pages instead of four. In the case of Big Dog no one noticed that terms like "foreground" were inappropriate for a paper that was going to a wider audience, and no one thought about changing the unreadable cumulative significance plots for plots that could be understood by outsiders—because no one was thinking too much about outsiders reading the paper. Simply, the scientists could not bring themselves to work and think as hard for Big Dog as they worked and thought about The Event. It was impossible when in the back of everyone's mind was the possibility that the work would be wasted and that no wider audience would ever read the paper. It was a hugely valuable rehearsal, but it was not quite the rehearsal that it was meant to be—just as war games are an immensely valuable rehearsal for war but are not war itself because nobody dies.

The moral seems to be that it is dangerous to overuse blind injection challenges and the like; it dissipates the best science and it introduces a moral corrosiveness into the scientific community because, once the early lessons from one or two such exercises have been extracted, more and more scientists will cheat so as to preserve their domestic lives. They are willing to sacrifice their domestic lives over and over again only for the sake of real science, and that is how it should be. When the major motivation for blind injections becomes deceiving the outside world in respect of whether a discovery has been made, the cost far outweighs the benefit, not to mention the intrinsic moral cost of the intention to deceive.

THE PHILOSOPHY OF DECEPTION

In my narrow experience the scientific community is too comfortable with deceit even though they try not to "lie." I first encountered this distinction during my very first piece of scientific fieldwork, on the TEA laser, conducted in 1971. I asked scientists what they said to competitors who wanted to learn from them how to build the device. I was told:

> If someone comes here to look at the laser the normal approach is to answer their question but ... although it is in our interests to answer their question in an information exchange, we don't give our liberty. (*Changing Order*, 55)

And:

> Let's say I've always told the truth, nothing but the truth, but not the whole truth.

One might argue, and these scientists obviously feel this to be the case, that lying and deceit are not the same.

There is a philosophy of lying that explores all the possibilities.[11] For example, it is possible to deceive by omitting to make certain statements, or by remaining silent. I like the short, popular treatment of lying by Joel Marks.[12] Marks argues that deceit is the basic moral category and that lying is a subset of it. Thus, one may deceive a person by telling the truth: "No, I

11. See http://plato.stanford.edu/entries/lying-definition/#TraDefDec.

12. Marks's treatment of lying can be found at http://ethicsessays.blogspot.co.uk/2006/01/truth-about-lying.html. This is a revised version of his original, which appeared in *Philosophy Now* 27 (June–July 2000). Thanks to Joel Marks for a short but interesting personal interchange on the topic; he himself no longer believes in morality nor the analysis in his paper and has not pursued the relationship between lying and secrecy as we do here.

didn't break that vase last Thursday," knowing that one did indeed break the vase on Friday. In philosophy such an act is described as a "palter." Clearly there is a distinction between keeping a secret and deceiving, but it seems to be a distinction that depends on context. Consider adultery: there is no question that hidden adultery involves deceit even though all the adulterous partner may do is conceal the act without actively lying or misleading. Here, of course, the deceived person is not a volunteer party to the deception, though it is possible for someone deeply in love or deeply committed in some other way to their partner to choose to be deceived and let their companionship continue rather than to know a truth that they will be forced to act on, thus destroying their relationship; that could be a kind of participatory deceit. In the last few months of the discovery process described here, it seems to me that the actions of the community strayed too far into deceit while mounting the defense that no actual lies were being told.

The two kinds of deceit promulgated to journalists were that the collaboration was unsure whether they had discovered anything because they were still completing their analysis and that even if there was something that looked good in the data it could be a blind injection. I was party to these deceits—I had to be. Twice a journalist contacted me—once around the time of the original Krauss rumors, and then again at the time of the second raft of rumors at the turn of the year. I was faced with what philosophers call "cross-cutting moral imperatives." On the one had I did not want to deceive the journalist, and on the other I did not want to break my implicit agreement with the community who were allowing me to share their world; I was in little doubt that the latter imperative should take precedence. So, though I told no lies, I allowed the journalist to continue to think that a blind injection was a possibility—I did not tell him that this was wrong. And when, second time around, he told me that the rumors said that the event that had been seen had taken place during an engineering run when there were no blind injections, I said "I have not heard those rumors," which was strictly true, as I had not seen them myself at that time, but it was clearly a palter. I did not like what I was doing. And I think many of the rest of the community were engaged in similar things. One senior member even let slip that he had said to journalists that it might be a blind injection—I thought everyone was at least prevaricating—but now having looked at the philosophical analysis I no longer think the distinction between lying and prevarication is anything like as sharp as the scientists think it is.

By now—the end of January 2016—more than a thousand scientists are lying, misleading, and paltering to keep the discovery secret. What is happening, as I have already suggested, is that as this episode reaches its denouement, a generation of scientists is being trained in the arts of deception! That is exactly the opposite of what science should be doing; it should be leading democratic societies in a demonstration of integrity. The most important product of gravitational wave physics is not gravitational waves, it is truth.

And what is all this deceiving for? At this point I am not sure that anyone really knows. An assiduous analyst of the rumors can more or less figure out what is going on anyway (see chapter 10). I wonder if it has become a ritualistic reflex in the way that bureaucratic rules become ends in themselves; everyone knows what they have to do but no one is asking why or whether the time for this kind of secrecy is past. Why is it still with us? Some reasons can be gathered from the email exchanges.

First, there is the matter of ownership of the result. As we have seen, some groups were concerned that they would be "scooped" by other scientists as soon as the news got out. It is worth noting that a suspicious number of scientists had already written papers predicting with remarkable prescience what the first event that would be seen by Advanced LIGO would be, and this in turn (see below) had led to a solid rumor. It is, then, astrophysics that is most vulnerable to being scooped in this way. As one respondent wrote to me on January 17 when I quizzed him about the need for secrecy:

> The trouble is, [with a partial revelation] masses and distances would leak out, and external theorists would put papers on arxiv [the physics preprint server] on 10 Feb.

I suppose that the groups of astrophysicists who have worked for decades for the interferometer groups do have a right to get those papers out first even if other astrophysicists could do just as good a job with some fairly vestigial information, such as knowing that it was a binary black hole and the masses of its components. And if that is true it puts an upper limit on what *should* be released to a wider audience before the discovery paper is written. That is the position taken here.

Barry Barish had said to me when I quizzed him during a telephone call about this need for secrecy (January 5):

> Complete openness is crazy—you can't do that; you have to be reasonably closed. Complete secrecy, on the other hand, seems to me is not essential. But how do you get a thing in between—I'm not sure.

I am going to make some suggestions about the path in between. I think that revealing a little would make concealing the parameters of The Event easier, because you could then say "but, assuming this holds up, we are not going to reveal everything until February 11." In most circumstances, and that includes these, if you explain that you are intending to keep a secret until a certain date there is no moral ambiguity and no deceit. Consider that the recipients of prizes, the details of examination results, the awards of honors, the substance of new fashions, the design of new models of cars and consumer durables—all are kept a secret until they are "unveiled" and everyone knows they are secret. No one thinks that keeping these things behind a veil is deceptive or morally reprehensible.

It does, however, go against a certain principle of secret-keeping, which was nicely expressed in a lighter moment when the emailers fell into swapping literary quotes about the nature of secrecy and rumors (with what, priggishly, I took to be a little too much pleasure). One quotation, taken from the British sitcom *Yes Minister* (April 6, 1981), is very much to the point. Sir Humphrey Appleby, the ruthless and utterly amoral—in the nicest possible way—senior civil servant, remarks: "He that would keep a secret must keep it secret that he hath a secret to keep." That sums up what the community is trying to do and, like most of what Sir Humphrey recommends, it is efficient but immoral. In contrast, my argument is that one should not keep it secret that one has a secret to keep—and I think that in

Table 13.1
Stages leading up to the "unveiling" of a scientific discovery

SIX STAGES OF UNVEILING		
	Pronouncement	Indication of Delay or Disappointment
1	There is enough going on to make us busier than usual.	But it was a false alarm.
2	We are analyzing data that might or might not lead in interesting directions.	It did not lead anywhere.
3	We are clearing diaries in case we need to make an announcement.	Unfortunately, we discovered a mistake.
4	A paper has been submitted and is being refereed.	The paper is being revised or has been rejected.
5	We'll be making an announcement on a certain date unless things go wrong.	Further scrutiny shows the result is not reliable. (But watch this space?)
6	UNVEILING	DELAY OR DISAPPOINTMENT

cases like this, contrary to Sir Humphrey's view, it would make it easier to keep the secret because no one would have to dissemble; it would also score much higher on the scale of moral action.

But do we want science to engage in unveilings, like the car industry or the fashion industry? Taking a very distant view, we might argue that the data has been generated with public money and the career structure of scientists is less important to the public than knowing what has been found out. In a tongue-in-cheek paper written in 1985, I argued that taking this kind of view one could argue that where pure science was concerned—science on which we were not desperately waiting on the results—the rewards should not go to the scientists who found things first but those who found them cheapest.[13] Still, let us assume that we want to reward the work of scientists and accept the notion of unveiling. In a case like this, the notion of unveiling would go along with a series of statements—I suggest six stages of increasing substance, as listed in table 13.1.

13. Collins, "The Possibilities of Science Policy."

The unveiling would comprise the statement that the scientists now believe in what they have found and include, for the first time, the details of what it was and the estimated parameters—the numbers. As far as I can see, this way of presenting scientific work to the public would satisfy all the needs of the scientists as expressed in this case: no one could be "scooped" because the nature of The Event and the parameters would not be made public until stage 6; the scientists would never make a claim until they had jumped all the hurdles they had set from themselves in reaching their criterion of publicly announceable certainty; the scientists would have their day on stage as would the funding agencies—no one thinks the unveiling of a new fashion or a the naming of a royal baby has been spoiled because people know there is a fashion show coming up or that a princess was pregnant.

But there would be no need to mislead anyone—no need to allow people to believe such things as that The Event could be a blind injection or, from the highest source, to refuse to answer a question about whether it could be (as in fact happened).

> ***Nature*, Jan. 12, 2016:** The LIGO collaboration declined to comment on whether there was any time when both interferometers were active but no blind injection was possible. (http://www.nature.com/news/gravitational-wave-rumours-in-overdrive-1.19161)

There would be no need to worry about rumors since the only rumors worth promulgating would be about the nature of what had been discovered and its parameters—and if they are leaking, there is a problem that cannot be solved anyway. There would be no need to train a thousand people in subterfuge. When someone asks a question that should not be answered one would simply say: "Well assuming this works out, that's a secret until the unveiling, on such-and-such a date"; that's not lying, that's keeping a secret.

Compare that with the message that came around on the last day of January:

> Although our target for resubmission [of the discovery paper] is early this week, we will not send a message when this is done, or if/when we receive a message from PRL with more revisions or with acceptance. We'll let you all know if/when we send a press advisory for a Feb 11 press conference, or on Feb 11 to let you know we had to re-schedule the press conference. This unusual and unfortunate measure is to allow all of you to truthfully say "The collaborations don't know whether our results have passed peer review."

In other words: "We're keeping you in the dark so that you can say you are in the dark with what passes as a clear conscience." That's a nice conclusion to a calculation about how best to keep a secret, but maybe it is time for good judgment to take over.

This does not resolve the problem of the EM partners because they, presumably, would have to be told about the nature of the discovery—for example, "It's a black hole inspiral"—so that they know what to look for, before the unveiling. They would not have to be told the statistical significance, but they ought to know that there would almost certainly be nothing to see in this case or they would risk making fools of themselves. It's a hard problem to solve without widening the community to include them, but it seems to be a problem unique to this field: one needs to tell members of another field something of what one is doing early on. Luckily, it is a problem that should go away, in this field anyway. Soon we should detect so many gravitational waves that there will be no point in this kind of secrecy. I am disappointed that the moment is not yet here and that the Boxing Day event is concealed from the community even beyond the American Physical Society meeting in the second half of April.

THE PUBLIC UNDERSTANDING OF SCIENCE

Everyone seems to agree that there is a problem concerning the public understanding of science and I think that the way the discovery of The Event has been handled exemplifies how things can go wrong. Science, in the name of publicity, public acclaim, and the funds that follow, is too ready to adopt the iconography of religion and similar revelatory enterprises. The scientists want to bring the truth down from the mountain and display its glory. The trouble is that science's truth and glory is essentially flawed; if it were not flawed it would not be the product of science but of some other kind of enterprise. As we have argued above, that science can be wrong is integral to its very nature. The whole secrecy and deceiving business is intimately related to a model of science as revelatory, not a model of science as craft-work with integrity. Let us give concrete form to this other model of science by giving it an acronym: craft-work with integrity is CWI. So we have three models of science in play so far: the revelatory model, which, it has been argued, fits much better with religion or magic; the relentless professionalism model, which has little harm in it and emerges from noble motives but reflects a lack of balance between calculation and judgment within the craft of science; and the CWI model, which will turn out to be important.

The revelatory model goes wrong when something that is presented as flawless—glorious truth—turns out to be flawed: the monopole, BICEP2, every gravitational wave finding prior to this one. If truth and glory are the warrant for science then when the flaws are revealed it invites a reaction of "If the idols have feet of clay then my opinion is as good as theirs."[14] CWI, in contrast, presents scientists as virtuous craftspeople, not priests or prophets. We do not want a science that is too true to be good. Virtuous craftspeople are people we can trust even if they have the occasional failure. The six-stage model offers a way to present discovery as CWI.

In retrospect, I can see why I am less excited by the post-Event world than the scientists. We are all suffering some feelings of anticlimax; we guided our gravitational wave vessel through shallows, skirting sharp rocks and avoiding the crocodiles, portaging across hard changes of technology, and then suddenly and thrillingly we shot over the waterfall of impossibility and are now tearing along on a short stretch of white water to the gravitational wave astronomy sea. There will never be another waterfall for

14. This is what Pinch and I called the "flip-flop" model of science (see Collins and Pinch, *The Golem: What Everyone Should Know about Science*).

gravitational wave physics; white water is the most excitement there will ever be. Maybe it is better to have traveled for so many years through an ever changing world than to arrive at the new normal. For me, however, there is something more troubling to amplify the disappointment of arriving. The detection of gravitational waves is one of the enduring crown jewels of science and one cannot resent the scientists glorying in it—and I gloried in it too. But science isn't all crown jewels; hardly any of science is crown jewels. To quote from the Envoi of *Gravity's Ghost*:

> I suspect that when the final announcement of the terrestrial detection of gravitational waves is confirmed it will be impossible for the scientists to resist the retrospective redescription of their field as one of heroic point-discovery, and I suspect they will be unable to resist the glorification of the enterprise in culturally derivative ways. At that point, as I have suggested, "the lame" (that is, the gradualist and uncertain description of discovery) will be trampled under the stampede; the "Drums of Heaven" will be said to have sounded and "Einstein's Unfinished Symphony" will be said to have been completed.[15] Perhaps more important to the mass of working physicists who have dedicated their lives to the project will be the right to face their critics from the other exact sciences with pride; making the most of the newly accomplished similarity between gravitational wave detection and the sciences of their peers will be irresistible. . . .
>
> Scientists' responsibility lies in making the best possible technical judgements, not in *revealing* the truth. It is to represent every judgment as a calculated certainty that is to abrogate *social* responsibility. To be a producer of certainties is, at best, to consign oneself to the nonexemplary sciences—the corner of the scientific world that has dominated, and distorted Western thought with examples of what it claims to be a perfect and, worse, *attainable* mode of knowledge-making. To strive too much for certainty is to abrogate the responsibility of taking that leading role in Western societies that only science as a cultural activity can assume. (160–161)

15. With apologies to David Blair, whose book on gravitational wave detection was given the first title, and Marcia Bartusiak, whose very good, popular introduction to the field was given the second.

So my disappointment is deepened by this moment when, understandably, the gravitational wave scientists want to exploit their glory and thus remove themselves from the role that I believe they have fulfilled for the last half-century—role models for the handling of the uncertain. Now they want to be role models for the discovery of the certain and therefore models for a tiny corner of science—the certainties of Newtonian and Einsteinian physics, the calculation of quantum theory—that represents almost nothing in the real world of weather, economics, climate change, and scientific judgment in general. For nearly half a century I have felt comfortable in this community of scientists because they represent the kind of society I want to live in, but now I am beginning to feel much less comfortable because they are starting to present themselves as representing nothing but themselves and a few other famous physicists. They are drawing on a fourth model of science—the crown jewels model.

This is what I thought would happen, but to see it happening still makes me uncomfortable; and, of course, it weakens the thesis, presented below, that scientists have the potential to take a leadership role in democratic societies.

The scientific substance has not changed following this shift from the model of CWI with relentless professionalism to one of discovering crown jewels, of course. We need only remember the argument about "rates" of different astrophysical happenings in the heavens to know this. The discovery moment aside, the science is still full of uncertainties. A little sociological sensitivity reveals that it is all a matter of how one presents the science—how one "socially constructs" the meaning of the science, how the scientists and society think about the science.

THE ROLE OF SCIENCE IN THE DEMOCRATIC WORLD

The reason all this priggishness is worthwhile is that science is, potentially, hugely important to democracy—it is much more important than its findings, even findings as astonishing and inspiring as the one discussed in this book. It is even more important than the testament to human ingenuity, enterprise, and determination of over half a century that this finding represents. It is even more important than the change in the order of things that this discovery will engender—or, more properly, be constituted by. In their book *Leviathan and the Air-Pump*, Steven Shapin and Simon Schaffer argue that Robert Hooke's experiments with air-pumps, the results of which were violently contested by Thomas Hobbes, constituted an argument about democracy. Hobbes's famous view was that society was a leviathan whose head and source of all sound knowledge was the King; Hooke was showing that an individual in a laboratory could make sound knowledge irrespective of the King. For Hobbes, science promised disorder. The centuries since have worked that argument round and round, and even Shapin and Schaffer notoriously conclude that "Hobbes was right" because science is itself a collective process as subject to the concatenation of opinion, influence, and power as any other enterprise. The contemporary view among social scientists, like that expressed in my book *Changing Order*, upon which Shapin and Schaffer drew to some extent, is that science is a collective enterprise. The foundational role of science in democracy, then, can no longer be seen as celebrating the isolated individual as someone wiser than kings, for science has its own kings—and it better have its own kings. The foundational role of science in society is not to celebrate the individual but to act collectively as a beacon of values. We desperately need science to rescue democracy from greed; we need that even more than we need gravitational wave astronomy.

CONCLUSION AND REPRISE: WHAT KIND OF PHYSICS DO WE WANT?

This has been an account of one the greatest scientific discoveries ever—a discovery that can fill people, not least me, with emotion when confronted with the power and beauty of physics. Much to my surprise, however, after four and a half decades of loving gravitational wave physics, I now feel a little disenamored of it. I think this is more than jealousy of a triumph in which I, the sociologist, cannot fully share. I think it is because over the decades I have enjoyed the company of a community engaged in a near impossible task, exemplifying the CWI model of science, but now I see a triumphant community pulling away from the travails that beset the rest of us. Who can begrudge the physicists their moment—but triumphalism cannot provide a model for utopia.

That is why I am unhappy with the moment-by-moment correlates of triumphalism—the secrecy that expresses power and the desire to amaze with magic tricks, the relentless professionalism that is determined to muffle the sound of the second shoe dropping, the unwillingness to think about the possible meaning of the "zero-delay excess" now that a raft of 5-sigma results are in prospect because the more secure certainties from the next observation run are felt to be more the business of physics. I wish the scientists would handle these things in a way that keeps the provisional nature of their work at the forefront and therefore enables them to represent so much more than a freakish moment of revelation. Freakish moments, it is true, are the way physics justifies itself when it is not claiming to support the economy and replace religion; and no one questions that freakish moments are one of the things that make physics what it is. The billions of dollars spent on the Large Hadron Collider were justified by the Higgs particle, and the billion dollars spent on the search for gravitational waves is justified by The Event. But how utterly meaningless these things are to nearly all of us! Perhaps we are too close for this to be obvious in the case of The Event—after all, it has made a huge difference to my life and that of the scientists—but just ask what difference to your life, dear reader, was made by the discovery of the Higgs.

To repeat, what kind of physics do we want? The physicists now have another jewel in the crown of their subject; only a few gems in the entire history of science have glittered brighter. But treating what has happened as a jewel takes physics away from its potentially more important leadership role in democratic societies.[16] The crown jewel moments are, to

16. For an earlier argument between myself and Steven Weinberg that turned on this point, see Labinger and Collins, eds., *The One Culture? A Conversation about Science*, and, in particular, my "Crown Jewels and Rough Diamonds: The Source of Science's Authority," in that volume. For a book-length development, see Collins and Evans, *Why Democracies Need Science*.

repeat, quite atypical of what most science does and quite atypical of what even physics does. The physicists' view of the purpose of fifty years of noble argument and perseverance with integrity may be the jewel, but the jewel is a small thing compared to the work and effort. One cannot learn from a crown, however shiny it is; it is the effort and the integrity that made it that sets an example for society. Maybe the theme that has run through my account of gravitational wave physics from the beginning—certainty versus judgment—has found its point.

14 THE BOOK, THE AUTHOR, THE COMMUNITY, AND EXPERTISE

My contact with the field began in 1972 at the peak of the controversy over whether Joe Weber had detected gravitational waves with his room temperature resonant bars. The controversy over gravitational waves made up about a quarter of my PhD project in the sociology of science. In those days I conducted my research by going from laboratory to laboratory and interviewing the scientists. I conducted eight interviews in Britain and America in 1972 and a further fourteen in Europe and America in 1975–76. The work I did in 1972 gave rise to what became a well-known paper that secured my career. In 1985 I wrote *Changing Order*, basing it on three of the four sets of interviews I carried out in those years, with the gravitational wave research as the central case study. A much deeper engagement with the field began in in the mid-1990s. Between the mid-1990s and the mid-2000s I spent more time with gravitational wave physicists than with any other group including my fellow sociologists. In those years I traveled to pretty well every conference and workshop the physicists held, often completing more than half a dozen flights a year, many of them long-haul. During this period I got to know the community well and made new friends and acquaintances among them. Crucially, I got to like them and like their project. I felt comfortable among these scientists and privileged to be close to this extraordinary enterprise.

Relevant to this book is the analysis of expertise that has been central to my work for more than ten years. The crucial and most successful concept within that stream of work is "interactional expertise."[1] Interactional expertise is best exemplified by my understanding of gravitational wave physics during the period of my intense involvement; in that time, over coffee, lunch, and dinner I would talk gravitational wave physics with my new friends and acquaintances and do a pretty good job of it even though I was not a physicist myself: I could not do a calculation; I could not contribute to papers; I could not help with building the apparatus. Nevertheless, I thought that someone like myself, with interactional expertise, but, as the language went, without "contributory expertise," could understand the field to the point of being able to make reasonable technical judgments. I pointed out that peer reviewers and managers of technological projects are in a not dissimilar position. This led me to try it out. I took part in an "imitation game" in which a GW physicist asked technical questions—he asked seven in all—of me and another GW physicist. The dialogue—seven questions and seven pairs of answers with identities disguised—was then sent to nine other GW physicists who were asked to identify the participants, knowing one of them was me. Seven said they couldn't work out who was who, and two said I was the real physicist. So I passed![2]

Interactional expertise degrades if it is not continually refreshed by constant contact with the changing field of science or technology, or other domain, to which it pertains. Because the intensity of my engagement with gravitational wave physics has diminished since the mid-2000s, my expertise has also begun to fade. It was given a couple of boosts as I wrote *Gravity's Ghost* and *Big Dog*, but I don't think I ever quite regained the wide level of knowledge I had earlier.

Since the end of the Big Dog analysis, another three or four years have passed, during which time I have not been to more than about one meeting a year. I've fallen back further in my understanding of the field, particularly the detailed workings of the instruments. Fortunately the loss of that kind

1. See Note XV in "Sociological and Philosophical Notes," 371.

2. See Note XVI in "Sociological and Philosophical Notes," 373.

of knowledge has not been fatal; if I were trying to write a second edition of *Gravity's Shadow* it would have been more troublesome.

Luckily this book is not about building detectors but about the way a detection is confirmed once the signals emerge from the completed devices. That means I need a narrow body of understanding that, fortunately, I refined to a pretty high level in writing *Big Dog*.

But to leave it like that would be disingenuous. So I undertook another imitation game to see if it would reveal the fading of my expertise. B. S. Sathyaprakash, professor of physics at Cardiff, once more helped by inventing a new set of questions more suited to 2015. He set eight questions as follows.

Q1. Advanced LIGO and Virgo data contain two signals from identical binary neutron star systems with their coalescence times separated by just 1 second. Explain if you think it would be possible to disentangle the two overlapping signals by matched filtering.
Q2. Einstein Telescope is a possible future 3rd generation gravitational wave detector. It is conceived to be an underground detector using cryogenic technology. What sources of noise are being mitigated by going underground and using cryogenics.
Q3. An alert from LIGO–Virgo analysis is sent to astronomers a day after a transient event was found. What sort of telescopes (gamma-ray, x-ray, infrared, optical, radio) should astronomers use for follow-up and why?
Q4. A continuous gravitational wave signal is observed by a **single** LIGO detector with a high confidence. Explain if you think it will be possible to obtain the sky position of and distance to the source?
Q5. It is said that the pulsar timing arrays (PTAs) and laser interferometers are essentially based on the same principle of detection of gravitational waves. What is this principle and how does it apply to PTAs and laser interferometers?
Q6. An experimentalist suggests using 10 times greater laser power to improve the sensitivity of an existing interferometer. Assuming that mirrors are able to withstand such an increase in the power and neglecting mirror thermal noise how do you think the strain sensitivity of the detector improves at different frequencies.
Q7. LIGO and Virgo take data for 5 years during which there is a galactic supernovae, 200 short, hard gamma-ray bursts and 4 pulsar glitches but fail

to detect any signals. Under the circumstances, what are the implications of non-detection?

Q8. Two teams, E and N, came up with detector designs both of which had the same distance range for binary coalescences; the E team had better strain sensitivity with a low frequency cutoff of 5 Hz while the N team had relatively poorer strain sensitivity but a slightly lower low frequency cutoff of 1 Hz. Explain if you think they will both be equally good in measuring the parameters of a coalescing binary.

These questions were answered initially by me and three other gravitational wave physicists. This time I decided to make the test more elaborate and asked other kinds of people to answer too so as to obtain some comparisons. The "bottom line" is shown in table 14.1.

As can be seen, four groups answered the eight questions—counting Harry Collins as a group. There were three GW physicists, three astronomers/astrophysicists who worked in the same department as the GW physicists (referred to as "savvy physicists"), and two social scientists who were acquainted with my work on gravitational waves and with imitation game tests of this sort (referred to as "savvy social scientists"). Four groups also marked the questions; there were 72 questions and answers in a list randomized at the level of the individual question.[3] Markers scored the answers according to the following four-point scale giving a maximum of 32 points:

Knows GW physics: 4
Understands: 3
Unconvincing: 2
Does not know GW physics: 1

Starting with the left-hand column, we see that Collins did pretty well when marked by the GW physicists—better than I expected—scoring 25 points as opposed to the 27 point mean which the three GW physicists achieved. The effective similarity of these two scores can be seen by noting the wide separation between them and the scores achieved by the non-GW

3. Thanks to Luis Galindo for help on this. See Note XVI in "Sociological and Philosophical Notes," 373.

Table 14.1
Overall outcome of new gravitational wave imitation game

		Who marked the answers?			
		4 GW physicists	2 Savvy social scientists	2 Social scientists	Harry Collins
Who answered the questions?	3 GW physicists	27	27	19	23
	3 Savvy physicists	19	23	17	13
	2 Savvy social scientists	17	20	19	11
	Harry Collins	25	27	20	28

specialist physicists and the social scientists. I believe, however, that my achievement in this test underestimates the erosion of my expertise over the last few years.

This erosion is a bit more evident if we look at the last column. Here I mark all the other answers and mark my own as well. I should add that the marking was done four months after I had answered the questions, and I recognized only two of my own answers. That I gave myself such a high score is, nevertheless, understandable, since one is bound to think one's own answers are right. Note also that I mark in a roughly similar way to the GW physicists. Recognizing what counts as a right answer is nearly as much of an indicator of understanding as providing the answer in the first place, so the fact that there is wide separation between my marks for the GW physicists and those of the physicists and savvy social scientists is a plus for me. The gap between the marks I gave myself and the marks I gave the GW physicists, however, reveals the erosion of my expertise. To reiterate, part of this gap is due to the fact that one tends to mark one's own questions well; if one didn't believe them one wouldn't have given them in the first place. Comparing the marks I gave myself and the marks the GW physicists gave me, I think I overmarked myself by two or three points. But I also investigated my low marking of the GW physicists very carefully, and it does show that in a couple of cases I did not understand the physicists' answers because of the weakening of my contact with the field. In the case of one answer, I just marked carelessly, but I can tie this down to the erosion of my understanding in the case of two other answers. If, over the last few years, I had been hanging around with the GW physicists as intensely as I did in the 1990s and early 2000s, I would not have made these mistakes, and being, as it turns out, an overall low marker—we all have different marking tendencies—I would have given the physicists a rounded average mark of 25 instead of 23 plus another point for careless grading of one question. So that is a measure of the erosion of my understanding as measured by this test.

I investigated the answers to question 2 the most carefully because I thought I had the answer right and the physicists had it wrong. But it turned out, after a lot of inquiries, that the account of gravity gradients that I had in mind, and in my account was the most important reason for building underground, was incorrect. It had been correct a decade earlier when I was more deeply embedded in the science, but I had missed the change in understanding that had happened over those ten years. The difference is that for me gravity gradients were mostly about changes in gravitational forces on the mirrors caused by the changes in air density associated with wind, and this can be mitigated by going underground to a shallow depth; as time went on, this effect was found to be small, and the serious problem has come to be seen as changes in gravitational attraction caused by minuscule ripples in the Earth's surface associated with seismic noise. To mitigate these one needs to go much deeper, and going that deep is not on many people's minds as there are other ways of compensating for the effect. We can see the difference in these two quotations from the literature from 2000 and 2011 respectively:

> At such low frequencies environmental effects, and particularly gravity gradients associated with tides and weather variations in the surrounding environment, create perturbations which greatly exceed the desired signal. (Ju, Blair, and Zhao, "Detection of Gravitational Waves," 1326)

> The dominant source of gravity gradients arise from seismic surface waves, where density fluctuations of the Earth's surface are produced near the loca-

tion of the individual interferometer test masses. (Pitkin, Reid, Rowan, and Hough, "Gravitational Wave Detection by Interferometry," 13)

To repeat, I'd missed this change and that is why I thought the physicists were all wrong and I was right. That is a kind of iconic example of what happens as one's interactional expertise begins to degrade and why it is bound to happen if one is not continually mixing with the community.

I found that apart from the gravity gradients problem, I was also not so good on the relationship between low frequency and determination of chirp mass and sky position. In retrospect, I can see that if I had not been following things closely since September I would have been out of date on the optimum length of time slides (see chapter 4) and the status of the idea of a freeze on detector states. But my background knowledge is such that in respect of all of these technical matters, I could have caught up with a short conversation.

Going back to the overall marks, we can see that the non-GW physicists, even when marked by GW physicists, did not do much better than the savvy social scientists, and the separation of GW physicists plus Collins on the one hand and non-GW physicists plus social scientists on the other is quite clear. This little test reveals how specialized science is—there are no "science experts in white coats." It is made more striking by the fact that the savvy physicists were pretty savvy. The one who did best wrote to me later:

> For what it's worth I based my answers on what I've learned over the years from seminars by GW people, papers that I've read as a member of peer review and policy committees, and reading semi-popular (Scientific American level) articles. But also the questions about noise and sensitivity made some sense to me as my own field is experimental (electromagnetic) astronomy.

In spite of this, the physicists' score was similar to that of the social scientists when marked by GW physicists. Furthermore, the two mean scores, given that there were eight questions, were 2.4 and 2.1—both hovering only just above "unconvincing" on the grading scale. So it was not so much that the savvy social scientists were good, but that the savvy physicists were not very good!

Looking at the middle columns, we see that the ability to distinguish between the classes diminishes as the markers grow less expert—this is exactly what we would expect; looking at the third column we see that ordinary social scientists mark everyone roughly the same. This shows that "the public" is very easy to fool; all "experts" look the same to them.

Finally, sometime after the main exercise was completed, I asked two physicists from a university where gravitational waves are not a central concern to mark the test. Their specialties were theoretical optical physics and particle physics, respectively. If presented as one of the columns in table 14.1 their average rounded scores would be 22, 18, 18, 18, which, in terms of discrimination between the experts and the nonexperts, is closer to the nonsavvy social scientists than any other group. The two sets of marks, however, were rather different, with the second mark being close to random; so it seems right to report them both. They were 25, 18, 18, 19, and 19, 18, 19, 17. The first set is the only one to make a marked separation between Collins and the GW physicists; this is puzzling, and it may be that stylistic features were playing a role.

Going back to the question of my fading understanding, expertise is multidimensional, and erosion in respect of one of the less obviously technical dimensions is much harder to rectify—it cannot be rectified by a few minutes of conversation. In *Rethinking Expertise*, Robert Evans and I pull out sixteen components of expertise or judgments about expertise which, however, we have found we have to modify. In later papers, we explain that the main components of technical understanding, interactional, and contributory expertise need further refinement. A vital component of these kinds of expertise is an understanding of the credibility of those who are making expert claims. This is easily seen in the "meta-expertise" rows of what we call "The Periodic Table of Expertises"—the rows that cover ability to judge between different experts and enable one to discount certain politicians and salespersons; judge between astronomers and astrologers; judge, or unfortunately fail to judge, between doctors and vaccine-scare mongers; and, most usefully, judge between scientists paid by the tobacco and oil industries to deliver certain results favorable to them and those who are driven by scientific values. But that kind of metajudgment also makes up a component of the "purely" technical expertises.

Our standard and oft-repeated example is, once more, taken from gravitational wave physics. In 1996, Joe Weber published a paper claiming to have a found a correlation between gamma-ray bursts and the gravitational waves he had found in earlier years. I went around the community asking scientists what they made of this paper. I discovered that I was the only person to have read it. The "technical" judgment being made was that Weber's credibility had now fallen so low that this, in spite of its perfectly respectable appearance, was a "nonpaper." The physics preprint server, arXiv, uses a computer algorithm to screen submissions. In 2015 we discovered that this paper, albeit twenty years old, still passed arXiv's screening process without problem. Indeed, to outsiders the paper has all the appearance of a potential Nobel-Prize-winning contribution to physics.[4]

In a 2011 paper, Martin Weinel and I call the kind of expertise needed to make the judgment that Weber's 1996 paper should be ignored "Domain-Specific Discrimination" (DSD) and define it as "the 'non-technical' expertise used by technical experts to judge their fellow experts."[5] Apart from what has been shown by my inadequate marking of the GW question test, I have lost the ability to make this kind of judgment in respect of a good proportion of the community as result of my years of distance from the field; this is where I *feel* the loss of expertise most strongly. Given what I already know, the technical losses can be patched with a few email inquiries or telephone discussions, but you cannot get to know the competencies and biases of a community like this—that takes years. Over the last few years many new people have entered the field or shifted from relative obscurity to positions in which they are making significant contributions.

4. The paper is Weber and Radak, "Search for Correlations of Gamma-Ray Bursts with Gravitational-Radiation Antenna Pulses." See Note XVII in "Sociological and Philosophical Notes," 375.

5. Collins and Weinel, "Transmuted Expertise: How Technical Non-experts Can Assess Experts and Expertise." See Note XVII in "Sociological and Philosophical Notes," 375.

Even when I wrote *Gravity's Ghost* and *Big Dog*, both of which depended heavily on my perusal of email traffic, I knew pretty well all the people behind the emails and I knew their political and technical stakes in the matter—I could see where people were coming from and assess their contributions accordingly. This time around I do not know what half the contributions mean in the sense of *domain-specific discrimination*—I do not grasp how seriously to take what is being said because I do not know the person saying it and *why* he or she is saying it. Thus, when I went through the 104 emails that I saved rather than deleted on the first Monday and Tuesday of the appearance of G150914, I found there were 45 separate emailers of whom I roughly knew only 20. A few years ago that number would have been, perhaps, 40. That makes a lot of difference to how one understands a field.

Here is an example when I remark to one of the scientists:

> I found the argument that [A is B because P is Q] very weird

to which came the response

> A very weird argument indeed from XXXX, although perhaps not so weird considering it was from XXXX ;-)

On the other hand, in another conversation I am told:

> I think before we're done we're are going to have to understand whether there is any credibility to that… and I think that's going to be a struggle because YYYY is a really smart guy and he's pretty self-confident and he will say he believes it and people have enough respect for him that they will not blow him off so I don't know how we're going to resolve that. (Interview with Peter Saulson, Sept. 28)

As it happens I know the person referred to in the second comment well enough to understand and appreciate what is being said. But I don't know the person who is the subject of the first comment; and in general this is the kind of thing that I am finding much harder to work out for myself regarding GW50914 than was the case for Big Dog—I am, to repeat, losing Domain-Specific Discrimination. It is easy to see how DSD affects face-to-face communication. In another place I show how lack of familiarity causes me to make a mistake over the body language of someone with whom I am heatedly discussing an issue, but what has become clear to me since the event is that the same applies to the interpretation of emails.[6] The written word is not just written: its meaning rests on a raft of previous social interactions. This is a very important point, especially these days when communication among the younger generation appears to rest more and more on social media.

6. The other place is a book I am writing now, to be titled "Artifictional Intelligence."

POSTSCRIPT
The Beginning of Gravitational Wave Astronomy

It is now May 2016. Here I describe the final, detection-related, scientific acts of the first observation period (O1) of Advanced LIGO. We are back with the first ripple! O1 lasted 106 days—up to the January 12, 2006. During this time the interferometers were in a suitable state to make scientific observations for around 1,100 hours, or nearly 46 days. I think the results of these observations have been presented the wrong way. We have a second discovery paper devoted to the Boxing Day event, GW151226, which will be submitted to *Physical Review Letters* as the confirmation of The Event and which, as I later learn, will be published on June 15; and we have a longer paper—"The O1 paper"—which will be promulgated on arXiv on the same date and will sum up the results of all the observations over the period of O1.[1]

I think the second major paper should not be a confirmation of The Event—I think that the fact that The Event has been confirmed by a second observation should have been said in the initial discovery paper, since that "second shoe dropping" was, in fact, what made everyone so confident that The Event was the equivalent of an "Omega-minus" rather than a "monopole." That the exact parameters were not yet worked out did not matter as far as that confirmation was concerned. Here the reporting

1. There are many companion papers being written at the same time (see https://www.ligo.caltech.edu/page/detection-companion-papers), but here I am concerned only with discovery papers.

sequence is falsifying history; the secrecy impulse is too strong. The second major paper published should be a summary of the discoveries made in O1 along with the parameters. As we will see, it is now a matter not of simply an event and a confirmation, but of the beginning of gravitational wave astronomy, with *three* events to be reported in detail, not two; the extra event is Second Monday, which has continued to grow in social significance. The crucial figure and the crucial data, which reveals and compares these three events that comprise the start of gravitational wave astronomy, are to be found in the O1 paper under preparation; but they should be in the second announcement paper.

The physicists who are still part of this project know that the only way to move big science forward is to argue, reach agreements one way or another, and then work with the agreements, whether or not they believe they are optimum (see *Gravity's Shadow*). That's what's good about being an outsider—I am not bound by the agreements and can still say where I believe things went wrong. I think things started to go wrong at the teleconference I monitored during the meeting in Birmingham on January 19 (see p. 207), when I was horrified to find that there was a growing movement to keep the Boxing Day event secret even though it had played such a large part in building confidence in the reality of The Event. To be charitable, this movement was driven by relentless professionalism; to be less charitable, it was driven by a sense of power—"This is our finding and no one is going to know about it until we are good and ready." If scientists would see their job as telling everyone as much as they can consistent with not being scooped, all the ambiguities and problems would go away. Let us now go back to the sequence of events as they are unfolding.

At the beginning of May the community is having a big argument about the figures in the Boxing Day Event paper—GW151226. Some are echoing my feelings of discomfort over the figures that I mentioned in chapters 8 and 11 (for example, see figure 11.7): people are asking about the scientific point of presenting a figure and using omega scans that shows only that nothing can be seen! In draft 6 of the paper, the omega-plot figure is still there, but now there are some waveforms too; these are reconstructions from templates and what they are meant to be telling us is far from

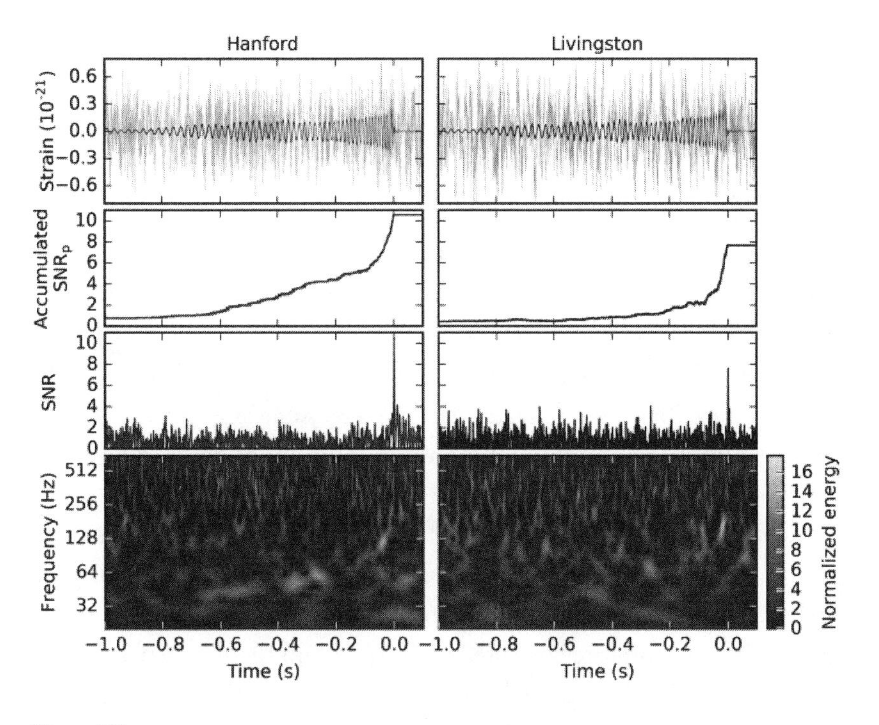

Figure 15.1
Figure 1 from the submitted GW151226 paper.

clear—at least to me. There is so much disagreement over how to present the figures that the community eventually decides to vote on it, but the vote, as usual, will not be decisive. This argument will go all the way to the wire—the moment that the paper is submitted to *Physical Review Letters.*

Figure 1 of the GW151226 paper as submitted is shown here as figure 15.1. The caption is long, but here is an extract:

> First row: Strain data from the two detectors, where the data are filtered. … Also shown (black) is the best-match template from a non-precessing spin waveform model reconstructed using a Bayesian analysis [20] with the same filtering applied. … Second row: The accumulated peak signal-to-noise ratio (SNRp) as a function of time when the template optimally overlaps with the signal waveform. The maximum cumulative SNRp corresponds to the peak in the next row. Third row: Signal-to-noise ratio (SNR) time series

produced by time-shifting the best-match template waveform and computing the integrated SNR at each point in time. The peak SNR is recovered when the template best overlaps in time with the signal waveform. Fourth row: Time-frequency representation [45] of the strain data around the time of GW151226.

The paper itself tells us: "The time-frequency representation of the detector data [top row] shows that the signal is not easily visible," which, indeed, it is not. The well-defined waveform that we see there is the template; it is the outer set of ill-defined spikes that is the signal. The second and third rows indicate that the cumulative signal-to-noise ratio is quite high, and that is where the statistical significance comes from. The omega plots in the final row serve no useful purpose that I can understand.

I wrote that last sentence toward the end of May. Since the relationship of my expertise to that of the scientists is a continuing methodological theme, let me say that in this case my judgment is confirmed by a referee; referees' reports are circulated on June 4. I am flipping between dates between June 4 and the end of May, while by the time I finished these passages I knew that the publication date was going to be June 15. I hope the changes of time and tense are not too confusing. Regarding the fourth row of the paper's figure 1, "Referee A" remarks (June 4):

> What are the authors' goals for this row of the figure? It appears that the authors hope to convey just how weak this signal is, especially when compared with GW150914. The figure caption only contains a terse sentence describing what these two panels show; the text of the paper just says that these panels show "that the signal is not easily visible." [And, referring to this row in particular, the caption adds: "In contrast to GW150914 [4], the signal is not easily visible."]
>
> I believe this row of Figure 1 could be omitted altogether. The top row already makes it clear this is a relatively weak signal; the second and third rows make it clear that the correct matched filter does an excellent job pulling this signal out of the noise. The fourth row is excellent fodder for a deeper discussion in a companion paper describing what a time-frequency

analysis shows (or fails to show), but does not add much to what is already described and discussed elsewhere in this paper.

As it happens, the fourth row is *not* removed from the version of the paper that is accepted for publication on June 8,[2] and I write to a member of the paper-writing team asking why. She replies (June 10):

Hi Harry, This point was raised by others in the collaboration. But just as many spoke out in favor of the panel. The reasons were that this figure was a key figure for GW150914 PRL and would serve as a nice comparison for those referring back. In addition, some outreach materials will use these spectrograms so it would be nice to also have them in the PRL. And finally, we are trying to teach our readers that we can't always see the signal in the data without matched filtering; thus, while we won't show this figure for future GW151226-like detections, we should show the figure for this first case.

So, it's not quite science that is at stake here.

The problem, of course, emerges from the relative weakness of the signal; it simply doesn't stand out, or even show up at all, when presented as a figure. That is what row 1 of the figure submitted for publication already shows. The way the template is matched to what looks like nothing but noise is as follows. First, the initial data is filtered to remove anything nonrandom and understood—such as the spikes caused by vibrations in the mirror-support wires—and filtered to restrict it to the waveband of interest. After this, what remains should be random if there is no signal in it. If, however, a template-like pattern has been impressed upon the data by some event in the heavens, the departure from randomness can be detected by overlaying a template on the data and multiplying the data by the template, then adding up all of the points in that product of data × template. If there's a small signal hidden in the data that matches the template, this

2. Abbot et al., "Observation of Gravitational Waves from a Binary Black Hole Merger," http://journals.aps.org/prl/abstract/10.1103/PhysRevLett.116.241103; http://arxiv.org/abs/1606.04855.

multiplying step will cause the signal to make a positive contribution to that sum of all points, while in the case of purely random data, everything in the sum of all points will cancel out on average. The procedure is carried out for the whole template bank at every point in the incoming signal. That is "matched filtering."

That weakness of the signal also reveals itself through varying statements about the length of the signal; in the first draft of the Boxing Day paper it was five seconds, whereas in the eighth draft it is only one second! Peter Saulson explains to me:

> Recall that a key feature of inspirals is that they evolve slowly at low frequencies, and more quickly at higher frequencies. That is because they have greater gravitational "luminosity" when the stars are moving more quickly. So that means that what we call the duration of the signal depends very sensitively on when we start counting. (The endpoint is dramatic and unambiguous, by comparison!)
>
> For our purposes, we consider our signal to have started when it reaches a frequency where the noise is "low enough." But that is a judgment call. For purposes of the search itself, in O1 we took the lowest frequency of interest to be 30 Hz. But for purposes of parameter estimation, the low end of the useful frequency band was taken to be 20 Hz. I think that is a rational judgment of the extra information that can be obtained from looking at a much longer template, even if it doesn't add much to the SNR.
>
> I'm pretty sure that the difference in the quoted durations depends on whether one is talking about the search or the post-detection parameter estimation calculations.

So this is a long weak signal in which the signal-to-noise ratio builds up and you can pretty well choose where it begins. The shorter choice is used to calculate the probabilities. But once the template has been chosen one can investigate the longer "tail"—a tail that in this case is attached to the front end of the animal—where the match between signal and template is good enough to allow the signal's properties to be estimated using much more information from a much longer stretch of data.

Referee A of the paper, the one who would be happy to see the omega plots dropped, spends some time trying to work out what the point of the paper is. The referee agrees that the paper confirms The Event, but his or her remarks, I believe, would apply better to a paper covering the whole of O1:

> The LIGO–Virgo Collaboration has demonstrated that the celebrated discovery event GW150914 was not a fluke: binary black holes merge sufficiently often and sufficiently nearby that we can expect multiple additional detections with further observing runs, and as the detectors' sensitivities continue to improve. The two high significance events GW150914 and GW151226 (supported by the lower significance event LVT151012) clearly demonstrate that these detectors have discovered a population of merging binary black holes. These two (or three) events already point to substantial dispersion in the properties of these merging binaries. Further observations and detections will teach us a tremendous amount about this population. As has long been promised, gravitational wave astronomy has enabled the study of an entirely new class of astronomical objects which cannot be studied in any other way.

Isn't this the proper topic of a paper that deals with the three events on the table?

Referee A also thinks the paper fulfills another very important purpose: to help readers to understand the kind of matched-filter analysis that is needed in the case of this kind of weak signal—the kind of signal that will be typical of future observations.

> The instantaneous amplitude of the signal is smaller than the noise, and can only be found with confidence by using a matched-filter analysis that cross-correlates the data with waveform models. As such, GW151226 is much more similar to what the community had long prepared for, and serves as an important prototype for much of the future of gravitational-wave astronomy. This event validates the data analysis and event characterization process for events that have low amplitude but many coherent cycles.

> It would be useful to note at this point that the strain amplitude is not only smaller than GW150914, but is in fact smaller than the noise amplitude at all frequencies [this is shown in row one of the figure]. This clarifies why a matched filter is needed: the analysis has to coherently integrate the signal's power over the 55 cycles spent in band in order to find a significant signal in the noise. ... This is such an important point (indeed, this point is a major driver of this paper's significance) that I would recommend finding a way to work it into the abstract as well.

Again, it seems to me that this kind of explanation would be better pursued in the context of the *three* events detected in O1, where it, and associated matters, could be applied to the even weaker—and therefore even more revealing, in data analysis terms—event of Second Monday as well as GW151226. What is needed, it seems to me, is a discussion of how signals of a whole range of different patterns and strengths are analyzed and assessed.

Going back to May, the community is also arguing fiercely about the title of the GW15126 paper. At one point they decide to take a vote though it does not quell the argument, which is still going on even as version 8, which is meant to be the submitted version, is being circulated. The argument turns on whether this paper should describe itself as confirming the original detection of gravitational waves; if that is what the paper is about, it calls for the title to include a phrase such as "second detection of gravitational waves." But it is pointed out that the Boxing Day event is probably not the second observation, it is the third—*if it was real,* Second Monday was the second observation!

The world of gravitational detection is changing before our eyes in the ways described in chapter 12. One of those ways is that the status of Second Monday is changing. If it had shown up before September 14, Second Monday would have hardly been noticed—it would have been noise. But its status is steadily growing as gravitational waves become more and more real and more detectable. This change is nicely brought out in an email from a respondent to whom I've written asking about the status of near noise events in the data:

Collins, May 22: I find myself out of kilter with you guys over the importance of weak signals—eg I was more excited by 2nd Monday than anyone else because you thought the significance was too low to make it interesting whereas I thought the "priors had changed" with the detection of The Event to such an extent that though 2nd Monday would not constitute a self-contained "discovery" it was still very interesting. But you guys reserved this degree of interest for the Boxing Day event.

Respondent: Yes, but it is interesting to see the ambivalence about Second Monday, isn't it? We hardly mentioned it in the GW150914 discovery paper, but whenever I give a talk and show the search result plots, I point it out and say that it is *probably* a real event, i.e. more likely to be real than to be a noise fluctuation. We just did not want to "claim" it as an event, to be conservative. But now, in the O1 BBH paper, it is analyzed and treated almost equally to GW150914 and GW151226. It certainly has an effect on the rate estimation. There is still some awkwardness about how we count events, e.g. with the new PRL title being "Observation of GWs from a second BBH coalescence." Did you follow the discussion thread of a number of people saying we should avoid calling it the second one in the title, and instead find another way to give it an identity (such as telling its mass)?

By the way, as predicted in chapter 10, the estimated rate of black hole inspirals in the heavens—the estimated outer limits of possible events per unit volume of space expressed as $Gpc^{-3}\ yr^{-1}$—has changed. Instead of the fiercely defended 2–400 of the discovery paper—fiercely defended against the *6*–400 that was nearly submitted—it has become 9–240 in the O1 paper! The huge argument that nearly "scuppered" the submission of the discovery paper was about a number that was, indeed, ephemeral. And it will continue to change as more detections are made.

As the respondent quoted above remarks, various ways of finessing the second/third problem are tried. Perhaps the best one is to remove "second" from the title and say something like "Detection of Gravitational Waves from a 20-Solar Mass Black Hole Inspiral," which is still telling us that there has been another observation of novel kind but without claiming it to be the second. But here is a contrary view in an email of May 23:

> When we were originally planning publications for GW151226 and the O1 BBH search, we decided on publishing a standalone PRL because the second clear detection is important confirmation of the first. As [XXXX] explained in her mail, that was the point the title is intended to emphasize. For most readers I can believe that this fact will be more significant than the mass.

In the paper submitted on May 29, the GW151226 paper does have a title that avoids the second/third issue and completely ignores the vote on what the title should be. It is:

> GW151226: Observation of Gravitational Waves from a 22 Solar-Mass Binary Black Hole Coalescence.

The adoption of this title, turning on the masses as it does, is explained as follows:

> We made this decision based on a number of email ... comments that suggested that, with this title, the fact that this was our second detection would be implicit. Additionally, the use of "second" seems not to be common in scientific discovery papers. Finally, the point that this is our second detection is made a number of times throughout the paper. We did not make this decision based on considerations of LVT151012 which we feel is properly represented as the third most significant *detection*. The selection of "22" came from rounding the initial m1+m2 numbers to 2 significant figures and our distaste for using a tilde in the title.

Going back to the central point, in the new era there are going to be many events that don't reach the 5 sigma criterion; but, as the social existence of gravitational waves grows stronger and stronger, these weak events will demand more and more attention, just as weak events are used in astronomy. It is, as we have seen, this transformation that is causing all the trouble for the *PRL* paper: in the new era, Second Monday won't go away, making GW151226 a second or a third event depending where your head is. When we watch the arguments caused by the term "second" in the title of the *PRL* paper we are watching the social construction of physical reality (of Second Monday) as "directly" as when we watch the changing strength of the mirror feedback signals in the interferometers and take them to represent the collision of black holes.

We don't know how things will play out in public relations terms, but in scientific terms it is the O1 paper that seems to me to be the really exciting one because it shifts us into GW astronomy. Figure 15.2 shows the wonderful figure 1 taken from an early draft (the final posting is Abbot et al., "Observation of Gravitational Waves," http://arxiv.org/abs/1606.04856).

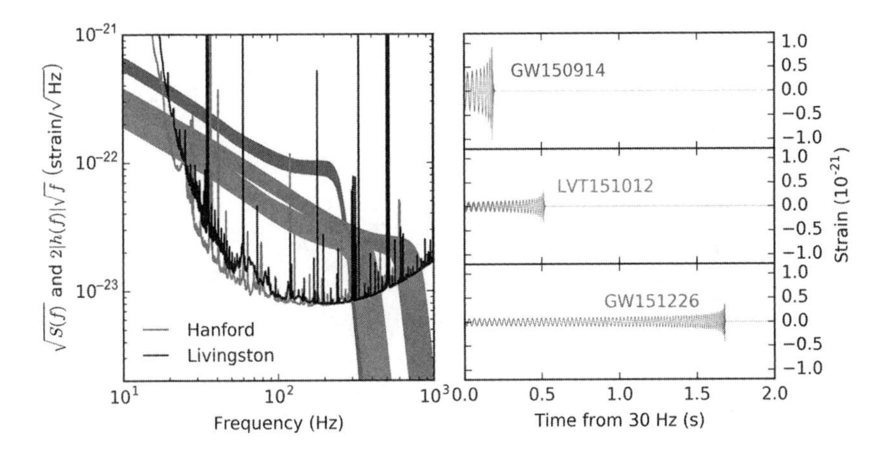

Figure 15.2
The wonderful figure 1 from the O1 paper.

Start with the panels on the right: they are the three waveforms of events detected so far. The top one is the short, energetic "Event" that stood out so clearly from the noise; the second one down is Second Monday, too weak to be counted as an event on its own but here grown into social acceptance; the third is GW151226, with its slow build-up of signal-to-noise ratio that gives it independent statistical significance. Figures like this are the future of gravitational wave astronomy—they show the different things that can be found in the heavens. In this case all three are inspiraling binary black holes, but the future will see more and more variegated waveforms; and many of them, like Second Monday, will be well below the 5 sigma needed for an independent discovery. To repeat, that is how the world is changing. Incidentally, relentless professionalism has stopped Second Monday being granted "GW" nomenclature and it is stuck with "LVT"—standing for "LIGO–Virgo trigger"—before its date. This is how it is described in the O1 paper:

> The significance of this event is such that we do not confidently claim this event as a gravitational wave signal. However, it is more likely to be a gravitational wave signal than noise based on our estimate for the rate of gravitational wave signals. . . . Detector characterization studies have not identified an instrumental or environmental artifact as causing this candidate event.

The left-hand panel shows the sensitivity curves for the two detectors at Hanford and Livingstone with all their noise spikes and shows the frequency evolution (from right to left—high to low frequency) of the three events, all standing well above the sensitivity threshold for a substantial frequency range. The thickness of the lines represents error bars, and the area between the lines and the sensitivity curves is proportional to signal-to-noise ratio. Second Monday is, of course, the bottom curve.

Table 15.3, extracted from the O1 paper, shows the parameters, with uncertainty intervals, for the three sources. As can be seen, Second Monday is a big event, radiating energy associated with conversion of a mass of around 1.5 suns.

It seems to me that all this material should be in the second paper published—the second major announcement.

Table 15.3

Event	GW150914	GW151226	LVT151012
Signal-to-noise ratio ρ	23.7	13.0	9.7
False alarm rate FAR/yr^{-1}	$< 6.0 \times 10^{-7}$	$< 6.0 \times 10^{-7}$	0.37
p-value	7.5×10^{-8}	7.5×10^{-8}	0.045
Significance	$> 5.3\,\sigma$	$> 5.3\,\sigma$	1.7σ
Primary mass $m_1^{\rm source}/{\rm M}_\odot$	$36.2^{+5.2}_{-3.8}$	$14.2^{+8.3}_{-3.7}$	23^{+18}_{-6}
Secondary mass $m_2^{\rm source}/{\rm M}_\odot$	$29.1^{+3.7}_{-4.4}$	$7.5^{+2.3}_{-2.3}$	13^{+4}_{-5}
Chirp mass $\mathscr{M}^{\rm source}/{\rm M}_\odot$	$28.1^{+1.8}_{-1.5}$	$8.9^{+0.3}_{-0.3}$	$15.1^{+1.4}_{-1.1}$
Total mass $M^{\rm source}/{\rm M}_\odot$	$65.3^{+4.1}_{-3.4}$	$21.8^{+5.9}_{-1.7}$	37^{+13}_{-4}
Effective inspiral spin $\chi_{\rm eff}$	$-0.06^{+0.14}_{-0.14}$	$0.21^{+0.20}_{-0.10}$	$0.0^{+0.3}_{-0.2}$
Final mass $M_{\rm f}^{\rm source}/{\rm M}_\odot$	$62.3^{+3.7}_{-3.1}$	$20.8^{+6.1}_{-1.7}$	35^{+14}_{-4}
Final spin $a_{\rm f}$	$0.68^{+0.05}_{-0.06}$	$0.74^{+0.06}_{-0.06}$	$0.66^{+0.09}_{-0.10}$
Radiated energy $E_{\rm rad}/({\rm M}_\odot c^2)$	$3.0^{+0.5}_{-0.4}$	$1.0^{+0.1}_{-0.2}$	$1.5^{+0.3}_{-0.4}$
Peak luminosity $\ell_{\rm peak}/({\rm erg\,s}^{-1})$	$3.6^{+0.5}_{-0.4} \times 10^{56}$	$3.3^{+0.8}_{-1.6} \times 10^{56}$	$3.1^{+0.8}_{-1.8} \times 10^{56}$
Luminosity distance $D_{\rm L}/{\rm Mpc}$	420^{+150}_{-180}	440^{+180}_{-190}	1000^{+500}_{-500}
Source redshift z	$0.09^{+0.03}_{-0.04}$	$0.09^{+0.03}_{-0.04}$	$0.20^{+0.09}_{-0.09}$
Sky localization $\Delta\Omega/{\rm deg}^2$	230	850	1600

THE RESPONSE OF THE FRINGE

At last, on June 15—six months too late as I see it—I am able to write to Reg Cahill, author of one of the viXra papers and longtime correspondent, and ask him what he makes of the replication of the initial Event result. I also write to Stephen Crothers, author of the other viXra paper, but with whom I have not corresponded before.

Crothers replies immediately (June 16) with a long email, from which I reproduce the first sentences:

> Dear Harry,
> There is no detection of Einstein gravitational waves or black holes by LIGO in this latest report, just as they did not detect them in their first report. They must make such reports in order to justify their existence. With such large sums of public money involved on "experiments" that cannot ever be independently replicated, they must always find what they wanted to find, and so they do, not by brains and laboratory, but by mass-media induced mass-hysteria, since they are bereft of science.

Reg Cahill wants to know the exact time of GW151226 so he can look for an occurrence of his alternative kind of gravitational wave event at the same time. I send full details and he replies on June 17:

> Dear Harry,
> I looked at the GCP data using diodes (Quantum Gravity Detectors). There are no strong correlations with the time of the 2nd LIGO event. So I have no useful comment to make about this LIGO claim. However it is very curious that the Australian Defence Dept. ordered me to stop my gravitational wave experiments one week before the 1st LIGO announcement.

THE "ZERO-DELAY EXCESS"?

Further interchanges with respondents indicates that the intriguing orange square or squares near the tail of "the line" of events, which I discussed on page 251 above as indicating a "zero-delay excess," is/are too weak to be counted as anything but noise. I would love this feature to be a Joe-Weber-type "zero-delay excess," and it just could be because the sheer number of events seems a little higher than it should be—at least according to one of the pipelines; the other pipeline shows no excess.

But because interferometery is "broadband"—it can plot the details of each excursion in detail—it was possible for the scientists to examine each of the eight "signals," numbered 4 to 11, that added up to the orange box or boxes. I should add that this happened late in the day and no one was very excited about it; it happened only after a lot of prodding from one or two people. The scientists are looking forward to the strong events that will be forthcoming in O2 and about which they will be able to say definite things—there is a strong feeling that close examination of very marginal events is not a good use of time. One scientist, however, gave what he described to me (personal correspondence, June 13) as "my (rather informal) reactions ... a first cut at assessing the events based on easily available information, which, if there was more interest in them, might be modified by other investigations." Given this, when each of these proto-events was examined in detail, it was found that 7 out of the 8 were "more likely than not" to be noise because there was little coherence between the two detectors or because they occurred during periods when a few glitches were seen too, though not always enough for a definite veto. One, however, "Event 9," had slightly better credentials. Here is an edited version of the brief technical report on this excursion:

> Event 9, 1128348574.48
>
> Both SNRs between 6–7, rchisq very close to 1, time delay 3ms, phase difference 3.9. Fairly close to pi so reasonably consistent with signal.
>
> Trigger series doesn't look particularly glitchy
>
> L1 omega scan looks somewhat suggestive of blips of power lying along a chirp trail: https://ldas-jobs.ligo-la.caltech.edu/~tdent/wdq/L1_1128348574.487/ (H1 omega scan is just blue with no visible features)

Event is basically too quiet to say anything definite (note that IFAR after trials factor is 0.016yr ~ 60/yr).

Later, the possible parameters of the proto-event were worked out and reported on June 10:

John actually ran PE on the remaining 1 (very quiet) event... aside from having a very wide range of m1 values (and a correspondingly large range of chi_eff) it looks fairly OK... for a quiet high-mass NSBH event.

Looking at the corresponding website, the masses of the components seem to be between 20 and 60 solar masses and at the lower end of 1 or 2 solar masses, respectively. The range of possibilities for mass 1 is very large given the weakness of the signal, if signal it was.

Returning to the discussion of "image and logic" (see chapter 7), there is no "logic" insofar as the statistical significance of this event is close to zero—noise coincidences of this energy level can happen often; there is, literally, no statistical significance "to speak of." Nevertheless, it would be entirely reasonable that the occasional very weak yet real event would show up in the noise; indeed, we can say that as gravitational wave astronomy unfolds into the future there are certain to be some very weak but genuine signals—this does not prove that this is one, but it could be. Furthermore, this one has an "image" that is consistent with its being real, the image of a neutron-star black-hole inspiral—something very much expected but not yet seen and therefore still more interesting. But this is a blurred "image"—something much closer to seeing pictures in fire than physicists like, especially this community. And bear in mind, of course, that this "image" is (a) constructed from numbers and (b) has been drawn from the data by a numerical process of fitting it to a bank of other images—even more like seeing pictures in the fire.

And yet, something like this must happen from time to time. For you readers who have been trying to absorb something of the regular perspective of the physicists described here—who are trying to think like "we (mainstream) physicists"—your decision about how much attention to pay to this event is almost as good as that of the physicists themselves; you are in as good a position to answer the question asked of the O1 paper coordinator by the scientists who worked through events 4–11:

> Is the investigation going to result in any specific statement being made in the paper or to the people who asked the questions? Or is it extra work to back up the usual boilerplate statement that I see on p. 4, i.e., "There were no other significant BBH triggers in the first advanced LIGO observing run. All other observed events are consistent with the noise background for the search."

You can see that, mostly, the physicists will want to stick with the boilerplate statement; they cannot *know* it was not noise, and in this field that is still a major sin. Eventually, a sentence is added reflecting the more detailed search of the excess coincidences:

> There were no other significant BBH triggers in the first advanced LIGO observing run. All other observed events are consistent with the noise background for the search. Follow up of the [eight near noise] coincident events ... suggests that they are likely due to noise fluctuations or poor data quality, rather than a population of weaker gravitational-wave signals.

This seems to me to be a bit more negative than it needs to be given the reservations of the person who analyzed the signals. But you can also see, maybe, that as gravitational wave astronomy develops, this kind of thing is going to crop up continually and the bounds of what counts as interesting is going to fall farther and farther into the noise. How far into the noise is something for the future: my guess is a long way, and that is because gravitational wave detection is changing from physics to astronomy—a science with a very different evidential culture—and because the new normal is being established. Under the new normal, an event that roughly matches the same template coincidentally and roughly coherently is going to seem more likely to be a gravitational wave than anything else (see the analogy discussed on page 271, above). For fifty years it has been the other way. Thus ends the first observational run of Advanced LIGO.

LAST WORD

I finished this postscript in June. By the time this book is published many more inspiral signals should have been detected in the course of O2 and, perhaps, reported. The concerns expressed in these final chapters will seem trivial, and they will be trivial, as far as the new gravitational wave astronomy is concerned. Nevertheless, this is how it was. We can see in this sequence how easy it is for retrospective history of science to go wrong. Very soon, without this contemporaneous account, there would be little trace left of the fact that so much of the scientists' confidence that The Event was the equivalent of an "Omega-minus" rather than a "monopole" came from GW151226. This fact would disappear because the scientists kept it resolutely hidden from the public record; the public record will show only the detection of The Event and then the detection of GW151226 as though these were separate happenings taking place in sequence, whereas the detection of GW151226 was part of the detection of The Event. That it happened this way is also important for the relationship of physics and society—if we want our physics to take the leadership role that I believe we desperately need, then physics has, always, to do things right. And doing things right for society may not always be doing things right for physicists.

HOW THE BOOK WAS WRITTEN AND THOSE WHO HELPED

I have been watching the science of gravitational wave detection since 1972. My first paper on the topic was published in 1975. Amusingly, Rai Weiss aside, I have been around this field longer than any other currently active member. As a sociologist of knowledge, I ask what makes scientists believe what they believe and disbelieve what they disbelieve, and what makes others believe what the scientists believe. Scientists have been trying to see gravitational waves for half a century, building generation after generation of more and more sensitive and more and more expensive machines, without a single success until September 14, 2015.

A subtheme to the story of the discovery has to be a methodological and personal one: how does the sociologist alternate between the native experience of sharing with the scientists this moment of overwhelming joyful love for the subject and its outcome and the lonely professional responsibility of maintaining a distanced perspective? The book reflects this tension. In most of what I have written I have assumed the role of a participant—a near-physicist asking, "Why do *we physicists* believe this?" Only after answering that could we switch to the sociologist's more distanced perspective and ask why *anybody* believes that a few numbers represent a massive event in space that happened 1.3 billion years ago—what a ridiculous idea and what a huge body of taken-for-granted, and extraordinarily forceful, reality it rests upon. But just say you don't believe and see where you wind up in the way people classify groups in society!

I wrote the larger part of this book in real time. As things unfolded I wrote sections. I wrote pretty well every day for five months, slowing down just a bit after the first press conference. Looking back, it would have been a very different book had I written it retrospectively. If I had been restricted to looking back after the discovery was confirmed, I don't think I could have recaptured the sense of dawning discovery and the dramatic change in the order of the universe that took place so fast. I try to convey my immersion in events by frequent use of the present tense.

The material on which the book is based comprises around 17,000 emails, mostly circulating around the gravitational community but with about another 450 or so direct exchanges between me and individual physicists, around 320 of them with Peter Saulson. Around 12,000 of these emails pertain to the discovery and the other 5,000 or so, much more sketchily analyzed, have to do with post-discovery events. In the discovery period there were also a half-dozen group-wide teleconferences and about fifteen teleconferences or telephone calls, an hour to an hour and a half in length, between me and the physicists; all these were recorded. I also recorded, or obtained recordings of, a number of events such as press conferences or broadcasts, collected newspapers, attended a couple of meetings or conferences, and did some questionnaire-type fieldwork so as to generate the material for chapter 14.

I should also explain why there is so much about me in this book when the story is the detection of gravitational waves. This is, first, because the method is what I call "participant comprehension." I try to become as like the group I am studying as I can so that I can write from my own experience as a member as well as from what the scientists tell me. That it is possible to become something that approximates an insider is made understandable by the relatively new concept of "interactional expertise," which shows how it is that one can understand technical and practical matters from long immersion solely in the spoken discourse of a specialist community.[1] That I can approximate an insider—as discussed at length

1. See Note XV in "Sociological and Philosophical Notes," 371.

in chapter 14—means that a lot of the time I am answering the question, "How is it that some momentary, minuscule oscillations can be taken to represent such a huge event so far away?" through my experience as much as that of the scientists: what made me come to believe in this event after nearly fifty years of not believing? The way my acceptance of the event developed is put together with that of "this" scientist and "that" scientist, with everyone's growth of belief following a slightly different trajectory.

And that trajectory varied from time to time—or at least the way it was presented varied. Thus, when the scientists were experiencing the happiness of finally discovering gravitational waves, they were believing; but when they were deceiving journalists about the fact of the discovery, they could tell themselves, and the journalists, that there was nothing to say because not all the tests that would confirm the discovery had yet been completed and the discovery paper had not yet been peer-reviewed. So a finding is a slippery thing—which will slip around for some people until the initial event has been confirmed several times, and for others will continue to slip around even after that.

Crucially, I need to be continually examining the extent to which my abilities are compromised by any deficiency in my own expertise. That is why I am always comparing what I think about certain episodes with what the scientists think. Furthermore, in my wider professional life I write a lot about expertise, including the expertise of sociologists, journalists, and so forth. I would like journalists who write about science to account for their own expertise—I want to avoid future occurrences of such things as vaccine scares based on ignorance of how science works. This means I have to account for myself and such mismatches in understanding between me and the scientists that I come across.

The form of the book is the unfolding of a discovery over time and distance. In time, it is about six months; in distance, it is the social space that reaches out from the specialist gravitational wave community to the wider scientific community and beyond. But within that two-dimensional framework the topics follow the pattern of my reading emails, listening to teleconferences, talking to scientists, and pulling together the analysis. Themes are introduced as they occur, but if it seems appropriate they may be treated again in more depth a bit later, just as they were in the discussions among the scientists. Of course, some themes, such as the tangled web of deceit, are my themes rather than theirs, but I indulge myself by introducing them whenever they were troubling me. Themes like the process of writing of the discovery paper for publication don't begin until we are well into the book because, in this case (unlike the case of Big Dog), the scientists didn't start thinking about the paper for several weeks.

Chapter 3 and the final chapters depart from the pattern of the discovery, publication, and reception chapters. Chapter 3 is an introduction to the field of gravitational wave detection for readers new to the subject. Without chapter 3 it would be hard for those new to the field to understand a lot of what is going on, and this book is meant to be self-contained. At the end of the book, in chapters 12–14, I take my professional role as a philosophically minded sociologist more seriously, revisiting certain themes of the book in more of my subject's depth, and reflecting on the way this book fits into my profession's concerns.

ACKNOWLEDGMENTS

Peter Saulson, Martin A. Pomerantz '37 Professor of Physics at Syracuse University in New York State, author of *Fundamentals of Interferometric Gravitational Wave Detectors* (1994; new edition being prepared), one-time spokesperson for the LIGO Scientific Collaboration, one-time chair of its Detection Committee, and my friend, is a kind of virtual coauthor of this book. In the past I have asked him if he wanted to be a coauthor proper but, sensibly, he has always refused. I should add that we don't always agree about the right way to present results, award prizes, and so on. He is not to blame for anything bad here but takes a good bit of the credit for anything good. He has read the entire manuscript line by line.

Many other gravitational wave physicists helped me write this book. I owe the whole community a debt that goes back forty-three years but, starting from September 14, a few of the physicists, knowing what I would be doing, sent me helpful unsolicited emails, concerned that I should understand what was going on. Many more responded generously to my requests for help, with the initial exchange often turning into extended strings: excluding Saulson, there are forty people named with whom I had considerably more than a hundred email exchanges altogether, a couple here, five or six there. I suffered only one "nonresponse" and one explicit refusal to respond on the reasonable grounds that the information was too personal. I had to send out only two prompts, and they elicited immediate replies with apologies. Saulson aside, the gravitational wave physicists who aided this project directly include: Bruce Allen, Stefan Ballmer, Barry Barish, Angus Bell, David Blair, Duncan Brown, Sarah Caudill, Massimo Cerdonio, Karsten Danzman, Tom Dent, Marco Drago, Matt Evans, Stephen Fairhurst, Peter Fritschel, Adalberto Giazotto, Mark Hannam, Jim Hough, Vicky Kalogera, Mike Landry, Albert Lazzarini, Sean McWilliams, Christopher Messenger, Laura Nuttall, Brian O'Reilly, Guido Pizzella, Fred Raab, Dave Reitze, Keith Riles, Sheila Rowan, B. S. Sathyaprakash, Peter Shawhan, David Shoemaker, Josh Smith, Patrick Sutton, David Tanner, Rai Weiss, Stan Whitcomb, Roy Williams, Graham Woan, and Mike Zucker. I apologize to anyone I have missed.

The forty-three years of research in this book have been generously supported by a series of grants from the UK Economic and Social Research Council, or "Social Science Research Council" as it was, and one from the US National Science Foundation. These are:

1971–1973, an SSRC PhD studentship with fieldwork expenses of around £240

1975, SSRC, £893, "Further Exploration of the Sociology of Scientific Phenomena"

1995–1996, ESRC (R000235603), £39,927, "The Life after Death of Scientific Ideas: Gravity Waves and Networks"

1996–2001, ESRC (R000236826), £140,000, "Physics in Transition"

2002–2006, ESRC (R000239414), £177,718, "Founding a New Astronomy"

2007–2009, ESRC (RES-000-22-2384), £48,698, "The Sociology of Discovery"

2010–ongoing, "To complete the sociological history of gravitational wave detection," funded out of US National Science Foundation grant PHY-0854812, $975,000, to Syracuse University, "Toward Detection of Gravitational Waves with Enhanced LIGO and Advanced LIGO," principal investigator: Peter Saulson. Open-ended.

I thank the MIT Press for pulling out all the stops and getting this book into print far more quickly than is normal for a university press, making it possible for the book to celebrate the one-year anniversary of the discovery announcement. I thank Katie Helke for her enthusiasm for the project from the moment it was first mentioned and pushing through an incredibly rapid decision process, and Judy Feldmann for assiduous copyediting with speed and sympathy.

SOCIOLOGICAL AND PHILOSOPHICAL NOTES

In these pages are assembled what would otherwise be twenty-two discursive footnotes referring to seventeen entries on matters of sociology and philosophy. Putting them together makes it possible to set them out in a less constrained way. Each "Note" can be read alone and refers back to the page number in the text given in parentheses.

The intention is that the Notes, as well as serving their original purpose, should also be readable as a free-standing, if somewhat episodic, section comprising a set of short essays on themes in the sociology and philosophy of science. Those primarily interested in the story of the discovery can safely ignore this section, but some might be interested in the perspective of a sociologist approaching a contemporaneous historical project like this and the professional concerns in which it is embedded.

The key is the revolution in outsiders' understanding of science that took place in the early 1970s. Before that time, this science was thought of as a kind of logic-machine run by brilliant attendants; it seemed insane to state that the unfolding of the very content of scientific knowledge was in any way a matter of the social. Nowadays, however, it is less unreasonable to discuss the social aspects of scientific discovery, and here these aspects are in continual focus.[1] That such a focus has become possible

1. That this shift in perspective may have affected more than narrow-minded sociologists is acknowledged in an essay by two erstwhile antagonists; Franklin and Collins (2016), in "Two Kinds of Case Study and a New Agreement," describe a rare instance of academic reconciliation.

begins with certain philosophical insights such as "the problem of induction." The problem of induction shows that we cannot predict the future from the past in a logically secure way; rather, predictions have to be based in agreements about how to predict and what to predict. Other important philosophical contributors are Wittgenstein's idea that rules do not contain the rules for their own application, which shows that we must find some conventional way of deciding what a rule means, and Duhem's notion that a scientific claim is embedded in a network of other claims so that which of these should be maintained and which sacrificed is a matter of agreement. The approach also grows out of the sociology of knowledge, which starts from the point that most of what we call knowledge is a matter of the society within which we were born and brought up; only after the 1970s was it understood that this applies to science as much as anything else. These underlying ideas are expressed in "the experimenter's regress" and its equivalents: what I call in this book "the proof regress," the idea that scientists have to agree to stop questioning at some conventional point if anything is to count as a scientific finding; the conventional nature of what counts as a secure level of statistical significance; the fact, explored here and in *Big Dog*, that the calculations that give rise to a finding are based on judgments, along with the tension between calculation and judgment in general. All these things are descriptions of science that show it to be embedded in social agreements and therefore social life. The scientists have to work within tacit agreements about how to go on.

Of course, as shown in chapter 11, there are scientists who do not share these social agreements and, therefore, can use the same data to reach different conclusions—exactly as we would expect. The acceptance of the detection of gravitational waves by the public—along with the Big Bang, black holes, the Higgs Boson, and anything said by Stephen Hawking—is purely a matter of social convention: these are the kinds of thing you have to believe if you are not to be accounted as mad or eccentric. The new understanding of science makes it possible to see how wider social and political forces can affect the content of science. There is not much of this kind of thing in this book except at the highest level, such as the very support for the funding of gravitational wave research and the willingness to believe its results, but we can see how there could easily be more immediate low-level impacts in "softer" sciences like economics or sociology.

As it turns out, *Gravity's Shadow* and its successors are currently the most detailed history of direct gravitational wave detection, and I am delighted that scientists recommend these books when someone needs to come to grips with the subject; scientists also find them useful as quick technical introductions. But I wrote in *Gravity's Shadow* that the books do not answer to the professional standards of historians. This is because the history I am primarily trying to set out is of changing ways we think about things—the changing ways in which scientists make the choices they must make, given the necessity of choice and judgment explained above. To understand this kind of change it is not necessary, or even particularly helpful, to provide documentary evidence of every historical claim; the evidence needed concerns the kind of things people were thinking about, and this evidence is obtained from talk. To give a simple example, at one point, under Robbie Vogt's regime, Ron Drever was locked out of his office (*Gravity's Shadow*, 575); Drever and many others thought it was an act intended to demean, while another group thought that the locks were changed as a matter of routine maintenance. I could have found out which it was by exploring documents in Caltech's maintenance department but I didn't bother; all I needed to know was that relations were such that it was possible to believe either account—exactly which one was true didn't really matter, so long as either could be believed. That is the kind of way in which the history in these books differs from some other kinds of history.

Many of the issues described in these introductory remarks are now dealt with at greater length in the notes that follow.

THE NOTES AND REMARKS

I (6). *Public understanding of science* There is a long-running debate about the nature of the public understanding of science. Much of the initial impetus came from scientific bodies anxious to justify the huge sums of public money spent on scientific research. The so-called deficit model took public resistance to science to arise out of ignorance about science:

if the lack of scientific understanding among them could be remedied, the public would be more ready to support science. Social scientists tend to ridicule this idea. After all, scientists themselves disagree about what scientific projects are worth supporting and which scientific findings are valuable, and scientists do not suffer from a deficit of scientific knowledge. Therefore, it seems unlikely that educating the public to some intermediate level of scientific knowledge is going to engender enthusiasm for science.

Rather, this kind of public understanding, though it is often called "outreach," should be seen as propaganda or advertising. There seems little doubt that there is a high level of enthusiasm for space exploration, astronomy, and astrophysics, with the most recent finding being a fine example. Of course, the most recent finding is also a fine example of a science that was not supported by what was probably the large majority of the knowledgeable scientific community for most of its lifetime.

More worrying is that the public, while supporting this kind of science, do not really understand it at all. Consider the best-selling books *Brief History of Time* by Stephen Hawking and *Elegant Universe* by Brian Greene. Almost no one of the wide body of readers of these books understands them; effectively they are purchasing and giving reverence to the modern equivalent of the Latin Bible. There is no real *understanding* of science going on here, there is only the touching of the robe. The same goes for the public "worship" of Stephen Hawking as a person—he is hugely admired for his brave battle to overcome crippling disease, but almost no one understands what he does *as a scientist*. And, to bring things closer to home, when Kip Thorne gives one of his immensely enjoyable lectures to the public covering black holes and the like, the graphics are wonderful and the corresponding Hollywood movie is terrifically entertaining, but none of it has anything to do with the day-to-day world of science such as is described in these pages. The public is not being scientifically educated in these cases; the public is being entertained.

It may be, of course, that this is good for the future of science, as it causes the public to favor it. On the other hand, as a prig when it comes to science (see chapter 13), I worry that it is dangerous to adopt the iconography of religion and the entertainment and advertising industries to win support for science: science is far too important for that (and see the "Envoi" chapter in *Gravity's Ghost*).

The motivation for this note was, however, something different. It was the belief, widely held among social scientists, that there is no deficit of scientific knowledge among the public, because the public can understand science as well or better than the experts. When social scientists make this kind of claim, they usually have a narrow range of scientific topics in mind though they seem reluctant to draw the boundaries; I have heard it claimed a decade or so ago that gravitational wave physics could escape from the problems that beset it at that time if only they would bring in some artists to help out. The usual topics that the social scientists have in mind, however, are medicine or environment related. The iconic cases are vaccine revolts, notably the revolt against the mumps, measles, and rubella (MMR) vaccine that was triggered by the remarks of medical doctor Andrew Wakefield in 1998 that linked MMR to autism. Though based on no scientific evidence, the idea of such a link was picked up by the press who quoted parents' experience of children exhibiting the symptoms of autism soon after an MMR injection. Parents were led to believe that their personal experiences could stand alongside the large body of epidemiological research that showed, for example, that there had been no increase in autism rates in countries where MMR vaccine had been newly introduced. The value of this kind of "lay expertise" was championed by some social scientists in spite of the lack of science. That there was no evidence of any type supporting the link between MMR and autism and that it was a statistical inevitability that some children would start to exhibit symptoms of autism shortly after administration of MMR—just as it was a statistical inevitability that some children would exhibit the symptoms shortly after eating their first banana—was not thought to count against the instinctual lay understanding of parents or the rights of parents to demand an investigation of a completely groundless possibility.

That certain members of the public can contribute to scientific matters is not in doubt. Medicine is a place where this would be expected, since in many cases it is patients upon whom medical scientists have to rely for a description of the symptoms that they are trying to understand and cure. At the same time, an understanding of the efficacy of treatments often starts with laypersons' descriptions of their bodily or mental states. Furthermore, the notion of interactional expertise (see Note XV below) allows for the creation of bodies of expert patients with substantial expertise in their own conditions and their treatment. A notable example of sufferers developing a sufficiently high level of expertise to make a contribution to the design of the testing of new treatments was the case of San Francisco–based AIDS treatments (see Epstein, *Impure Science: AIDS, Activism, and the Politics of Knowledge*). It is only where cases such as this—in which small groups of uncertified persons become experts in a subfield as a result of assiduous reading and spoken interaction with the experts—come to be taken as signifying a generally high level of expertise among laypersons that problems arise. I address this issue in *Are We All Scientific Experts Now?* (2014).

II (11). *First sighting of optical pulsar* Harold Garfinkel, Michael Lynch, and Eric Livingstone (1981; hereafter GLL) report on the first sighting of an optical pulsar. In this case a tape-recorder was left on by chance as the scientists examined their observational data in real time. GLL obtained the recording and analyze it. Unfortunately the paper is not easy to read, as the authors are mainly trying to establish a new and esoteric disciplinary style. Here is an example of the prose:

> The conversational record of the inquiry exhibits in argumentative dialogue, in expressed troubles, and in interactively voiced conjectures, that "first time through" the object was tied to an unfinished and developing course of embodiedly "lived" procedure, procedure enacted as plan lived-through in its contingent and recalcitrant worldliness.

The content of the tape-recording is very interesting, and various extracts are presented in the article. Excerpts are also readily available at https://www.aip.org/history/exhibits/mod/pulsar/pulsar1/05.html, from where the following is taken.

McCallister: This next observation will be observation number 18.

Disney: We've got a bleeding pulse here.

Cocke Hey. Wow, you don't suppose that's really it, do you? Can't be.

Disney: It's right bang in the middle of the period, look. I mean, right bang in the middle of the scale. It really looks something from here at the moment.

Cocke: Hmm!

Disney: It's growing, too. It's growing up the side a bit, too!

Cocke: God, it is, isn't it. Hmm!

Disney: Good God, you know that looks like a bleeding pulse.

[laughter]

It's growing, John.

McCallister: It is.

Disney: Look.

McCallister: It is. Hey, you're right.

Cocke: God, I hate to believe it right now, though.

[laughter] Well, we're up to 2,000 counts. We're now at 750,700.

[inaudible]

Disney: It's really building up. Look at that.

Cocke: It is, isn't it. Yeah.

Disney: There's not one left behind now. See, look, not one of those dots left behind.

Cocke: By God, yeah, uh huh.

[crosstalk]

Disney: There's a second something over here.

Cocke: Well, we expect two: a small pulse and a larger pulse, remember?

Disney: Uh huh. Right. I wasn't too sure of this one, but—.

That's a bleeding pulse.

Cocke: It is, isn't it. God. I don't believe it.

[laughter]

Disney: I don't believe it—I won't believe it until we get a second one.

Cocke: I won't believe it until we get the second one and until the thing has shifted somewhere else.

Disney: God, just come and look at it down here.

[laughter]

This is an historic moment. Hmm!

Cocke: I hope it's an historic moment. We'll know when we take another reading. And, if that spike there is again right in the middle—that spike is right in the middle and that scares me.

The sociologists' analysis—my understanding of which comes from discussion with Mike Lynch—includes two points, among others: (a) there is something special going on when an observation is made for the first time when a new reality is being created, and (b) some of what the scientists are thinking about is the potential reception of what they are seeing.

The last point is especially important in the context of competing styles of sociological analysis of science. Thus, the style of what became known as "controversy studies" took the establishment of a new scientific fact as a matter of interaction between separate, and often competing, groups (e.g., as exemplified by the papers collected in Collins 1981), whereas the style of "laboratory studies" took it that one could understand how new facts came into being by examining work within a single laboratory—e.g., Latour and Woolgar (1979), or Knorr-Cetina (1981). GLL was neither a controversy study nor a laboratory study since it was simply based on a tape-recording, a copy of which the authors came to possess some time later as a result of fortuitous circumstances. Lynch's point, as expressed to me, is that the anticipation of the reception of the finding brings other laboratories into the discussion even though the only piece of data is this single recording. We have to assume, nevertheless, that GLL would have preferred to have richer data covering a greater cross-section of the activities that eventually led to the optical pulsar being widely accepted as a scientific finding.

III (33). *Rules regress and double blind tests* A great deal of the way sociologists of scientific knowledge understand the world, and a great deal of wisdom, can be extracted from of the philosopher Ludwig Wittgenstein (1953), crucially, that "rules do not contain the rules for their own application." Thus attempts to solve problems by dogged application of rules are doomed to failure and frustration because the way rules should be applied is understood differently by different people. Mindless bureaucracy can be one result; it is not so much a dogged application of rules as a dogged application of one particular interpretation of the rules. The "airplane event" is a classic example of the problem and so is the tension between relentless professionalism and judgment that is discussed from chapter 10 onward.

So-called evidence-based medicine, or policy, if interpreted as meaning that nothing can be trusted unless it is supported by experimental work based on double-blind testing, is another example of the problem. In chapter 1 of *Dr. Golem: How to Think about Medicine* (Collins and Pinch 2005), double-blind tests used as safeguards against "experimenter effects" are discussed at length. Blinding is a good default position in experimentation but becomes ridiculous when taken to extremes. Collins and Pinch (2005, 32–33) show how ridiculous double-blind testing would be in the case of using casts for curing broken limbs, while Smith "Parachute Use to Prevent Death and Major Trauma Related to Gravitational Challenge: Systematic Review of Randomized Controlled Trials" (2003), satirizes the pedantic use of such tests by complaining about their absence in the case of parachutes.

IV (97). *Science, humanities, and the social studies of science* In spite of the fact that it is discussed only rarely these days, C. P. Snow's *The Two Cultures* (1959) is still highly germane in the field of social studies of science. The early 1970s revolution in social studies of science grew out of history and philosophy of science with its practitioners believing that a reasonable understanding of the science in question was a necessary foundation to a critical analysis. In less than a decade, however, ways of analyzing science emerged that turned on treating it as a subject for ethnographic description, with the perspective of the stranger being taken as yielding an analytic advantage: under such an approach understanding the science was taken to be, if anything, a disadvantage. Along with this came an approach that treated science as essentially a literary enterprise—after all, what did scientists do if not produce "inscriptions" (written records of their findings) and assemble them into publications? Once science studies were amalgamated into the existing tradition of literary criticism and semiotic analysis, there was a huge expansion of the subject among humanities scholars with little constraint imposed by the technicalities of the science. This lack of constraint opened up some of the activity to accusations of being "fashionable nonsense." The barb was driven home by such events as the Sokal hoax (see Sokal 1994, 1996; https://en.wikipedia.org/wiki/Sokal_affair) and a more recent, if less elaborate, counterpart (http://prospect.org/article/academic-drivel-report).

The difference in culture can, I believe, be tracked back to a difference in what Evans and I call (Collins and Evans 2007) "The Locus of Legitimate Interpretation" in the sciences and the arts. In the sciences the acknowledged assessors of the value of a piece of scientific work are other scientific experts—the closer, professionally, to the originators of the work, the better: this is the basis of peer review. In the arts, however, producers of the work count much less as assessors of value—both artistic and financial—than the audience for the work. The value of a work of art is mainly judged by the public or its representatives—for example, newspaper critics. Works of art are therefore often legitimated more by their quality as "performances," intended to satisfy an audience, than by their intrinsic value as judged by the artist's peers. The humanities and social sciences are pulled in two different directions by these competing understandings of the world of knowledge.

For example, during the nasty period known as the "science wars," when scientists attacked social scientists who criticized science (Sokal was a central figure), even the scientists, now acting as philosopher- or sociologist-critics, slipped into artistic mode as far as locus of legitimate interpretation is concerned. After all, Sokal's hoax was a performance intended to convince, not its targets, but a wider audience, and Sokal exploited the hoax in that way. Most of the science-wars arguments were of that sort, whichever side you were on—performances intended to convince the public rather than anything that would have a chance of convincing an expert opponent. That this style can be pervasive in the arts and humanities accounts for the lack of felt duty to clarity; if one is to convince an opponent, or if one is to open oneself up to the kind of criticism that might cause one to change one's own ideas, one must present one's work as clearly as possible so that everyone can understand it, and one must begin one's criticisms by thoroughly understanding one's opponent and then producing counterarguments of such clarity that they cannot be gainsaid. If, however, one is interested in performing for an audience, there is a lot to be said for misrepresenting one's opponent's position so as to make it seem less convincing to outsiders and presenting one's own work in an obscure form that is hard to criticize or that will make criticism seem muddled. Here, then, is the key to at least some of the differences between scientific and artistic culture are reflected in the difference between scientific and humanistic approaches to social analysis of science.

One difficulty the social sciences face is that the notion of science is often confounded with technique. It is thought that a scientific social science must be experimental or survey-based with results supported by complex statistical analysis. It is tempting for social scientists to reject any association between social sciences and "science" because they find the statistical and experimental approach uninteresting. But a scientific approach is not a matter of technique; it implies only that the researcher is willing to claim that his or her results are robust and could be repeated by any other researcher with the same level of technical or native understanding of the field. It may be that gaining an appropriate level of native understanding is a "subjective" process, but it does not follow that the results of analyses based on that understanding are anything other than objective or that they cannot be expressed at a sufficiently general level to be applied to situations outside of immediate subject of investigation. This is illustrated by the discussion with Gary Sanders concerning riding on buses on Baton Rouge set out in footnote 7 in chapter 6 (97). But not even all ethnographers share the aim of producing objective knowledge; some want to provide an essentially personal description of a local society that has no implications for societies in general.

The analysis of gravitational wave detection carried out here, and in the other books listed, is executed in a scientific spirit. Though the method is deeply "subjective," were it not to lead to findings that are robust, reproducible—and this will often mean that the findings are readily recognizable to scientists as a reflection of the reality they inhabit—and were the high-level findings of the study not applicable to other fields of science, it would be a failure.

V (97). *Millikan oil-drop experiment* The story of the famous Millikan oil-drop experiments that established the integral charge on the electron is told by Holton (1978). Millikan, though he interpreted his experiment as showing that the charge on the electron was integral, had a number of entries in his notebook that implied the existence of fractional charges; he chose to ignore these on the grounds that, in his judgment, he thought they were experimental artifacts. They looked genuine enough in the 1970s, however, to encourage Bill Fairbank, the pioneer of cryogenic resonant bar gravitational wave detectors, to claim to have found a fractional charge corresponding to a free quark. This story is told in *Gravity's Shadow*, 216–217.

VI (98). *Networks of beliefs and experimenter's regress* Another important idea that drives the sociology of scientific knowledge is that scientific findings are not isolated but exist only within a network of other beliefs and claims. For a finding to be supported, many related "subhypotheses" have to be accepted, and a disputed finding can always be saved by sacrificing this or that related hypotheses or belief. This idea is the basis of chapter 5 in *Gravity's Shadow*, which talks about what Joe Weber would have had to sacrifice to keep his high-flux gravitational wave claims alive. Indeed, in the end he did sacrifice his own theory of the sensitivity of resonant bar detectors, replacing it with another theory that few people believed. That story is told in chapter 19 of *Gravity's Shadow*. I also argue, in chapter 10 of

Gravity's Shadow, that Weber could have kept his claims alive by refusing to accept the general view of how resonant bars should be calibrated, but he did not refuse—that was one subhypothesis he did not exploit.

The idea of subhypotheses, or "auxiliary hypotheses," probably begins Pierre Duhem (1914/1981) and what subsequently became known as the Duhem–Quine thesis (see https://en.wikipedia.org/wiki/Duhem%E2%80%93Quine_thesis). Imre Lakatos's (1970) theory of research programs is related to the idea, but Lakatos separates the subhypotheses into a "core" and a "protective belt," or periphery. According to Lakatos, the periphery can more readily be sacrificed to save the core—giving up the core means abandoning the theory.

The experimenter's regress (see p. 263 and *Changing Order*) has been said to be a version of the Duhem–Quine thesis, but, if it is, it shows how it is a day-to-day problem for contemporary science rather than a theoretical idea. In this book we see the basic idea drawn upon in the form of what is called the "proof regress," in order to mount the argument that scientists have to stop questioning at some point that is a matter of social convention within a scientific community. Where the conventions differ—as in the case of fringe science—the questioning will stop at different points; what is a secure finding in one place will turn out to be less than secure in another place.

VII (134). *The nature of scientific papers* The main text describes how what Latour and Woolgar (1979) call "modalities" are stripped from accounts of scientific work as tentative findings are transformed into publishable papers. The modalities are descriptions of the details of experimental activity—who carried it out, when they carried it out, and so forth. As the modalities are removed, the entities described shift from being unique events to regular features of the world that are found everywhere rather than specific events occurring at a particular time and place. It is not that some electrical excitations that might be an indicator of a gravitational wave occurred at this time and place and were observed by this person but rather that "gravitational waves were detected." The shift in literary style gives rise to what Shapin (1984) calls a "literary technology." By means of this technology the reader is placed in the position of a "virtual witness" of the experiment or observation—it is as though reading the paper enables him or her to make the same observation as is described within it. To engender the illusion, the paper is written in a passive voice—things, engineered by Nature, happened to the observers rather than the observers making things happen through their active intervention. These things would happen under similar circumstances whenever and wherever the observers placed themselves, because the phenomena are not a historical event but universal. In the text I point out that the style of the discovery paper departed from these conventions because a historical event was being described. The historical event was the first direct detection of gravitational wave, the first more or less direct observation of black holes and the first detection of an inspiraling binary black-hole system. These are the first observations of what are expected to be many but right now we are celebrating this historical moment—only later will sightings like this become routine and universal.

VIII (238). *"Beer mat knowledge" and other types of expertise* For a discussion of the full range of different types of expertise, organized in what we call "The Periodic Table of Expertises," see Collins and Evans 2007 (*Rethinking Expertise*). "Beer mat knowledge" is one of three categories (these three should really be called "information") that can be acquired by exercising no more than "ubiquitous tacit knowledge"—that is, the tacit knowledge that is gained by living in society as an ordinary citizen. This tacit knowledge includes fluency in the native language and literacy, along with an understanding of books and other written sources and an understanding of their reliability and where they are to be located. Beer Mat Knowledge is the lowest level of these three. The next level up, when it comes to knowledge of technically specialist matters, is "popular understanding"; this comes from reading popular science books. Next is "primary source knowledge," which is what ordinary citizens can garner from finding their way into specialist libraries or specialist locations on the Internet. To garner primary source knowledge is a demanding exercise, and, partly as a result of this, it is easy to mistake this kind of *information* for one of the higher forms of *knowledge*. But these higher forms are different. There are two higher forms: "contributory expertise"—which is just the standard form or expertise possessed by those who work in a specialist field—and "interactional expertise"—which is based not on working within a field but on immersion in the field's spoken discourse; it is discussed at length in Note XV. These two kinds of expertise depend on "specialist tacit knowledge": the tacit knowledge that can be acquired only by long immersion in the discourse of the specialist community. Thus both kinds of expertise depend on immersion in the specialist discourse, with practice being properly executed and coordinated through the shared practice language—or common interactional expertise (Collins 2011).

The important difference between the three categories of specialist information that rest only on ubiquitous tacit knowledge and the two that rest on specialist tacit knowledge can be illustrated with an anecdote. Joe Weber published a paper in 1996 (Weber and Radak 1966—see also Note XVII) that claimed to see a correlation between his early sightings of gravitational waves and gamma-ray bursts. The paper was carefully written and claimed that the correlation was associated with a statistical confidence level of 3.6 standard deviations. But my investigations indicated that the paper was not read by anyone (*Gravity's Shadow*, 366); by this stage Weber's credibility in respect of sightings gravitational waves was so low that the paper—though it had all the appearances of a scientific paper, with all the right literary technology (see Note VII) such as would make it appear perfectly sound to a reader with only ubiquitous tacit knowledge—had, as it were, no scientific "soul." It would require specialist tacit knowledge to know this; it is not something you could learn from the paper itself. I argue elsewhere (e.g., Collins 2014) that failure to understand this kind of distinction is responsible for serious and potentially fatal problems such as vaccine scares.

IX (159). *Kuhn, paradigms, and forms of life* For an analysis of the different ways in which Thomas Kuhn's famous (1962) book *The Structure of Scientific Revolutions* can be interpreted, see Pinch 1997. Many of Kuhn's ideas were anticipated by Ludwik Fleck (1935/1979). Fleck was an active medical doctor and researcher who developed his ideas about science by reflecting on his own practices. He was imprisoned in a concentration camp during the World War II where he prepared typhoid vaccine. His book, *Genesis and Development of a Scientific Fact*, was not translated into English until 1979, and that is why it had little effect on what I have called the revolution in the understanding of science that took place in the 1970s. His version of the notion of paradigm, or form of life, was the "thought collective" characterized by a certain "thought style"; he understood the way scientific thinking is developed among groups of agreeing scientists. Kuhn acknowledges a debt to Fleck in the opening remarks of his book, and it is Kuhn who would set the scene for the new kind of socially based understanding of science even though the detail was generally provided by other philosophers—for example, in my case, Winch and Wittgenstein.

Kuhn's book—said to be the best-selling academic book of the twentieth century—was published to a storm of criticism. Working within the model of science as a quasi-logical machine with empirical reference, critics described Kuhn's idea that scientific "paradigms" underwent periods of revolution when concepts and practices changed en masse, as driven by "mob psychology." Kuhn, it seems, was eventually convinced by the relentless criticism that his idea of paradigm was too simple. In a second edition he described it as two separate ideas, one to do with scientific concepts and one to do with iconic experiments and the experimental style to which they gave rise. Here, it seems, he failed to understand paradigm as an exemplification of the deeper idea of form of life in which the very way that we think and define the objects and concepts in our world is just the other side of the coin of the way we act in the world. This is explained in the text (259) with Winch's (1958) example of the germ theory of disease. The way scientists make judgments, the places at which they stop arguing, and all the other social aspects of science mentioned in the introduction to this chapter are to be understood as part and parcel of the way scientists inhabit a form of life; a major scientific change, as I try to explain in chapter 12, is a change in a form of life.

X (261). *Automating science with computers* Given the ambitions of the artificial intelligence (AI) community, the idea of science as a quasi-logical machine was bound to lead to attempts to build computers that could displace scientists. Of course, computers already displace scientists in lots of roles, and their domain is continually increasing. For example, as I understand it, whereas once upon a time the apogee of brilliance among scientists would be the ability to integrate difficult functions, this kind of task can now be executed by the computer program Mathematica. And, of course, the stabilization of LIGO's interferometers along with the collection and analysis of data would all be impossible without computers. But the ambition of the AI community was to replace all the creativity of scientists. A famous case is that of the program known as BACON. It was said that, given the data, BACON could deduce Kepler's laws of planetary motion—no doubt it could if the data was perfect (see, e.g., Langley et al. 1987). The trouble is that this idea of science is impoverished (Collins 1991a, 1991b; Simon 1991). There is no "data"; there is only data plus noise, and the art of science is separating them. Separating data from noise depends on an interplay between minimizing noise and considering what seems reasonable. In Note V, we discussed Millikan's oil-drop experiment. We saw that to establish the integral nature of the charge on the electron, Millikan did not simply feed raw data into a calculating machine, not even a very clever calculating machine, because if he had he would have reached a different conclusion. Instead, Millikan used his scientific judgment to throw out some of the data and retain that which supported his hypothesis. The art of science is to make judgments of this sort that will withstand criticism. That means anticipating the judgments of the wider community and how certain judgments will be received—essentially a kind of sociological wisdom. That is why, even though computers are taking over more and more scientific procedures, there is little likelihood of them replacing scientists en masse in the foreseeable future.

XI (261). *Studies that established the sociology of scientific knowledge* Though my 1975 study of the Joe Weber controversy was probably the first of its kind, it was preceded by the theoretical work of David Bloor (1973), which used a similar Wittgensteinian (1953) starting point to argue for the possibility of a sociology of scientific knowledge. The opposition came from philosophers such as Laudan (1983), who said that the only job for a sociology of scientific knowledge was explaining why some supposed scientific knowledge failed to adhere to the normal, rational course of things that would lead to the truth; the only job was a sociology of failure, since true scientific knowledge was self-explanatory. This view is rarely expressed these days, though nothing in philosophy ever seems to die entirely.

But philosophy is a form of life just like any other kind of knowledge-making, and what triumphs has as much to do with what people practice as what they think and say. Just as Joe Weber's actually building a gravitational wave detector put an end to the debate about whether gravitational waves could be detected in principle (or so I would argue, but see Kennefick 2007), and just as, according to Peter Winch (1958), surgeons scrubbing their hands and the like establish the reality of germs even as they remove them, the case studies of controversies and the social aspects of their closure established the sociology of scientific knowledge. A collection of these can be found in Collins 1981a, which is a special issue of the journal *Social Studies of Science* entitled *Knowledge and Controversy: Studies in Modern Natural Science.* They include: Collins 1981b, which deals with the way the Weber controversy was closed down; Harvey 1981, which looks at experimental confirmations of non-locality in quantum theory; Pickering 1981, which is concerned with the debate over the quark versus the color theory of subatomic particles; Pinch 1981, which looks at the controversy over the detection of solar neutrinos; and Travis 1981, which documents the debate over the chemical transfer of memory in worms and rats. Other important early case studies in this style include Pickering 1984 and Pinch 1986, which are extended treatments of the debates in the earlier articles; Shapin and Schaffer 1987, which is a historical study of the conflict concerning the Robert Hooke's air-pump between him and Thomas Hobbes; and Mackenzie 1990, which is a detailed analysis of claims and tests of the accuracy of intercontinental ballistic missiles. I do not pretend that this list is exhaustive.

As explained, this kind of analysis shows how it is possible for wider societal concerns to seep into the very content of scientific knowledge.

What has become known as science and technology studies (STS) has become more and more concerned with public issues at the expense of analysis of the nature of knowledge. In that sense, this book reflects the formative period of the subject rather than its modern face. The case studies described above seem, nevertheless, to have left their mark on history of science (Franklin and Collins 2016), and the future of studies of knowledge might lie in a combination of the sensibilities and technical skills of historians and the questions and methods of sociology. These include a stress on the social texture of scientific knowledge-making supported by contemporaneous research as well as or instead of the exploration of archives. While historians Steven Shapin and Simon Schaffer belong to the generation that helped to found the sociology of scientific knowledge, it is encouraging that a sociological sensitivity also informs the work of historians including, among others, Daniel Kennefick (2007), David Kaiser (2009, 2011), and, nowadays, Allan Franklin (2013).

XII (282). *History of particle physics and multiple authorship* Peter Galison has written the definitive histories of particle physics, notably his *Image and Logic* (1997), which has been discussed in the text (p. 126). He has also written on the way the authorship of scientific papers has changed as particle physics has grown—"The Collective Author" (2003)—explaining how author lists now number in the thousands. Galison's account includes many contemporaneous quotations from scientists working out how to decide who should count and who should not count as an author in a large collaboration and how the pattern has changed over time. One of the problems that Galison tries to deal with, as the title of his piece implies, is the collective nature of scientific work:

> The team replaced the individual because the individual did not (could not) *know* the length and breadth of the experimental problem. ...
>
> The high-energy physics collaboration functioned not at all like a collection of homogeneous agents of which one could be a spokesman just because of his typicality. Indeed it was precisely because of the heterogeneity of the collaboration that the fundamental practical paradox of authorship arises. Each subgroup is necessary precisely because its special function is needed. ...

> The answer to "Who Is We?" in the context of a two-thousand-strong fluid collaboration must always remain unstable, oscillating between the desire to make scientific knowledge the issue of a single conscious mind and the desire to recognise justly the distributed character of the knowledge essential to any demonstration. (2003, 351–353)

We cannot speak with authority on high-energy physics, but in light of what we know about the division of labor in gravitational wave physics, this seems unnecessarily mysterious. There will always be conflicts about the role of different contributions to a joint activity, as illustrated by, for example, the disagreement over the amount of space and importance of the "rates" calculation to the discovery paper (p. 214), but there is no conflict between a division of practical labor and common understanding of the entire project by every, or nearly every, individual within it. This is because of the existence of interactional expertise and a shared practice language (Note XV). Multiple authorship does reflect the complex division of labor in this field, but it also reflects a shared understanding, or potential shared understanding, of every aspect of the work. Indeed, without this possibility it would be hard to understand how the division of labor or the management of the overall project would work. Therefore, in principle, each of the 1,011 authors, though each contributes to the project in a different way, can still be a full author of the paper.

XIII (291). *Popper and falsifiability* Popper's philosophy was at one time very popular with scientists, who believed it correctly described their work. Its influence lingers on—particularly the notion of falsifiability. We argue elsewhere (Collins and Evans 2017) that Popper describes the way scientists work correctly, even though his philosophy does not deliver what he thought it did. We argue that many philosophical ideas about science are correct descriptions of the way scientists live their scientific lives, but this does not mean they are a quasi-logical account of science. Thus, in the ordinary course of their scientific lives, scientists try to corroborate their findings and also, in the case of good science, try to set out the conditions under which their findings could be falsified. Popper treated corroboration and falsification as if they were in tension, but it makes perfect sense to try to do both. Here I will try to explain, very briefly, what Popper was trying to do and what he achieved.

Popper was writing at a time when it was thought that the job of philosophy of science was to explain the mechanics of scientific discovery—expressed in the 1959 title of one of his books: *The Logic of Scientific Discovery*. Any attempt to build a logic of scientific discovery was beset with "the problem of induction." A favorite version of this is Nelson Goodman's (e.g., 1973) "new riddle of induction." Goodman invents a new color, "grue," which means, very roughly, green today and blue tomorrow. All the evidence we have for the greenness of grass is equally evidence for the "grueness" of grass, so all the observations of green grass we have ever made should just as much lead us to expect that grass will be blue tomorrow as that it will be green tomorrow. The problem is an old one but amounts to the fact that we cannot predict the future from the past. Popper's example is the well-known impossibility of making the claim "All swans are white" irrespective of how many white swans have been observed. (For a discussion of these matters in the context of scientific replication, see *Changing Order*.)

Popper believed he had solved the problem of induction by replacing corroboration with falsification. While no finite number of observations of white swans could lead in a logical way to the claim "All swans are white," he believed that a single observation of a black swan could lead in a logically inviolable way to the claim "All swans are not white." Hence the popular view that sound scientific knowledge is made by falsification rather than verification.

Evans and I argue in *Why Democracies Need Science* that the whole problem is misconceived because scientific knowledge is not logically inviolable and what is needed is an assembly of the best procedures rather than some unbreakable formula; one can see from the remarks in the introduction to this chapter why we should believe this to be case. But, in any case, Popper's logic does not get him where he thinks it does because, as argued in the introduction, a claim, including an observational claim, can only be established by accepting a network of other hypotheses. Thus, if one sees a black swan there is always the possibility that the swan is covered in paint or soot, that its feathers have become discolored as a result of something the swan ate, or that there is something wrong with the light or with your eyes. Poppers central idea is that there is an "asymmetry" between corroboration and falsification, the former being insecure whereas the latter is secure, but there are so many possibilities that stand in between an observation of a black swan and the claim "Not all swans are white" that the asymmetry disappears, and falsification comes to look very like corroboration as far as the logic of the matter is concerned (see Lakatos 1970).

The solution is simple—stop worrying about "the logic" of scientific discovery because there isn't one. Instead, start putting together ideas about what constitutes a set of procedures to which one should aspire as a scientist; this set of procedures will include both attempts to corroborate and attempts to falsify. We can say that a scientific finding is stronger if it can be corroborated and, with Popper, more scientific if it is possible to explain the circumstances under which it could be shown to be false.

XIV (291). *The scientific form of life and formative intentions* This Note follows directly on from the previous one about Popper and falsification. In the last Note I argued that the search for the logic of scientific discovery is best abandoned and replaced by the idea that science is defined by a set of what are considered to be best practices. These are not rigidly assembled according to an overall logic but link scientists into a community like any other social community or "form of life." Forms of life are discussed in the text around page 259 and in Notes IX and XV. "Formative intentions" are things you can intend to do in different societies or forms of life (Collins and Kusch 1998). Thus, if you belong to the Zande people you can intend to divine a witch by administering poison to a chicken but you cannot intend to take out a mortgage; if you are a British person you can intend to take out a mortgage but, unless the circumstances are very unusual, you cannot intend to divine a witch by administering poison. That is to say, the nature of society is constituted by the kind of things members can legitimately intend—by their "formative intentions." Studies of radical scientific change are explorations of the way formative intentions change. And one can also see, though this is a very minor example, that once it had been established that it could reasonably be believed that Robbie Vogt *might* have intended to lock Ron Drever out of his office—see above—the sociological job had been done whether Vogt actually intended to lock Drever out or not. Note how extraordinary this possibility is, and how revealing of the level of strain in the group, but note that no one was ever going to accuse Ron Drever of being a witch because this is something that we simply cannot imagine being part of our form of life however strained the relations. The argument put forward in *Why Democracies Need Science* is that the formative intentions that constitute science overlap heavily with the values of democracy.[2]

XV (187, 314, 344). *Interactional expertise* The idea of interactional expertise brings language into central focus. It is argued that by spending enough time taking part in the spoken discourse of a specialist group—by acquiring their "practice language" (Collins 2011b)—one can learn to understand their world of practice without taking part in the practice itself. This idea flies in the face of a long tradition in philosophy that stresses the central importance of practice to understanding.

The origin of the argument is debates in artificial intelligence (AI). The influential philosopher Hubert Dreyfus, drawing on the philosophers Heidegger and Merleau-Ponty, argued in a famous paper (Dreyfus 1967) and subsequent books (Dreyfus 1972, 1992) that computers would never have human-like intelligence unless they had bodies with which they could move around and experience the world in the ways that humans experience it. AI enthusiast Doug Lenat argued that this must be wrong because of the capacities of persons who did not possess ordinary human-like bodies.

2. Under the form of life model of science, people acting in ways that do not follow the guidelines are not fatal in the way they would be to a theory based on the logic of scientific discovery. A logical account, like a mathematical proof, is vitiated by a single exception. (That this is not a universal characterization of mathematics is, however, made wonderfully clear by MacKenzie in his 2001 *Mechanizing Proof: Computing, Risk, and Trust.*)

His example was "Madeleine"; Madeleine was severely disabled from birth yet was entirely fluent with her fluency attained through conversation not physical interaction (Sacks 2011). The response to this by philosophers (e.g., Selinger 2003; Selinger, Dreyfus, and Collins 2007) was that Madeleine had a body with front, back, and so on and could work from this. But this answer allows that even if we need some kind of vestigial body to have intelligence (the argument seems doubtful, but let us allow it), we don't need much of a body to learn everything practical we need to learn; we can learn from conversation.

I (e.g., Collins 1996) tried to resolve the problem by splitting it into two. Human societies, or specialist groups within those societies, would not be human-like unless they had bodies, but individuals can learn from those societies without sharing their bodily form.

> Wittgenstein said that if a lion could speak we would not understand it. The reason we would not understand it is that the world of a talking lion—its "form of life"—would be different from ours ... lions would not have chairs in their language in the way we do because lions' knees do not bend as ours do, nor do lions "write, go to conferences or give lectures." ... But this does not mean that every entity that can recognize a chair has to be able to sit on one. That confuses the capabilities of an individual with the form of life of the social group in which that individual is embedded. Entities that can recognize chairs have only to share the form of life of those who can sit down. We would not understand what a talking lion said to us, not because it had a lion-like body, but because the large majority of its friends and acquaintances had lion-like bodies and lion-like interests. In principle, if one could find a lion cub that had the potential to have conversations, one could bring it up in human society to speak about chairs as we do in spite of its funny legs. It would learn to recognize chairs as it learned to speak our language. This is how the Madeleine case is to be understood; Madeleine has undergone linguistic socialization. In sum, the shape of the bodies of the members of a social collectivity and the situations in which they find themselves give rise to their form of life. Collectivities whose members have different bodies and encounter different situations develop different forms of life. But given the capacity for linguistic socialisation, an individual can come to share a

form of life without having a body or the experience of physical situations which correspond to that form of life.[3]

To put this in terms of more familiar examples, one cannot have a tennis language to learn from unless there are groups of humans with the physical ability to play tennis, but one can learn what it is to play tennis and, in principle, all its practical nuances, just by talking to tennis players. The same, of course, goes for gravitational wave physics.

It is not easy to acquire interactional expertise—it takes a long time—but once acquired it is much more than the ability to "talk the talk." It is better described as being able to "walk the talk." It has much in common with the kinds of knowledge managers of technical projects possess (Collins and Sanders 2007). The idea of interactional expertise seems necessary if we are to understand many features of the way societies work, how they support the division of labor in technical specialties, and the way subgroups interact with society as a whole. A useful discussion of the concept in the context of a classification of different types of expertise is Collins and Evans 2007; the most recent and complete review is Collins and Evans 2015.

As explained in the text, it is the idea of interactional expertise that makes it possible to contemplate an outsider without a physics degree undertaking a project like this one. It is also the idea of interactional expertise that gives impetus to imitation game exercises such as that discussed in chapter 14 and Note XVI.

XVI (314, 316). *Imitation games and Turing tests* As mentioned in Note XV, imitation games are associated with interactional expertise. Imitation games were the precursor to the Turing test. Turing based his test on a parlor game in which hidden men and women pretended to be each other while responding to written questions from a judge. Turing believed that a computer should be called "intelligent" if, say, it was as good at pretending to be a woman as was a man pretending to be a woman; to reiterate, what

3. From a review of Hubert Dreyfus's book, *What Computers Still Can't Do* (Collins 1996), which was published in the journal *Artificial Intelligence.*

was supposed to be indistinguishable according to the original description of the test (Turing 1950) was a computer pretending to be a woman versus a man pretending to be a woman (or vice versa), not a computer pretending to be a human compared to a human—the way so-called Turing tests are conducted these days.

We use imitation games, that is, Turing tests with humans, to test for the possession of interactional expertise. A classic example is an experiment in which blind persons pretending to be sighted are compared to sighted persons, with sighted persons asking the questions. The questioning can be mediated via interlinked computers. This configuration is compared to one where sighted persons pretend to be blind while being compared to blind persons, with blind persons asking the questions. In these pairs of tests, the blind persons pretending to be sighted do much better that the reverse because blind persons are immersed in the spoken discourse of the sighted whereas the sighted are not, generally, immersed in the discourse of the blind. Thus blind persons have many opportunities to acquire interactional expertise in the world of the sighted. We have run such tests at various scales and on various topics (Collins and Evans 2014).

The original gravitational wave imitation game test was run over email, with a gravitational wave physicist setting seven questions that were answered by me along with another gravitational wave physicist. The completed dialogues were sent to nine other gravitational wave physicists who were asked: "Which is the real gravitational wave physicist and which is Harry Collins?" Seven of these said they could not tell the difference, and two said Collins was the real gravitational wave physicist. An account was written up as a news item in *Nature* (Giles 2006). It is important to understand, however, that contrary to what the *Nature* story can be read to imply, this exercise was not a hoax but a demonstration of genuine understanding as exhibited by a display of interactional expertise. I describe a more elaborated version of this test applied to my slightly eroded level of interactional expertise in chapter 14. Interactional expertise is, of course, closely related to the concept of tacit knowledge—see Notes VIII and XV. Collins 2010 is an analysis of tacit knowledge.

XVII (236, 238, 321). *Domain discrimination, specialist expertise, and the fringes of science* As explained in the text (224, 226), the sociologist must studiously avoid short-circuiting the process of inquiry into the social factors that feed into scientific belief. This implies that truth, or rationality, or similar cannot be allowed to be part of the explanation of why something came to be believed. It thus becomes very difficult for the sociologist to distinguish between mainstream science and what we can call "fringe science." There is a large, organized fringe with its own journals and its own annual conferences (Collins, Bartlett, and Galindo 2016). One concern of many fringe scientists is a rejection of the theory of relativity; some claim the theory is a massive conspiracy, even, given Einstein's support for Israel, a Jewish conspiracy! Whatever the case may be, rejection of relativity implies rejection of The Event. Chapter 11 gives some examples of such rejections.

The self-denying ordinance of the sociologist makes the partition of science into mainstream and fringe a much more interesting problem than it is when analysts simply allow themselves the luxury of being parasites on the opinions of the scientists themselves—"Scientists say relativity is right so people who do not believe in it are irrational." Instead, one must try to find sociological demarcation criteria, a much more demanding task. It is an important task, not so much for the future of scientific knowledge where what counts as the truth will emerge over the decades, but for the decisions that have to be made by policy makers. For example, in 1992 (see *Gravity's Shadow*, 361) Joe Weber wrote to his congressional representatives indicating that his new theory of the sensitivity of resonant bars meant that the much more expensive interferometers were a waste of money. This presented no problem to the mainstream community, but how was an outside decision maker to judge the claim? The only answer seems to be that that, in a democracy, decision makers will have to base their decisions on the opinions of the mainstream institutions, and that is why demarcation criteria that are more than the opinions of scientists are needed. Strangely enough, arXiv has a similar problem in that it receives many submissions from fringe papers, and the sheer logistics of the operation demands some automation. In collaboration with Paul Ginsparg, the founder of arXiv we

(Collins, Ginsparg, and Reyes-Galindo 2016) have shown that the automated methods used by arXiv, though they represent the state of the art, do not recognize papers such as Weber and Radak 1996 as anything out of the ordinary. What is needed in those cases is what we call "Domain-Specific Discrimination" or "Domain Discrimination," for short (Collins and Weinel 2011, 407).

In Collins, Bartlett, and Reyes-Galindo 2016, we establish a series of characterizations of the fringe. Indicative is the difference between the fringe and the mainstream in respect of what Thomas Kuhn (1959, 1977) called "the essential tension." The essential tension is that between the need to preserve the right of the individual to make novel claims, setting him- or herself outside of the consensus, and the need to accept a degree of regulation of scientific thinking and acting if science is to move forward. In the normal way, science is always balancing these two needs. We find that in the fringe the balance shifts markedly to the side of the individual, with consensus being thought dull or suspiciously authoritarian; this shift in the balance is even visible in fringe scientific conferences, with each delegate's pet theory being given space to be expressed, resulting in a general lack of organization. In this book, of course (chapter 11), given the uniform lack of criticism among the mainstream, we find ourselves drawing on the fringe to provide criticism of the first detection of gravitational waves; it is the fringe, in refusing to accept the social consensus, that allows us to see the extent to which acceptance of The Event is a matter of social consensus.

APPENDIX 1
Procedure for Making a First Discovery

STEP 0: GETTING READY FOR A DISCOVERY

To make the whole process as smooth and fast as possible, it is important that the needed procedures be in place and validated before an observation run starts. These procedures and the responsible parties include:

- Carrying out thorough reviews of search pipelines, at least for the main search types. This will be taken care of by the search groups review committees.
- Establishing detection checklists, in order to verify the sanity both of data analysis results and of the state of the instruments. The checklists will be prepared by the search groups and detector characterization groups and will be reviewed by the DC [Detection Committee].
- Establishing procedures to ensure that state of the observatories and the activities going on at the time of a possible event can be documented and recorded rapidly in response to a possible (transient) detection. The LIGO Observatory Heads and a person to be designated by the Virgo Spokesperson are responsible for defining and implementing these procedures.
- Establishing a policy about the level of significance required to claim a detection. This policy will be proposed by the DAC [Data analysis committee] and the DC for approval by the collaborations.
- Establishing procedures for folding the results of any electro-magnetic follow-up observations of a candidate in the assessment of the significance of that event. These procedures will be proposed by the EM forum for approval by the collaborations.

- Preparing detection paper outlines for the main expected first detection source types. These paper outlines should be prepared by the search groups and be reviewed by the search group review committees, the DC and the editorial boards. This review should examine whether or not key issues likely to arise in a detection claim are addressed in the outlines, to ensure that necessary expertise has been engaged prior to the start of Step 1. The collaborations should agree on a target journal for the discovery paper.
- Reviewing, and revising if necessary, the procedures for communication with the media and the general public about the discovery as established by the two collaborations. These procedures are the responsibility of the LSC spokesperson, the LIGO executive director, the GEO Data analysis coordinator, and the Virgo spokesperson.
- Establishing criteria for cross-checking a detection claim with a second independent pipeline, as required by the multiple-pipelines policy (provide reference). The DAC and the DC will review the cross-check procedures described by the search groups in their search plans.

In the weeks leading up to an observing run, the DC chairs will ask each of the responsible parties listed above to report on the status of each preparation and will report the status of each to the collaborations.

STEP 1: INITIAL STEPS TOWARD A POSSIBLE DISCOVERY CASE

As soon as hints for a discovery are observed in any of the search groups, the co-chairs of the group(s) will inform the LSC and Virgo Data Analysis Coordinators, the DC chairs, the Observatory Heads, the Detector Characterization and Instrument Leads, the LSC spokesperson, the LIGO executive director and the Virgo spokesperson.

The search group continues work to build the case, interacting with its review committee, using its protocols and detection checklists. The co-chairs of the search groups will remind all involved scientists that confidentiality must be observed.

During this stage:

- The search groups and review committees initiate an in-depth *technical* check of the correctness of the analysis.
- The search group performs cross checks/consistency checks with other search groups, under the supervision of the DAC.
- The search groups define if there is a case, and what this case is, for a detection.
- The Observatory heads activate the process for capturing the complete state of the observatories and the activities there for the time of the event (assuming a transient event).
- The detector characterization groups and instrument scientists collect and bring forward complete information on instrument state and data quality at the time of the possible detection.
- The search groups and detector characterization groups put the case in writing through technical notes and maintaining an up-to-date web site for the event(s).
- The EM follow-up group will seek the results of any follow-up observations that might have been performed in response to alerts issued by L–V [LIGO–Virgo].
- The DC is informed about the case. The DC begins its work by reviewing the work of the detector characterization groups and instrument scientists in documenting the state of the interferometers and in assessing data quality.

STEP 2: CONSOLIDATING THE CASE

Based on the results of Step 1, the LIGO/Virgo leadership decides whether to continue to proceed with the case. If their assessment is favorable, then:

- The Review Committee will complete its review of the search and its result.
- The DC will complete its review of the data quality and the instrument state.
- The Spokespersons will appoint two co-chairs of a team to coordinate the preparation of the detection case to the collaboration and to coordinate the writing of the detection paper; one of the co-chairs will be a Chair of the relevant Search Group, and the other will be a leader of

the instrumental work. The co-chairs will work with the Spokespersons to engage the contributions of appropriate experts in all relevant areas, including data quality, significance estimation, parameter estimation, and the EM follow-up (if appropriate). The Spokespersons should also take steps to ensure the highest level of clarity and style in the paper. Drafts of the paper will be shared with all interested members of the Collaborations, who are in turn invited to comment and suggest improvements.

- The LSC Spokesperson and the Virgo Spokesperson notify the collaborations of the possible detection via a spectrum of methods, including:
- an L–V Spokesperson e-mail to the collaborations
- convening a telecon open to all LSC and Virgo members to present and discuss the initial information about the event
- maintaining up-to-date information on the web, including up-to-date drafts of the detection paper.
- The outreach committees are informed and charged with preparing material for communicating with the general public about the detection

STEP 3: PREPARING FOR THE DECISION

When the outcomes of Step 2 are available, the LSC/Virgo leadership will convene a meeting including the chairs of the relevant search group(s), the chairs of the DAC, the chairs/representatives of the Detector Characterization/instrument teams, the chairs of the Paper Coordinating Team, and the DC chairs. This group will discuss the case as it is understood at that time, and decide whether to take it forward and make the case to the DC and the collaborations.

At this time, the Paper Coordinating Team must provide the draft of the paper that describes the discovery and detailed material supporting the evidence for detection.

The LSC/Virgo leadership will formally charge the joint DC to review the claim. The DC plays the role of an independent investigator, asking broader and different questions than the people doing the analysis may have asked themselves and examining the case with fresh eyes/perspective different from the previous stages. In essence, the DC plays the Devil's Advocate, providing an opportunity for the wider Collaboration to form their own opinion. This stage may be *somewhat* adversarial but will certainly involve multiple and closely coupled interactions with the search group, detector characterization groups, and instrument scientists.

During this stage, the search groups and their review teams, Detector Characterization and instrumental team, the DAC, the Paper Coordinating Team, and the DC *may* carry out the following activities:

- Deeper and/or broader review of the detection case and the draft detection paper
- Consideration of analyses that may take longer to complete than allowed by the previous steps (e.g., parameter estimation)
- Additional analysis with different parameters to assess the robustness of the data analysis
- Audit of the detector hardware and/or software to test that the configuration is known
- Repeat measurements of detector calibration and response to hardware injections
- Literature searches to assess the astrophysical predictions concerning sources of this type
- Any further checks and actions, if necessary

Input from the Collaborations concerning any aspects of the detection claim and the paper will be welcomed and considered.

STEP 4: MAKING THE DECISION

Once the DC has substantially completed its evaluation, its assessment of the candidate is communicated to the collaborations' managements and the collaborations. Here, the DC provides the spokespersons its evaluation of the proposed discovery. The Spokespersons will circulate the DC report executive summary and final draft of the paper to the collaborations. The spokespersons will remind the collaborations that confidentiality is paramount during this process.

At the earliest possible time after the collaborations have had time to read and evaluate the paper, an L–V collaboration meeting will be held, either in person or by teleconference. During this meeting, the search group will summarize the case for the candidate, and the DC will present their recommendation to the collaborations. Following this, a discussion by the members of the LSC and Virgo Collaboration will take place on the merits of the case. At the end of that discussion, the members of the collaboration will formally consider whether *or not to submit the detection claim paper for publication.* A strong consensus from both the LSC and Virgo Collaboration to submit the paper is necessary for the paper to go forward. If necessary, a vote may need to be performed, but a decision to claim a detection will require that the vote be overwhelmingly positive.

If the collaborations' decision is positive, a blind injection check is performed. If a blind injection was present for this type of event, the spokespersons/director opens the "envelope." If no blind injection occurred, then the hardware injection channels are crosschecked for unrecorded or malicious injections to ensure that the event is truly valid.

If the event is certified, the paper is submitted and the discovery announcement goes out.

APPENDIX 2

First Draft of the Discovery Paper without Author List or Bibliography

Direct Observation of Gravitational Waves from a Binary Black Hole Merger

The LIGO Scientific Collaboration
LSC

The Virgo Collaboration
Virgo

On September 14, 2015 at 09:50:45 GMT, the two interferometers of the Laser Interferometer Gravitational-wave Observatory (LIGO) observed a strong gravitational wave signal matching the waveform expected from the coalescence of a binary black hole system. The Advanced LIGO interferometers at both the Hanford and Livingston observatories detected the signal with a time difference of 6.9 msec. It was observed chirping upwards in frequency from 30 Hz to 250 Hz with a peak gravitational wave strain of 1×10^{-21}. The signal is easily visible in whitened data, and is recovered by matched-filtering with a signal to noise ratio of 23.5 for the combined detections. Compared with backgrounds estimated empirically from the data, the false alarm rate is below 1 event per 11000 years, equivalent to a Poisson significance of 5 σ. The two interferometers were operating well, and an exhaustive investigation revealed no known or suspected instrumental cause for the signal. The signal shows no significant deviation from the best-fit waveform computed within general relativity using post-Newtonian methods for the inspiral, numerical relativity for the merger, and perturbation theory for the ringdown of the final black hole. The best-fit parameters of the binary black hole system are: chirp mass, 30.6 ± 1.4; the individual black hole masses, 42.1 ± 2.7 and 29.6 ± 3.1 (all in solar masses). The distance is estimated to be 500 ± 100 Mpc, uncertain mainly because the sky location is not well defined with only a two-detector observation. This is the first direct detection of gravitational waves and the first direct observation of the dynamics of black holes.

PACS numbers: 04.80.Nn, 04.25.dg, 95.85.Sz, 97.80.-d

This early draft results from the collection of inputs received from the various working groups based on the request in the proposed outline. The paper coordinating team did not have the chance to edit this raw material yet. This will be done in the coming days.

INTRODUCTION

One year after the final formulation of the field equations for General Relativity in 1915, Albert Einstein predicted the existence of gravitational waves. He found that the linearized weak field equations had wave solutions, showing them to be transverse waves of spatial strain that travel at the speed of light, generated by time variations of the mass quadrupole moment of the source. Also in 1916, Karl Schwarzschild published a solution for the field equations that describes a black hole. At that time Einstein understood that gravitational wave strains would be remarkably small, and expected that they would have no practical importance for physics. With the technology available in 1916 to explore the universe as then understood, this was a sound judgment.

The steady advances of astrophysics, especially the discovery of compact objects such as neutron stars and black holes, and the remarkable advances in the technology of measurement have changed the prospects. The observations by Hulse and Taylor of the energy loss to gravitational radiation by the binary pulsar system PSR 1913+16 provided the first observational demonstration of the existence of gravitational waves.

The direct detection of gravitational waves has been a long held aim, it allows new tests of general relativity, especially in the strong field regime, at the source and opens up a completely new way of exploring the universe by listening to the gravitational waves emitted by relativistic systems, many of which are electromagnetically dark. The idea of applying modern experimental methods to the direct search for gravitational waves of astrophysical origin began with Weber and his resonant mass detectors in the 1960s. In the 1970s and 1980s long-baseline broadband laser interferometric detectors were proposed with the potential for significantly better sensitivity. These latter techniques have resulted in a developing worldwide network of detectors: LIGO, consisting of two 4 km long instruments separated by 3000 km in the United States, GEO600, a 600 meter interferometer near Hannover, Germany, and VIRGO, a 3km long system in Cascina, Italy. These instruments reached an initial plateau of performance by 2005; through 2010 they carried out observations for a wide variety of signals but with no detections.

As a result of the Advanced LIGO project, the LIGO interferometers' sensitivities have been substantially increased. The interferometers are now able to make strain measurements at frequencies ranging from 10's of Hz to a few kHz at levels of strain $h \sim 10^{-22}$ and smaller, just at the levels needed to intercept the gravitational waves from compact sources, with their short dynamical times and relativistic velocities5

On September 14, 2015 at 09:50:45 GMT, the two LIGO Hanford and Livingston gravitational wave interferometers detected remarkably strong signals 6.9 msec apart. The initial detection was made with a search for coincident excitations above noise in both detectors. It was observed chirping upwards in frequency from 30 Hz to 250 Hz with a peak gravitational wave strain of 1×10^{-21}. The signal is easily visible in whitened data. The waveforms match well the signal expected from a binary black hole system with best-fit component masses of 42.1 ± 2.7 and 29.6 ± 3.1 and a chirp mass of 30.6 ± 1.4 (all in solar masses). The combined matched-filter amplitude signal to noise ratio is 23.5. With just two detectors, the sky position is not well determined, and because of this the distance estimate of 500 ± 100 Mpc has large uncertainty. After estimating the coincidence background empirically, we have concluded that an event this strong with such a good match would be expected to be caused by instrumental effects less often than once every 11,000 years. Moreover, a rigorous examination of the auxiliary data of both interferometers shows no evidence of a possible instrumental cause for the signal. The best-fit filter is a waveform built from general-relativistic computations of the inspiral, merger, and ringdown phases of the system, and the residual when this waveform is subtracted from the unfiltered time-domain data is consistent with instrumental noise. Now 100 years after the fundamental predictions of Einstein and Schwarzschild, we report the first direct detection of gravitational waves, as well as the first direct observation of an astrophysical black hole through the gravitational waves it emitted as it was being formed.

At the time of the reported event only the LIGO detectors and GEO600 were operating. GEO600 is not designed to have adequate sensitivity at the low frequencies required to detect this event. VIRGO was still being upgraded as part of the Advanced VIRGO project to be completed in 2016.

THE DETECTORS

Gravitational wave observatories were first envisioned in the early 1970 [1]. They became a reality towards the end of the millennium, when the initial LIGO [2] and VIRGO [3] observatories went online. The VIRGO observatory is located in Pisa, Italy, whereas the two LIGO observatories are located in the United States near Livingston, LA and Hanford, WA, respectively. Since then, these observatories went through a major upgrade to install Advanced LIGO [4] and Advanced VIRGO [5]. For the first observational run in the advanced detector era both LIGO observatories were operational and participated in the run.

Even in this early stage, the sensitivity of the Advanced LIGO detectors is significantly better than during the initial era. During the first observational run the horizon distance for binary neutron star inspirals was approximately 150 Mpc, or approximately four times the distance compared to initial LIGO. For binary black systems with individual masses of $30 M_{\odot}$ the improvement was a factor of five to six—being visible to a distance as far as 2.5 Gpc ($z \approx 0.4$). Since the rate scales with the cube of the distance, a week at current sensitivity surpasses all previous runs combined.

The first observational period lasted from August 17, 2015 to Jan 12, 2016, with the first 4 weeks designated as an engineering run. The first week of the engineering run was used to tune the instrument, the second week was used for an extensive calibration, and weeks two and three were used to shake down the online analysis code and the hardware injections. The event GW150914 was observed during the forth week. But, it is important to recognize that the instrument was in its final configuration and running at nominal sensitivity. As a matter of fact, no changes to the detector configuration were allowed for the following 22 days to make sure that we understood the rate of background events with sufficient statistics.

Figure 1 shows a simplified layout of the experimental setup. The Advanced LIGO detector comprises of a 4 km long Michelson laser interferometer. Optical resonators are deployed in the arms to enhance its sensitivity by about a factor of 300 [6]. The antisymmetric port of the Michelson interferometer is held near a dark fringe, so that the majority of the power reflected from the arm cavities is sent back in the direction of the laser. Power recycling uses a partially transmissive mirror in the input laser path to form a third optical resonator enhancing the power incident on the beam splitter by a factor of 35–40 [7]. Ideally, the transmission of the power recycling mirror is adjusted to the losses of the interferometer, so that no light returns to the laser. Such an arrangement optimizes the optical power in the arm cavities, and hence the sensitivity to differential length changes induced by gravitational waves. During the first observational run the light stored in the arm cavities reached 100 kW. A forth optical resonator with a partially transmissive mirror at the antisymmetric port is used to optimize the extraction of the gravitational wave signal [8].

The light source is a pre-stabilized laser followed by a high power amplifier stage [9]. The maximum available power is around 200 W, but only about 20–25 W were used during this run. An electro-optics modulator is used to impose RF modulation sidebands onto the laser light. These RF modulation sidebands are used to sense the auxiliary degrees-of-freedom of the interferometer utilizing the Pound-Drever-Hall reflection locking technique [10]. A triangular optical resonator of 16 m length is placed between the laser source and the

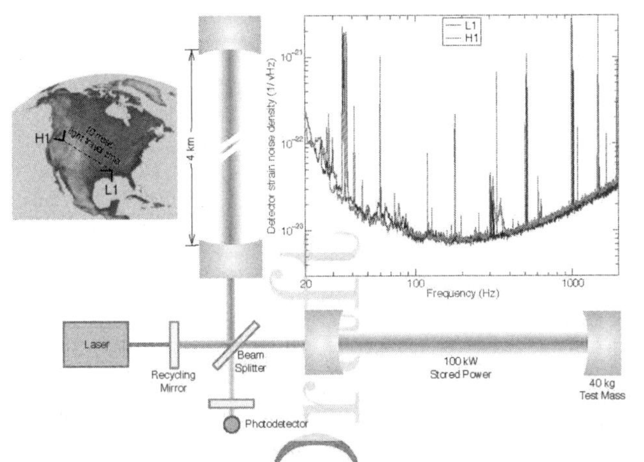

FIG. 1. Simplified setup and sensitivity of the Advanced LIGO detector. H1 and L1 are used to represent the LIGO detectors at Hanford, WA and Livingston, LA, respectively.

Michelson interferometer to clean up higher-order transverse optical modes and to further stabilize the laser frequency [11].

The LIGO test masses are high quality fused silica mirrors of 34 cm diameter and 20 cm thickness. These 40 kg optics are suspended by multi stage pendula to isolate them from ground vibrations [12]. The suspensions themselves are mounted on active seismic isolation platforms to reduce the absolute motion and provide further isolation [13, 14]. Both systems reside inside an ultra-high vacuum system to prevent acoustic couplings and to keep the Rayleigh scattering from the residual gas at a minimum. The seismic isolation system provides 3–4 orders of magnitude suppression of ground motion above 10 Hz. The suspension system provides another 6–7 orders to eliminate ground motion from affecting the test masses in the sensitivity band. A thermal compensation system consisting of ring heaters and CO_2 lasers is used to apply selective heat to the test masses [15]. This allows to simultaneously correct for intrinsic curvature mismatch, to compensate for thermal lensing by the main laser beam, and to mitigate parametric instabilities [16].

The optical resonators are locked on resonance with a servo controls system [17]. Servo controls are also required by the laser source to stabilize both intensity and frequency, by the the seismic isolation systems to lock the platform to an inertial frame, by the suspension system for local damping and external actuation, and by the auto-alignment system to control angular drifts [18]. Most of these servos are implemented in digital. The sensor signals are digitized at 64 kHz, then digitally filtered to compute the controls signal, before converted back into analog. The digital controls computers also serve as the front-end of the data acquisition system which continuously writes $\sim$ 6 MB/s of time series data to disk. A state based automation controller provides hands-free operations during running.

The gravitational wave signal is extracted at the anti-symmetric port using an output mode cleaner and DC offset locking [19]. The output mode cleaner is a small fixed spacer optical resonator used to reject unwanted light from the Michelson contrast defect and from the RF modulation sidebands. A gravitational wave will slightly offset the phase of the light in each arm and produce a differential signal which will be seen as a change of the average light level by the main photodetectors.

The strain sensitivity is shown in the inset of Figure 1. The sensitivity is limited by shot noise at higher frequencies and residual ground motion at lower frequencies [20]. Several line features are visible in the spectrum which are well understood, such as 60 Hz harmonics, the roll and bounce modes of the suspended test masses, the violin modes of the suspension fibers and their harmonics, some acoustic resonances of optics mounts, and the calibration lines. These lines do not significantly degrade the sensitivity to detect the merger of inspiral binary systems.

The interferometers are calibrated using the photon recoil of separate calibration laser beams that are directed onto the end test masses. These laser beams can be amplitude modulated to create a known time varying force on the test mass, which translates to a precise knowledge of the arm length variation using the mass's inertia.

The calibration beam power is measured using integrating sphere detectors that are regularly calibrated against absolute references at NIST, to sub-percent levels of precision and accuracy.

When the interferometers are in observation mode, their calibration is continuously monitored with the photon calibrator beams by applying sine wave modulations to them at select frequencies. These calibration lines are visible in the noise spectra shown in Fig. [?]. The calibrator beams can also be used to inject forces that mimic the waveforms of specific gravitational wave sources. Outside of observation mode, a swept-frequency modulation of the calibrators is periodically performed to verify the interferometer calibration across a large fraction of the detector frequency band.

The calibration procedure also accounts for the action of the servo controls that hold the arm cavities on resonance. This is done using a detailed model of the servo loop, which is verified with a collection of tranfser function measurements that characterize all components of the loop. With this system model and the absolute calibration provided by the photon calibrator, the output data stream can be accurately calibrated across the full detection frequency band. The 1σ statistical calibration uncertainty is less then 10% in amplitude and 10 degrees in phase over the band of 10-2000 Hz. To check for systematic errors in the absolute calibration, we have compared the photon calibrator to two alternative displacement references; one of these is based on the laser wavelength and the other is derived from a known modulation of the laser frequency [? ?].

At each observatory, timing signals are synchronized to Global Positioning System time to better than 1 microsecond; the timing signals are distributed to all data acquisition computers via optical fiber [?]. The overall timing uncertainty at each interferometer is less than 10 microseconds.

OBSERVATIONAL RESULTS

In this Letter, we report the first direct observation of gravitational waves, detected at 2015-09-14 09:50:45 UTC in coincidence by the two Advanced LIGO detectors at the Livingston (LLO) and Hanford (LHO) observatories. Our observation period began on 12 September, 2015 00:00 UTC and concluded on 20 October, 2015 13:32 UTC. We analyzed all times when both LIGO detectors were operating, for a total of 16 days of coincident data. This time period represents data from the first observing run (O1) of Advanced LIGO, in addition to some time of high quality data before the nominal O1 start time. During our observation time we were sensitive to binary black hole mergers with total mass between 50–100 $M_\odot$ at a distance of $\gtrsim 400$ Mpc.

GW150914 was first discovered on September 14, 2015 09:53:51 UTC by a real-time search for unmodeled gravitational-wave transients, approximately three minutes after data was collected. The discovery was promptly confirmed by a second burst search, also operating in low latency [?]. The signal waveforms and the estimated chirp mass of 27.6 $M_\odot$ calculated by the unmodeled search suggested that the signal was from a compact object merger with at least one component large enough to be a black hole. The observatories were notified to freeze the state of the instruments and start the data validation process.

m_1	m_2	s_1	s_2	$\mathcal{M}$	ρ_H	ρ_L
$48 M_\odot$	$37 M_\odot$	0.96	−0.90	$36 M_\odot$	20	13

TABLE I. Inferred parameters of GW150914. FILL IN WITH FULL PE RESULTS, GIVE RANGES AS WELL AS MAX LIKE VALUES

At the time, modeled searches for compact binary coalescence were also running in real-time. LIGO-Virgo collaborations perform a low-latency search for modeled compact binary signals aiming to send alerts to astronomical observers [21, 22]. At the time of the event, two low-latency pipelines were targeting binary systems with maximum total mass of 15 $M_\odot$, significantly smaller than the estimated total mass of the Event. Due to this limitation, neither pipeline detected it. On September 15, 2015 14:28:03 UTC, GW150914 was first circulated to observing partners for electromagnetic follow-up observations.

Subsequent searches of archived data around the time of the event confirmed the detection of a high amplitude signal. Two independent templated analyses searching for GW from compact binary mergers each recovered an event with a false alarm probability of $< 1.6 \times 10^{-6}$, corresponding to 4.5σ significance (see Section X.X). Offline searches for unmodeled GW transients found the signal to have a false alarm rate of less than 1 in 28,000 years, or a false alarm probability of $< 1.5 \times 10^{-6}$.

The observed signal agrees extremely well with the theoretical predictions gravitational waves radiated in the final moments of a black-hole binary merger [] forming a more massive black hole. No detectable residual remains after subtracting the best-matching GR template from the detector data. The waveforms, amplitudes, and arrival times in the two LIGO detectors are consistent with having arrived from a single astrophysical source (See Figure 2). Estimates for the parameters of the binary merger are shown in Table I, and suggest a final black hole with a total mass of $\sim 85 M_\odot$.

INSTRUMENTAL AND ENVIRONMENTAL VALIDATION

In evaluating the detection candidate, investigations were made into mechanisms that could generate spuri-

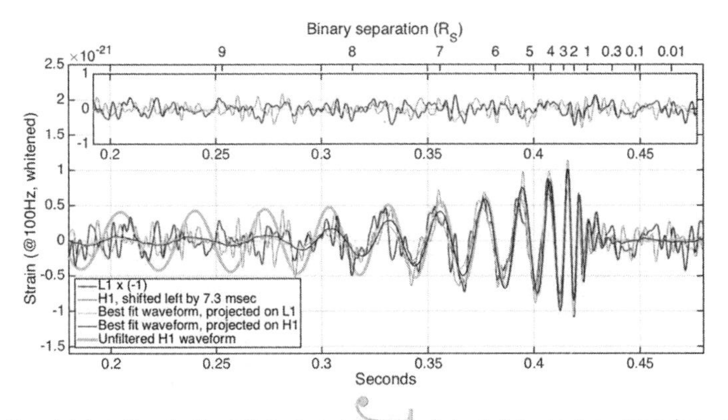

FIG. 2. Josh's updated caption, should get Stefan to upda Observed signals (after bandpass filtering) compared with the best fit theoretical models obtained from General Relativity. Time series of the detector data are filtered with a zero-phase and a-causal band-pass filter with nearly flat magnitude response in the band from 20 to 400Hz and notches for strong lines. The time axis is Livingston arrival time, offset from September 14, 2015, 09:50:45 UTC. Hanford is shifted early by 7.3 milliseconds. The best-fit template and versions processed with the same filter and projected on to the Livingston and the Hanford detectors are also plotted. Inset: Template-subtracted time series, filtered and processed the same way. No residual signal is visible above thedetector noise. LHO in red, LLO in green. Further simplification of this figure: remove unfiltered template?

ous signatures in the interferometers. Instrumental artifacts, excitations produced in the surrounding environment, and the subsystem that injects digital excitations into the detectors were considered. We also validated site timing system performance and the integrity of the data acquisition file system.

The detectors have a baseline of stationary Gaussian noise dominated by seismic and photon shot noise. In both detectors the band between about 20 Hz and 120 Hz shows a stationary noise floor slightly exceeding the predicted noise. Typical output data also contain short transient noise events (glitches) with a higher amplitude than would be expected from Gaussian processes alone [23]. These transients arise from a number of mechanisms and thus have various morphologies. Many have well-understood causes, e.g. loud bursts from overflows of actuator control signals, 50 and 60 Hz glitches caused by motors switching on, or fluctuations in the radio frequency modulation used for control of auxiliary degrees of freedom. These transients can be identified by instrumental monitoring channels. In addition, L1 experienced transients resulting from the beat of two radio frequency signals sweeping through the measurement band. These features however can be predicted using the readback of the laser frequency control and are also witnessed more strongly by auxiliary channels than by the gravitational-wave channel, so this mechanism can be ruled out. In addition to these, both detectors experienced short transients of unknown origin consisting of a few cycles around 100Hz. While these are in the same frequency band as the candidate event, they have a very characteristic time-symmetric waveform with no clear frequency evolution. The search included an algorithmic test to reject most transients with that morphology, and any that survive are accounted for in the background estimate.

To assist in validating a candidate signal the sites are equipped with Physical Environment Monitoring (PEM) sensors: seismometers, accelerometers, microphones, magnetometers, radio receivers, and a cosmic ray detector (pem.ligo.org). Injections of magnetic, radio frequency, acoustic, and vibration signals, as well as correlation studies, indicate that the SNR is higher in these sensor channels than in the GW channel for events produced by environmental signals. These injections are used to quantify the coupling between the environment and the GW channel [24, 25]. The PEM sensors did not record anything that could account for the event. We also checked that environmental signal levels at our observatories, and at external electromagnetic weather observatories and cosmic ray detectors were typical of normal times.

This paragraph needs updating for aLIGO hardware, or should be struck: The hardware and software operating the two LIGO interferometers are nearly identical and their timing is synchronized to Global Positioning System time. This precise timing has resulted in low-amplitude combs of spectral lines (e.g. 1 and 16 Hz) that are coherent between the two sites, produced by slight cyclical corruption of the data [?]. We do not see synchronized transient data corruption events because we do not normally synchronize processes between the Hanford and Livingston sites.

As a means of validating the LIGO instrument response and calibration, actuators on the interferometer test-masses are used to apply small forces which simulate the effect of a gravitational wave. These end-to-end tests, referred to as "hardware injections", require that the digital control systems be capable of emulating gravitational waves from astrophysical sources. As such it is important to verify that the observed signal was not produced by the digital system.

After the observation of the candidate event, hardware injections of binary-black hole waveforms were performed. Digital signals were coherently added to the control stream of the two interferometers, and recovered by analysis codes, within the known calibration error (xx-yy%).

A record of all injected signals is kept, and no injection was recorded at the time of GW150914. To further rule out this possibility, the signal which entered the digital system (labeled XXX in figure YYY) was used as input to a standalone model of the interferometer response. This model contained only the known filters, gains and responses of the real detector. At all points along the signal path they were found to match (in particular at the output, labeled ZZZ in figure YYY) thereby excluding the possibility of a hardware injection.

The Advanced LIGO Timing System [26] has a hardware defined timing accuracy and overall clock synchronization precision of better than one microsecond. At each LIGO site the synchronization reference is derived from GPS satellites that ultimately ensure both relative and absolute synchronization over the entire detector and between the sites.

The timing system has self-diagnostic information implemented in hardware. Additional GPS synchronized timing witness channels are also recorded along with the aLIGO datastream, and independent atomic clock timing signals provide an additional redundant check [?]. Analysis of available timing diagnostic data indicated that the timing performance of the aLIGO detectors around GW150914, was according to specifications.

We are uncertain if this paragraph is warranted: An audit was performed of all of the control and data systems on each interferometer [?]. This audit included verification of the real-time software running on control computers; a dump of the running kernel object build information, provided by the running software itself; trace and verification of all build information back to the source code; a visual check of all user provided source code used in real-time code objects, and a recompilation of all code against traced sources in order to verify that the newly compiled code matched the running software. In addition, CRC checksum tests were performed on all frame data files. Once generated these frame files are rapidly distributed both to on-site and remote systems. No problems were found with any of the audited systems, software or data. Each system was also checked for evidence of remote login activity at the time of the event and nothing unusual was found.

DATA ANALYSIS

To search for a broad range of transient GW sources [] **[references to S5S6 papers]** we identify coincident events in the time-frequency (TF) data from the two LIGO detectors in the frequency band 48 − 1024 Hz. Figure ?? shows the TF plot of the whitened strain data around the time of the event. A "chirp" signal is clearly visible in the LHO and LLO detectors as a TF cluster of the excess power above the baseline detector noise.

The burst search coherentWaveBurst (cWB) combines the TF data[] from both detectors and selects the coincident clusters. Due to non-stationary detector noise, the initial rate of selected events is high: 10^{-2} Hz or more. To reduce the false alarm rates (FAR) the coherent events are identified. First, the likelihood analysis is performed [] to reconstruct the detector responses to a GW signal and the residual detector noise. Then, the coherent waveforms and source sky location are obtained by maximizing the likelihood statistic c_cE_r over the sky, where E_r is the total energy of the reconstructed signal. The network correlation coefficient $c_c = E_c/(|E_c| + E_n)$ depends on the energy of the residual noise E_n and the coherent energy E_c, which is proportional to the cross-correlation between the reconstructed GW waveforms in LHO and LLO. For highly correlated GW signals $c_c \sim 1$ and for spurious events (glitches) $c_c << 1$. Events with $c_c < 0.7$ are excluded from the analysis. The cWB detection statistic $\rho_c = (2c_cE_c)^{1/2}$ is a biased estimator of the network signal-to-noise ratio (SNR), which approaches the true SNR value for highly correlated GW signals. The event candidate GW150914 was detected by the burst search with $\rho_c = 20.0$. The statistic ρ_c ranks events by their significance above the background due to the random coincidence of glitches in the LHO and LLO detectors. We re-run the same burst search multiple times on time-shifted data to measure the accidental background rates. Large number of independent (multiple of 1 s) time-shifts were performed to accumulate 28,000 years of the equivalent background time. The event candidate GW150914 has passed all selection cuts. It is louder than any background event expected in xxx years of the equivalent observation time.

Two additional unmodeled methods were used to estimate the significance of this event. The oLIB pipeline found GW150914 in an online search, and later evaluated it to be more significant than all background events found in XXX years of equivalent background time. A Bayesian follow-up pipeline (BayesWave Burst) processed the detection candidate and all background triggers identified by cWB with $\rho_c > 11.3$ and computed the evidence ratio (Bayes factor) between hypotheses that the candidate event is due to instrument noise or a gravitational wave signal. The evidence in favor of the hypothesis that GW150914 is a gravitational wave signal is higher than any background event from XXX years of data.

Having found this event to be highly significant in searches for unmodeled transients, any hypothesis for a non-astrophysical source would be forced to explain the "coincidence" that the most significant transient found in LIGO data so far, also displays a waveform consistent with the predictions of general relativity. This qualitative evidence strongly supports the compact object merger scenario described in this letter.

Binary systems containing two compact objects, e.g., black holes and/or neutron stars, are a canonical source of gravitational radiation. As the objects orbit one another, they lose energy [27–29] in a predictable way eventually causing the system to merge. The merger timescale and phase evolution of a compact binary system depends on the intrinsic parameters such as the mass and spin of each binary component [30–36]. The binary's physical parameters lead to a unique time evolution of the binary's decaying orbit, which in turn leads to a unique gravitational waveform, which can be precisely modeled to extract the signal from the detector data [37, 38].

Because the parameters of the binary are not known *a priori*, searches for compact binary coalescence correlate a set of waveform models with detector data in an attempt to cover the full plausible parameter space of detectable binaries with minimal loss in signal-to-noise ratio [39–45]. The dynamics of the early phase of binary coalescence, known as the *inspiral* phase, is well modeled by perturbative methods for solving the Einstein equations. However, describing the late stage dynamics involving the *merger* of the binary and eventual *ring-down* of the final black hole, requires full numerical solutions [46–51]. For systems with total mass above $\sim 10\ M_\odot$, the merger occurs at low enough frequency to be visible with the LIGO detectors and therefore requires all three phases of inspiral to be accurately modeled [52].

Modeled searches for compact binary systems have been ongoing throughout initial LIGO [53–68]. Previous searches targetted various aspects of binary coalescence, but were not as complete as the search presented here. Early searches only modeled the inspiral or ringdown phases of binary coalescence separately. Searches starting with LIGO's fifth science run used waveform tempates that captured the complete inspiral, merger and ringdown of binary black hole system. With one exception, all previous LIGO and Virgo searches neglected spin effects in waveform models [64]. Advances in waveform modeling and analysis techniques [57, 69, 70] have made it possible to effectively search for binary black holes accounting for component spin.

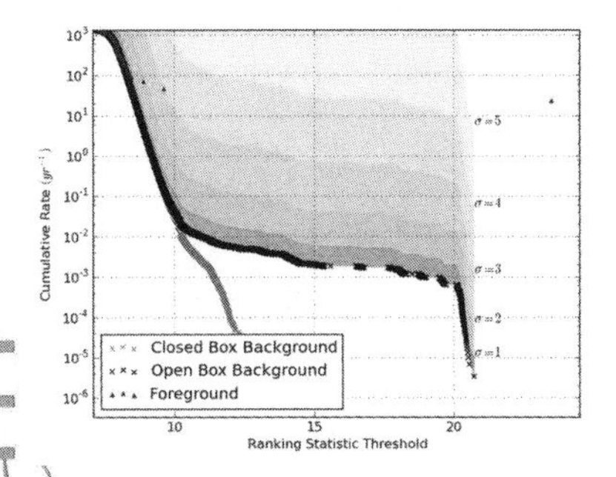

FIG. 3. Placeholder for the significance plot(s) (here showing CBC significance plot obtained by pycbc)

The search described here targeted binaries with component masses between 1 and 99 $M_\odot$, with a total mass less than 100 $M_\odot$. The waveform models incorporated spin for each component object along the direction of the orbital angular momentum. The magnitude of the dimensionless spin parameter varied between 0 and 0.04 for components below 2 $M_\odot$, and between 0 and 0.98 for components above 2 $M_\odot$. Although the waveforms did not explicitly model spins that are misaligned with the orbital angular momentum, the models used are known to effectively recover systems with misaligned spin [71].

Modeled searches for compact binary coalescence begin with matched filtering [72, 73], which correlates each waveform template with the strain data. The matched filter output gives the signal-to-noise ratio (SNR) of each waveform model as a function of time for each detector. Peaks, known as triggers, are identified every second for each template. Due to the non-stationary nature of gravitational wave detector data, a simple correlation of the data with waveform models is not sufficient to determine if a signal is present. Additionally, signal consistency checks that perform a χ^2 fit to the expected correlation are used to rule out noise transients [74]. Triggers with a high signal-to-noise-ratio and a low χ^2 value which are found to be temporally coincident in two or more detectors form the set of candidates. The candidates are ranked by the likelihood that their parameters are caused by a signal rather than noise, with signal like parameters generally having a high signal-to-noise ratio and a low χ^2 in two or more gravitational wave detectors. The probability of obtaining the candidate parameters from noise is calculated and determines the significance of each.

SOURCE DISCUSSION

The transient signal from GW150914 is consistent with the general-relativistic predictions for the late stages of the coalescence and subsequent merger of a binary black hole system with parameters given in Table II. In what follows we briefly discuss some implications resulting from the detection of this source.

GW150914, with individual masses of at least $20\,M_\odot$ and potentially as high as $45\,M_\odot$, provides the most reliable evidence to date for the existence of massive stellar black holes. Such massive stellar-mass black holes are predicted, within models of stellar evolution, to form in environments with metallicities of a tenth or less of the solar value.

GW150914 not only provides the cleanest evidence for the existence of stellar-mass black holes, but clearly demonstrates that *binary* black holes can form in nature, and in addition that they can form with physical properties that lead to their merger within a Hubble time. The formation of binary black holes has been predicted by a wide range of astrophysical models, since the discovery of binaries with two neutron stars about forty years ago (e.g., see the reviews [75, 76] and references therein. They have been predicted to form both in galactic fields [77?] need cit for Clark et al 1979 and in globular clusters [77] through different pathways and at varying rates, most recently [78, 79]. Regardless of the specific formation channel, GW150914 allows us to constrain the rate (per co-moving volume and proper time) of binary black hole coalescences to be $10^{-1.40^{+0.83}_{-2.70}}\,\mathrm{Gpc}^{-3}\,\mathrm{yr}^{-1}$ (90% CL). This assumes a population uniformly-distributed in proper time and co-moving volume, an estimated false-alarm rate threshold of $10^{-4}\,\mathrm{yr}^{-1}$ and an Advanced LIGO sensitive time-volume of $25\,\mathrm{Gpc}^3\,\mathrm{yr}$, FIXME: see RATES PAPER for further details. The measured physical properties of GW150914, see Table II and the inferred merger rate for similarly massive systems appear broadly consistent with the wide range of possibilities allowed by theoretical predictions at low metallicities, both from isolated binary evolution and from dynamical interactions of black-hole systems in dense stellar environments. Careful examination of the system properties combined with its occurrence in the relatively local universe ($z \lesssim 0.2$) will provide concrete constraints on theoretical models. These astrophysical implications are further explored in the FIXME: Collaboration-Astro-paper.

The discovery of GW150914 also has profound implications for direct studies of the strong-field dynamics described by General Relativity. It offers us a direct probe of whether Nature's black holes correspond to the objects predicted by Einstein's theory. By subtracting the best-fit waveform model from the data, we find that the residual is consistent with Gaussian noise, indicating that the observed signal is consistent with the predictions from general relativity FIXME: at what CL?. In addition, from the FIXME: ≈ 6 cycles of the inspiral portion of the coalescence that were observed in the LIGO band, we find agreement with the post-Newtonian expansion with bounds on the individual coefficients that are up to $\sim 10^8$ times more constraining than previous limits (such as those obtained through the timing of the double pulsar FIXME: ref?). These bounds directly translate into an improved limit on gravitational-wave Compton wavelength describing a non-standard dispersion relation for the propagation of gravitational radiation of $\lambda_g \gtrsim 10^{16}$ m, FIXME: need exact number and CL. Finally, within general relativity a Kerr black hole is fully described by its mass and spin. If such a black hole is the end-product of the coalescence of two Kerr black holes, there exists a specific relation between the masses and spins of the initial and final objects. We find that these relations are satisfied by GW150914 at the 68% confidence level. FIXME: See TestCG paper for full details

TABLE II. Parameters for GW150914. We report the median value as well as the range of the 90% credible interval. Masses are as measured in the detector frame unless otherwise noted. The source redshift and source-frame component masses assume standard cosmology.

Primary component mass	$38.9^{+5.4\pm z}_{-3.2\pm y}\,M_\odot$
Secondary component mass	$32.8^{+3.3\pm z}_{-5.1\pm y}\,M_\odot$
Source-frame primary component mass	$35.4^{+5.1\pm z}_{-3.1\pm y}\,M_\odot$
Source-frame secondary component mass	$29.8^{+3.2\pm z}_{-4.5\pm y}\,M_\odot$
Luminosity distance	$456^{+171\pm z}_{-176\pm y}$ Mpc
Source redshift	$0.099^{+0.03\pm z}_{-0.04\pm y}$

IMPLICATIONS/BIGGER PICTURE

Since the relative strain sensitivity of Initial LIGO to BBHs with component masses similar GW150914 is a factor of ~ 4 lower than aLIGO, this event would not have been detected had it occurred in Initial LIGO. The sensitive time-volume of the search described here for mergers with comparable masses is 150 $\mathrm{Mpc}^3\,\mathrm{Myr}$. The observation of GW150914 is consistent with the Initial LIGO 90% confidence upper limit on the merger rate of binaries with similar component masses to GW150914 of 0.1 $\mathrm{Mpc}^{-3}\,\mathrm{Myr}^{-1}$ [54].

The rate density of sources with component masses similar to GW150914 is $X\mathrm{Gpc}^{-3}\,\mathrm{yr}^{-1}$ [?].

Besides the individual detection of the loudest events at relatively close distances, the superposition of all the faint unresolved sources at high redshift create a stochastic background, that we expect to see in the near future, depending on the rate and the average chirp mass of the population. Predictions for different models are presented in the companion paper [?]. The detection of a stochastic background will complement individual detections and will have a profound impact on our understanding of the evolution of binary systems properties over the history of the universe.

the bigger picture of O1: more data, reference other searches. Responsible to gather input and provide text for this bullet point are the DAC chairs. Laura Cadonati will liason with the paper-writing coordinating team.

Over the next five years, continued commissioning of the LIGO instruments [80] and the addition of Virgo [81] and Kagra [82] will increase the accesible volume of the Universe by about an order of magnitude. Based on GW150914, the global gravitational-wave detector network will deliver tens to hundreds of similar events per year. These observations will revolutionize our understanding of stellar evolution and black hole formation [?]. The most significant events, which might have SNR as large as ~ 70, will also enable unique tests of strong-field general relativity through the measurement of quasi-normal modes, higher harmonics of the inspiral waveform, and searches for additional polarizations [?].

In the future, the global gravitational wave detector network will significantly expand its reach, as described in [83]. Advanced LIGO is expected to reach its design sensitivity in 2019, with a factor of three increase in sensitivity across a broad band and an extension of the sensitive band to 10 Hz [80]. Advanced Virgo [81] will begin observations in 2016, extending the network and significantly improving the position reconstruction of sources, and will reach design sensitivity early next decade. The KAGRA detector [82] is currently under construction and is expected to begin observations in 2017-18. A proposal to install a LIGO detector in India [84] is in the final stages of consideration by the Indian government. If approved, LIGO India will join the global network around 2022. In addition, the Einstein Telescope [85] is a proposed future detector that will be an order of magnitude more sensitive than the existing detectors and extend the sensitive band down to 1 Hz [6].

Further details about these results, including pointers to companion papers, supplementary information, and associated data releases are available at this URL: `http://losc.ligo.org/GW150914`.

CONCLUSIONS

LIGO has made a direct detection of gravitational waves from an astrophysical source. The event waveform shows the three phases of a binary black hole merger: the radiation from the inspiral before the merger, the radiation from the merger itself signaling the formation of the new more massive black hole and the final stage, the oscillation of the space-time geometry as the new event horizon forms. The waveform is consistent with the predictions of General Relativity and provides evidence for the dynamics of General Relativity in the strong field limit. The observation also indicates that black hole binaries may be more numerous and more massive than had previously been believed.

The next near term steps in the research are to improve the detector sensitivity to the Advanced LIGO design to increase the rate of these events by searching further into the universe and being able to observe lower mass systems. In the next runs it is expected that Advanced VIRGO will be operating and the gravitational network will provide better localization for the sources allowing a connection to traditional astronomy. In the longer term we expect KAGRA in Japan and (hopefully) LIGO in India to join the network. With this we truly begin the new field of gravitational wave astrophysics.

[1] R. Weiss, Quarterly report of the Research Laboratory for Electronics, MIT, https://dcc.ligo.org/LIGO-P720002/public (1972).

[2] The LIGO Scientific Collaboration, Reports on Progress in Physics **72**, 076901 (2009).

[3] The Virgo Collaboration, Journal of Instrumentation **7**, P03012 (2012).

[4] The LIGO Scientific Collaboration, Classical and Quantum Gravity **32**, 074001 (2015).

[5] The Virgo Collaboration, Classical and Quantum Gravity **32**, 024001 (2015).

[6] R. W. P. Drever, *The Detection of Gravitational Waves*, edited by D. G. Blair (Cambridge University Press, 1991).

[7] B. J. Meers, Phys. Rev. D **38**, 2317 (1988).

[8] J. Mizuno, K. A. Strain, P. G. Nelson, J. M. Chen, R. Schilling, A. Rdiger, W. Winkler, and K. Danzmann, Phys. Lett. A **175**, 273 (1993).

[9] P. Kwee, C. Bogan, K. Danzmann, M. Frede, H. Kim, P. King, J. Pöld, O. Puncken, R. L. Savage, F. Seifert, P. Wessels, L. Winkelmann, and B. Willke, Opt. Express **20**, 10617 (2012).

[10] A similar approach was used in initial LIGO, see P. Fritschel, R. Bork, G. González, N. Mavalvala, D. Ouimette, H. Rong, D. Sigg, and M. Zucker, Appl. Opt. **40**, 4988 (2001).

[11] C. L. Mueller, M. A. Arain, G. Ciani, R. T. DeRosa, A. Effler, D. Feldbaum, V. V. Frolov, P. Fulda, J. Gleason, M. Heintze, E. J. King, K. Kokeyama, W. Z. Korth, R. M. Martin, A. Mullavey, J. Pöld, V. Quetschke, D. H. Reitze, D. B. Tanner, L. F. Williams, and G. Mueller, to be published in Review of Scientific Instruments (2015).

[12] S. M. Aston, M. A. Barton, A. S. Bell, N. Beveridge, B. Bland, A. J. Brummitt, G. Cagnoli, C. A. Cantley, L. Carbone, A. V. Cumming, L. Cunningham, R. M. Cutler, R. J. S. Greenhalgh, G. D. Hammond, K. Haughian, T. M. Hayler, A. Heptonstall, J. Heefner, D. Hoyland, J. Hough, R. Jones, J. S. Kissel, R. Kumar, N. A. Lockerbie, D. Lodhia, I. W. Martin, P. G. Murray, J. ODell, M. V. Plissi, S. Reid, J. Romie, N. A. Robert-

APPENDIX 3
Rules for Author Lists

Oct. 20, 00:01: The collaboration is moving rapidly towards a paper about GW150914. The paper will need to have a carefully assembled author list and this message is about that list. There may be two papers; the procedures here apply to either one or two coordinated papers. Any other companion papers are expected to use the regular procedures for assembling their author lists.

The procedures used by the P&P committee to attach the author list to a paper are described in LIGO-M060334. (See [1] below.) The list for this paper is expected to be the "August 2015 LSC and August 2015 Virgo author list" (M1500255 at https://dcc.ligo.org/LIGO-M1500255) with additions as sketched below.

The EMC [Elections and Membership Committee] has the task of assembling the LSC and LVC lists twice each year following the LSC Publication Policy (T010168). [2] This document also describes procedures for adding authors to any paper. [3,4] After discussion among the Elections and Membership Committee, Publications and Presentations Committee, and Spokesperson, we will adopt the following procedure. All requests for adding LSC authors should be sent to the LSC Executive Committee (lsc-excomm@ligo.org). The requests can come either from the Analysis group or from the PI of any LSC group. The requests can be for current LSC members who made significant contributions to the results presented in the detection paper or for past members who contributed in a significant way to bringing the LIGO detectors to the condition they are in today. Please detail the contributions of the nominated member.

All requests must be submitted by Sunday, November 15 and will be decided upon by November 30th. The EMC will produce a special author list (in the DCC) for this paper by December 7.

----

[1] The relevant sentences are: "The composition of the author list is specified in the LSC Publication Policy (T010168, latest revision). ... LSC policy states that the correct author list is the one in place when the paper is first circulated to the collaboration by the P&P Committee."

[2] "The August [2015] list will contain the names of current LSC members who joined the LSC prior to Dec 15 of the previous year [2014] and who have devoted more than 50% of their research effort toward LIGO since that date. It will also contain the names of past LSC members who had earned authorship but have left the collaboration (or whose research effort fell below the 50% level) after Aug 15 of the previous year [2014]."

[3] For this paper, the most relevant paragraph is "Individuals who have made significant contributions to a particular observational paper, but who are not on the LSC author list, may be added to the author list of that paper. To add an author, his/her name should be proposed for authorship by the Analysis Group when presenting the paper to the LSC Executive Committee for final approval. Consent of the LSC Executive Committee is required."

[4] A second paragraph says "Any special arrangements or conflicts concerning authorship should first be brought to the attention of the author contact from the relevant group (typically the PI) who can bring them to the EMC. The EMC will make a recommendation to the LSC Spokesperson, who will make the final decision, consulting with others as needed. Any conflicts on authorship on LIGO publications will be resolved by the Spokesperson in consultation with the LSC Executive Committee and the Laboratory Directorate."

--

David Tanner

REFERENCES

Abbott, B. P., et al. 2016a. Observation of gravitational waves from a binary black hole merger. *Physical Review Letters* 116:061102.

Abbott, B. P., et al. 2016b. GW151226: Observation of gravitational waves from a 22-solar-mass binary black hole coalescence. *Physical Review Letters* 116:241103. http://arxiv.org/abs/1606.04855.

Abbott, B. P., et al. 2016c. Binary black hole mergers in the first advanced LIGO observing run. http://arxiv.org/abs/1606.04856.

Barnes, V. E., P. L. Connolly, D. J. Crennell, et al. 1964. Observation of a hyperon with strangeness minus three. *Physical Review Letters* 12 (8): 204–206.

Berger, P. L. 1963. *Invitation to Sociology*. Garden City, NY: Anchor Books.

Bloor, D. 1973. Wittgenstein and Mannheim on the sociology of mathematics. *Studies in the History and Philosophy of Science* 4:173–191.

Castelvecchi, D. 2016. Gravitational-wave rumours in overdrive. *Nature*, January 12, http://www.nature.com/news/gravitational-wave-rumours-in-overdrive-1.19161.

Collins, H. M. 1975. The seven sexes: A study in the sociology of a phenomenon, or The replication of experiments in physics. *Sociology* 9 (2): 205–224.

Collins, H. M., ed. 1981a. *Knowledge and Controversy: Studies in Modern Natural Science: Special Issue of Social Studies of Science* 11 (1).

Collins, H. M. 1981b. Son of seven sexes: The social destruction of a physical phenomenon. *Social Studies of Science* 11 (1): 33–62.

Collins, H. M. 1985. *Changing Order: Replication and Induction in Scientific Practice*. Beverley Hills: Sage. (2nd ed. 1992, Chicago: University of Chicago Press.)

Collins, H. M. 1991a. AI-vey! Response to Slezak. *Social Studies of Science* 21:201–203.

Collins, H. M. 1991b. Simon's Slezak. *Social Studies of Science* 21:148–149.

Collins, H. M. 1996. Embedded or embodied? A review of Hubert Dreyfus' *What Computers Still Can't Do. Artificial Intelligence* 80 (1): 99–117.

Collins, H. M. 2001. Crown jewels and rough diamonds: The source of science's authority. In *The One Culture? A Conversation about Science*, ed. J. Labinger and H. Collins, 255–260. Chicago: University of Chicago Press..

Collins, H. M. 2004. *Gravity's Shadow: The Search for Gravitational Waves*. Chicago: University of Chicago Press.

Collins, H. M. 2010. *Tacit and Explicit Knowledge*. Chicago: University of Chicago Press.

Collins, H. M. 2011a. *Gravity's Ghost: Scientific Discovery in the Twenty-First Century*. Chicago: University of Chicago Press.

Collins, H. M. 2011b. Language and practice. *Social Studies of Science* 41 (2): 271–300.

Collins, H. M. 2013. *Gravity's Ghost and Big Dog: Scientific Discovery and Social Analysis in the Twenty-First Century*. Chicago: University of Chicago Press.

Collins, H. M. 2014. *Are We All Scientific Experts Now?* Cambridge: Polity Press.

Collins, H. M. In preparation. *Artifictional Intelligence: Human and Computer Understanding*.

Collins, H. M., A. Bartlett, and L. Reyes-Galindo. 2016. The ecology of fringe science and its bearing on policy. http://arxiv.org/abs/1606.05786.

Collins, H. M., and R. Evans. 2007. *Rethinking Expertise*. Chicago: University of Chicago Press.

Collins, H. M., and R. Evans. 2014. Quantifying the tacit: The imitation game and social fluency. *Sociology* 48 (1): 3–19.

Collins, H. M., and R. Evans. 2015. Expertise revisited I—Interactional expertise. *Studies in History and Philosophy of Science* 54:113–123.

Collins, H. M., and R. Evans. 2017. *Why Democracies Need Science*. Cambridge: Polity Press.

Collins, H. M., P. Ginsparg, and L. Reyes-Galindo. 2016. A note concerning Primary Source Knowledge. *Journal of the Association for Information Science and Technology*. http://arxiv.org/abs/1605.07228.

Collins, H. M., and M. Kusch. 1998. *The Shape of Actions: What Humans and Machines Can Do*. Cambridge, MA: MIT Press.

Collins, H. M., and T. J. Pinch. 1982. *Frames of Meaning: The Social Construction of Extraordinary Science*, vol. 5. London: Routledge & Kegan Paul.

Collins, H. M., and T. J. Pinch. 1993. *The Golem: What Everyone Should Know about Science.* Cambridge: Cambridge University Press. (New ed., 1998, subtitled *What You Should Know about Science*, reissued as Canto Classic in 2012.)

Collins, H. M., and T. J. Pinch. 2005. *Dr. Golem: How to Think about Medicine.* Chicago: University of Chicago Press.

Collins, H. M., and G. Sanders. 2007. They give you the keys and say "drive it": Managers, referred expertise, and other expertises. In *Case Studies of Expertise and Experience: Special Issue of Studies in History and Philosophy of Science*, ed. H. M. Collins, vol. 38 (4): 621–641.

Collins, H. M., and M. Weinel. 2011. Transmuted expertise: How technical non-experts can assess experts and expertise. *Argumentation: Special Issue on Rethinking Arguments from Experts* 25 (3): 401–413.

Dreyfus, H. L. 1967. Why computers must have bodies in order to be intelligent. *Review of Metaphysics* 21 (1): 13–32.

Dreyfus, H. L. 1972. *What Computers Can't Do.* Cambridge, MA: MIT Press.

Dreyfus, H. L. 1992. *What Computers Still Can't Do.* Cambridge, MA: MIT Press.

Duhem, P. 1914/1981. *The Aim and Structure of Physical Theory.* Trans. P. P. Wiener. New York: Athenaeum.

Epstein, S. 1996. *Impure Science: AIDS, Activism, and the Politics of Knowledge.* Berkeley: University of California Press.

Fleck, L. 1935/1979. *Genesis and Development of a Scientific Fact.* Chicago: University of Chicago Press. (First published in German in 1935 as *Entstehung und Entwicklung einer wissenschaftlichen Tatsache: Einführung in die Lehre vom Denkstil und Denkkollektiv.*)

Franklin, A. 2013. *Shifting Standards: Experiments in Particle Physics in the Twentieth Century.* Pittsburgh: University of Pittsburgh Press.

Franklin, A., and H. M. Collins. 2016. Two kinds of case study and a new agreement. In *The Philosophy of Historical Case Studies*, ed. T. Sauer and R. Scholl, 95–121. Boston Studies in the Philosophy of Science. Dordrecht: Springer.

Franzen, C. 2016. Listen to the sound of gravitational waves: The "chirp" would make a great ringtone. *Popular Science*, February 11, http://www.popsci.com/listen-to-sound-gravitational-waves.

Galison, P. 1997. *Image and Logic: A Material Culture of Microphysics.* Chicago: University of Chicago Press.

Galison, P. 2003. The collective author. In *Scientific Authorship: Credit and Intellectual Property in Science*, ed. M. Baglio and P. Galison, 325–355. New York: Routledge.

Garfinkel, H., M. Lynch, and E. Livingston. 1981. The work of discovering science construed with materials from the optically discovered pulsar. *Philosophy of the Social Sciences* 11:131–158.

Giles, J. 2006. Sociologist fools physics judges. *Nature* 442:8.

Goodman, N. 1973. *Fact, Fiction, and Forecast.* 3rd ed. Indianapolis: Bobbs-Merrill.

Hall, S. 2016. About the LIGO gravitational-wave rumor… *Sky and Telescope*, January 13, http://www.skyandtelescope.com/astronomy-news/about-this-weeks-gravitational-wave-rumor.

Harvey, B. 1981. Plausibility and the evaluation of knowledge: A case study in experimental quantum mechanics. *Social Studies of Science* 1 (11): 95–130.

Holton, G. 1978. *The Scientific Imagination.* Cambridge: Cambridge University Press.

Ju, L., D. G. Blair, and C. Zhao. 2000. Detection of gravitational waves. *Reports on Progress in Physics* 63:1317–1427.

Kaiser, D. 2009. *Drawing Theories Apart: The Dispersion of Feynman Diagrams in Postwar Physics.* Chicago: University of Chicago Press.

Kaiser, D. 2011. *How the Hippies Saved Physics: Science, Counterculture, and the Quantum Revival.* New York: W. W. Norton.

Kennefick, D. 2007. *Traveling at the Speed of Thought: Einstein and the Quest for Gravitational Waves.* Princeton: Princeton University Press.

Knorr-Cetina, K. 1981. *The Manufacture of Knowledge: An Essay on the Constructivist and Contextual Nature of Science.* Oxford: Pergamon Press.

Kuhn, T. S. 1959. The essential tension: Tradition and innovation in scientific research. In *The Third University of Utah Research Conference on the Identification of Scientific Talent*, ed. C. W. Taylor. Salt Lake City: University of Utah Press.

Kuhn, T. S. 1977. *The Essential Tension: Selected Studies in Scientific Tradition and Change.* Chicago: University of Chicago Press.

Labinger, J., and H. M. Collins, eds. 2001. *The One Culture? A Conversation about Science.* Chicago: University of Chicago Press.

Lakatos, I. 1970. Falsification and the methodology of scientific research programmes. In *Criticism and the Growth of Knowledge*, ed. I. Lakatos and A. Musgrave, 91–196. Cambridge: Cambridge University Press.

Langley, P. W., G. Bradshaw, H. A. Simon, and J. M. Zytkow. 1987. *Scientific Discovery: Computational Explorations of the Creative Process.* Cambridge, MA: MIT Press.

Latour, B., and S. Woolgar. 1979. *Laboratory Life: The Social Construction of Scientific Facts.* London: Sage.

Laudan, L. 1983. *Progress and Its Problems: Towards a Theory of Scientific Growth.* Berkeley: University of California Press.

Lyons, Louis. 2013. Discovering the significance of 5 sigma. arXiv:1310.1284 [physics.data-an].

MacKenzie, D. 1990. *Inventing Accuracy: A Historical Sociology of Nuclear Missile Guidance.* Cambridge, MA: MIT Press.

MacKenzie, D. 2001. *Mechanizing Proof: Computing, Risk, and Trust.* Cambridge, MA: MIT Press.

Marks, J. 2000. The truth about lying. *Philosophy Now* 27 (June–July): 51.

Mermin, D. N. 2005. *It's about Time: Understanding Einstein's Relativity.* Princeton, NJ: Princeton University Press.

Overbye, D. 2016. Gravitational waves detected, confirming Einstein's theory. *New York Times,* February 11, http://www.nytimes.com/2016/02/12/science/ligo-gravitational-waves-black-holes-einstein.html.

Pickering, A. 1981. Constraints on controversy: The case of the magnetic monopole. *Social Studies of Science* 1 (11): 63–93.

Pickering, A. 1984. *Constructing Quarks: A Sociological History of Particle Physics.* Edinburgh: Edinburgh University Press.

Pinch, T. J. 1981. The sun-set: The presentation of certainty in scientific life. *Social Studies of Science* 1 (11): 131–158.

Pinch, T. J. 1985. Towards an analysis of scientific observation: The externality and evidential significance of observational reports in physics. *Social Studies of Science* 15 (1): 3–36.

Pinch, T. J. 1986. *Confronting Nature: The Sociology of Solar-Neutrino Detection.* Dordrecht: Reidel.

Pinch, T. J. 1997. Kuhn—The conservative and radical interpretations: Are some Mertonians "Kuhnians" and some Kuhnians "Mertonians"? *Social Studies of Science* 27 (3): 465–482.

Pitkin, M. S. Reid, S. Rowan, and J. Hough. 2011. Gravitational wave detection by interferometry (ground and space). http://arxiv.org/pdf/1102.3355.pdf.

Popper, K. R. 1959. *The Logic of Scientific Discovery.* New York: Harper & Row.

Pretorius, F. 2005. Evolution of binary black hole spacetimes. *Physical Review Letters* 95:121101.

Sacks, O. W. 2011. *The Man Who Mistook His Wife for a Hat.* London: Picador.

Selinger, E. 2003. The necessity of embodiment: The Dreyfus–Collins debate. *Philosophy Today* 47 (3): 266–279.

Selinger, E., H. L. Dreyfus, and H. M. Collins. 2007. Embodiment and interactional expertise. *Studies in History and Philosophy of Science* 38 (4): 722–740.

Shapin, S. 1984. Pump and circumstances: Robert Boyle's literary technology. *Social Studies of Science* 14:481–520.

Shapin, S., and S. Schaffer. 1987. *Leviathan and the Air-Pump: Hobbes, Boyle, and the Experimental Life*. Princeton: Princeton University Press.

Simon, H. A. 1991. Comments on the symposium on "Computer Discovery and the Sociology of Scientific Knowledge." *Social Studies of Science* 21:143–148.

Smith, G. C. S. 2003. Parachute use to prevent death and major trauma related to gravitational challenge: Systematic review of randomised controlled trials. *British Medical Journal* 327 (1459).

Smolin, L. 2006. *The Trouble with Physics: The Rise of String Theory, the Fall of a Science, and What Comes Next*. Boston: Houghton Mifflin Harcourt.

Sokal, A. D. 1994. Transgressing the boundaries: Towards a Transformative hermeneutics of quantum gravity. *Social Text* 46–47 (spring–summer): 217–252.

Sokal, A. D. 1996. A physicist experiments with cultural studies. *Lingua Franca* (May).

Staley, K. 1999. Golden events and statistics: What's wrong with Galison's image/logic distinction. *Perspectives on Science* 7:196–230.

Travis, G. D. 1981. Replicating replication? Aspects of the social construction of learning in planarian worms. *Social Studies of Science* 1 (11): 11–32.

Turing, A. M. 1950. Computing machinery and intelligence. *Mind* 59 (236): 433–460.

Twilley, N. 2016. Gravitational waves exist: The inside story of how scientists finally found them. *New Yorker*, February 11.

Weber, J. and B. Radak. 1996. Search for correlations of gamma-ray bursts with gravitational-radiation antenna pulses. *Il Nuovo Cimento B Series 11* 111 (6): 687–692.

Winch, P. G. 1958. *The Idea of a Social Science*. London: Routledge and Kegan Paul.

Wittgenstein, L. 1953/1999. *Philosophical Investigations*. Trans. G. E. M. Anscombe. Upper Saddle River, NJ: Prentice Hall.

INDEX

Note: In general I have not indexed the gravitational wave scientists who helped me write this book. Most of them are anonymized and it seemed arbitrary to include just those who aren't. Exceptions include where the physicist is a topic as much as a source of help—e.g., the ideas of Adalberto Giazotto about detection and the roles of Peter Saulson and Joe Weber. Help from gravitational wave physicists is recognized in the acknowledgments.